Viruses: Biology and Applications

Viruses: Biology and Applications

Edited by Harvey O'Brien

⬜SYRAWOOD
PUBLISHING HOUSE
New York

Published by Syrawood Publishing House,
750 Third Avenue, 9th Floor,
New York, NY 10017, USA
www.syrawoodpublishinghouse.com

Viruses: Biology and Applications
Edited by Harvey O'Brien

© 2019 Syrawood Publishing House

International Standard Book Number: 978-1-68286-728-0 (Hardback)

Cataloging-in-Publication Data

Viruses : biology and applications / edited by Harvey O'Brien.
    p. cm.
Includes bibliographical references and index.
ISBN 978-1-68286-728-0
1. Viruses. 2. Virology. 3. Biology. I. O'Brien, Harvey.
QR360 .V57 2019
579.2--dc23

TABLE OF CONTENTS

PREFACE

Over the recent decade, advancements and applications have progressed exponentially. This has led to the increased interest in this field and projects are being conducted to enhance knowledge. The main objective of this book is to present some of the critical challenges and provide insights into possible solutions. This book will answer the varied questions that arise in the field and also provide an increased scope for furthering studies.

Viruses are submicroscopic parasitic and infectious agents that replicate inside the living cells of other organisms. Virology is the branch of science concerned with the in-depth study of all types of viruses. Viruses are generally studied under an optical microscope. They affect all organisms including bacteria, fungi, plants and animals. Some viral diseases in humans include common cold, influenza, chickenpox, AIDS, etc. Virology has played a significant role in the study of genetics, protein transport and immunology. Some of the diverse topics covered in this book address the varied aspects of virology. Different approaches, evaluations, methodologies and advanced studies on viruses have been included herein. Scientists and students actively engaged in this field will find this book full of crucial and unexplored concepts.

I hope that this book, with its visionary approach, will be a valuable addition and will promote interest among readers. Each of the authors has provided their extraordinary competence in their specific fields by providing different perspectives as they come from diverse nations and regions. I thank them for their contributions.

Editor

Viral Protein Kinetics of Piscine Orthoreovirus Infection in Atlantic Salmon Blood Cells

Hanne Merethe Haatveit [1], Øystein Wessel [1], Turhan Markussen [1], Morten Lund [2], Bernd Thiede [3], Ingvild Berg Nyman [1], Stine Braaen [1], Maria Krudtaa Dahle [2] and Espen Rimstad [1,*]

[1] Department of Food Safety and Infectious Biology, Faculty of Veterinary Medicine, Norwegian University of Life Sciences, 0454 Oslo, Norway; hanne.merethe.haatveit@nmbu.no (H.M.H.); oystein.wessel@nmbu.no (Ø.W.); turhan.markussen@nmbu.no (T.M.); ingvild.nyman@nmbu.no (I.B.N.); stine.braaen@nmbu.no (S.B.)

[2] Department of Immunology, Norwegian Veterinary Institute, 0454 Oslo, Norway; morten.lund@vetinst.no (M.L.); maria.dahle@vetinst.no (M.K.D.)

[3] Department of Biosciences, University of Oslo, 0316 Oslo, Norway; bernd.thiede@ibv.uio.no

* Correspondence: espen.rimstad@nmbu.no

Academic Editors: Corina P.D. Brussaard and Mathias Middelboe

Abstract: *Piscine orthoreovirus* (PRV) is ubiquitous in farmed Atlantic salmon (*Salmo salar*) and the cause of heart and skeletal muscle inflammation. Erythrocytes are important target cells for PRV. We have investigated the kinetics of PRV infection in salmon blood cells. The findings indicate that PRV causes an acute infection of blood cells lasting 1–2 weeks, before it subsides into persistence. A high production of viral proteins occurred initially in the acute phase which significantly correlated with antiviral gene transcription. Globular viral factories organized by the non-structural protein µNS were also observed initially, but were not evident at later stages. Interactions between µNS and the PRV structural proteins λ1, µ1, σ1 and σ3 were demonstrated. Different size variants of µNS and the outer capsid protein µ1 appeared at specific time points during infection. Maximal viral protein load was observed five weeks post cohabitant challenge and was undetectable from seven weeks post challenge. In contrast, viral RNA at a high level could be detected throughout the eight-week trial. A proteolytic cleavage fragment of the µ1 protein was the only viral protein detectable after seven weeks post challenge, indicating that this µ1 fragment may be involved in the mechanisms of persistent infection.

Keywords: *Piscine orthoreovirus*; PRV; non-structural protein; µNS; µ1; expression kinetics; proteolytic cleavage; pathogenesis; blood cells; Atlantic Salmon

1. Introduction

Piscine orthoreovirus (PRV) belongs to the genus *Orthoreovirus* in the family *Reoviridae* [1,2]. The orthoreoviruses are ubiquitous in various animal species, but only found to be of pathogenic significance in poultry and recently in fish [3–7]. PRV is abundant in farmed Atlantic salmon (*Salmo salar*), detected both in apparently healthy and diseased fish [8–11]. The infection causes heart and skeletal muscle inflammation (HSMI) and is associated with melanised foci in white muscle in Atlantic salmon [1,7,12]. HSMI is a prevalent disease and melanised foci is a quality problem; both conditions are of major economic importance to salmon aquaculture. The pathogenesis of HSMI is not completely elucidated. Outbreaks of the disease are primarily observed in the seawater phase and last for several weeks in the population [13], after which the PRV infection becomes persistent [9,11,14]. In experimental cohabitant infection trials, disease onset occurs after 8–10 weeks [15].

The study of molecular mechanisms linked to PRV infection has been limited by the lack of susceptible cell lines. Studies of the viral infection have therefore been performed in vivo or by infecting erythrocytes ex vivo [16]. Piscine erythrocytes are nucleated and contain the transcriptional and translational machinery necessary for expression of mRNA and proteins [17]. Erythrocytes are important target cells for PRV and the infection activates an innate antiviral immune response typical for RNA viruses in these cells [18]. During the peak phase of infection, more than 50% of all erythrocytes may be infected [19]. Interestingly, severe anemia has not been reported from HSMI outbreaks in the seawater phase, indicating low or no virus-induced lysis of infected erythrocytes [20]. Recently, a variant of PRV was demonstrated to be the etiologic agent of erythrocytic inclusion body syndrome (EIBS), a condition associated with anemia and mass mortality in juvenile Coho salmon (*Onchorhynchus kisutchi*). The level of anemia in EIBS affected fish corresponded with the level of viral replication in blood [7]. In addition, infection of rainbow trout in fresh water by yet another PRV variant is also associated with anemia and an HSMI-like disease [5].

The *Orthoreovirus* genome consists of ten double-stranded RNA (dsRNA) segments enclosed in a double protein capsid. The genomic segments are classified according to size with three large (L), three medium (M) and four small (S) segments encoding the λ, μ and σ class proteins, respectively [3,21]. In mammalian orthoreovirus (MRV), the species type of genus *Orthoreovirus*, the viral transcription machinery is located in the inner core and consists of $\lambda 1$, $\lambda 2$, $\lambda 3$, $\mu 2$ and $\sigma 2$ [22]. The outer capsid proteins $\mu 1$, $\sigma 1$ and $\sigma 3$ are involved in cell attachment and membrane penetration during the initial stages of infection [23–25]. The two non-structural proteins μNS and σNS participate in the formation of viral factories where viral genome replication and particle assembly occur [21,26,27]. Although some important amino acid motifs are conserved between MRV and PRV, sequence identities between homologous proteins are generally low [2]. MRV enters the cell by receptor-mediated endocytosis. The outer capsid is largely removed and $\mu 1$ is cleaved at two positions that generate, in addition to the full-length protein, five different fragments [24,28]. The N-terminal autolytic cleavage site, which produces $\mu 1$N and $\mu 1$C, seems conserved across orthoreoviruses, including PRV [2,29,30]. Further cleavage of $\mu 1$C, mediated by exogenous proteases, generate fragments δ and ϕ [24].

Structures resembling viral factories have also been observed in PRV-infected erythrocytes, and recombinant expression of the protein in fish cell lines indicate that PRV μNS has an analogous role in factory formation [16,19,31]. The majority of virus-encoded proteins localize completely or partially within these viral factories [2,3,21]. The viral factories in PRV-infected cells resemble the globular structures observed for the MRV type 3 Dearing (T3D) strain, in contrast to the filamentous-like viral factories generated by MRV Type 1 Lang (T1L) [19]. The latter is considered the most common morphology type of orthoreoviral factories [21,32]. Gene segment M3 in MRV and avian orthoreovirus (ARV) are reported to produce two isoforms of the factory forming μNS protein in infected cells [33–35]. The second isoform is produced by different mechanisms in the two viruses; in MRV, μNSC is expressed by a second in-frame AUG (Met$_{41}$) while in ARV, post-translational cleavage in the N-terminal region releases μNSN [33,35,36]. In ARV, only full-length μNS interacts with σNS in infected cells, suggesting that the two isoforms play different roles during ARV infection [34].

Considering the emerging occurrence of HSMI, PRV exhibits a considerable risk for the aquaculture industry and proper disease control is highly desired. To understand the association between PRV infection and disease outcome, and also to limit further disease outbreaks, more information regarding PRV protein kinetics is essential. In the present study, the kinetics of viral RNA, viral protein and antiviral immune response in blood cells from experimentally PRV-infected Atlantic salmon were investigated. We hypothesized that PRV causes an acute infection in blood cells correlating with innate antiviral gene expression, before the infection subsides to a low persistent level.

2. Materials and Methods

2.1. Construction and Expression of Recombinant Piscine orthoreovirus (PRV) μNS

Following the supplier's protocol, the BaculoDirect™ Baculovirus Expression System (Invitrogen, Carlsbad, CA, USA) was used to generate recombinant μNS. The μNS open reading frame (ORF) (acc. no. KR337478) was obtained by polymerase chain reaction (PCR; primers listed in Table S1) of the plasmid construct pcDNA3.1 μNS N-FLAG [31] and cloned into the pENTR™ TOPO® vector (Invitrogen). The pENTR μNS construct was used in a recombination reaction to generate the recombinant baculovirus DNA. Sanger sequencing (GATC Biotech AG, Konstanz, Germany) confirmed the sequence of the construct. Spodoptera frugiperda (Sf9) insect cells (BD Bioscience, Erembodegem, Belgium) cultured in Grace Insect Medium (Invitrogen) supplemented with 10% heat inactivated fetal bovine serum (FBS, Life Technologies, Paisley, Scotland, UK), 100 U/mL Penicillin, 100 μg/mL Streptomycin and 0.25 μg/mL Fungizone (Life Technologies), were transfected with recombinant baculovirus DNA. Passage 1 (P1) viral stock was harvested 11 days post transfection and used to produce high titer viral stocks according to the supplier's protocol. The BacPAK quantitative PCR (qPCR) Titration kit (Clontech, Mountain View, CA, USA) was used to determine the viral titer. Finally, Sf9 insect cells were infected with Passage 2 (P2) or higher passage of recombinant baculovirus stock ($>1 \times 10^8$ copies/mL) and incubated at 27 °C for 96 h for expression of the recombinant μNS protein containing a C-terminal 6xHis-tag.

2.2. Construction and Expression of Recombinant PRV λ1

The ORF of PRV structural protein λ1 (acc. no. KR337475) encoded by gene segment L3 was amplified (primers listed in Table S1) using cDNA originating from a HSMI outbreak [31] as template. The PCR product was cloned into pET100/D-TOPO (Invitrogen) and the sequence verified by Sanger sequencing (GATC Biotech AG). The pET100-λ1 plasmid was transfected into E. coli (BL21 DE3 strain, Invitrogen) and expressed with a N-terminal 6xHis-tag, following the manufacturer's instructions. Protein expression was monitored by sodium dodecyl sulfate polyacrylamide gel electrophoresis (SDS-PAGE).

2.3. Protein Purification

The Sf9 insect cells and the E. coli cells expressing recombinant PRV μNS and λ1 proteins, respectively, were pelleted by centrifugation at $5000 \times g$ for 10 min, then dissolved and washed in phosphate-buffered solution (PBS). Purification of recombinant proteins was carried out using ProBond Purification System (Life Technologies) following the manufacturer's instructions. The recombinant μNS protein was eluted with an elution buffer containing 8 M Urea, 20 mM $Na_2H_2PO_4$ (pH 4.0), and 500 mM NaCl. The purity of the recombinant protein was monitored by SDS-PAGE using a 4%–12% Bis–Tris Criterion XT gel (Bio-Rad, Hercules, CA, USA). To purify λ1, the Ni-NTA agarose was run on a SDS-PAGE where a band matching the size of λ1 was excised. The gel sample containing λ1 protein was solubilized in 250 mM Tris-HCl with 0.1% SDS, pH 6.8, sonicated 3×5 s and incubated at 4 °C with shaking overnight. The sample was centrifuged at $10,000 \times g$ for 10 min and the supernatant was dialyzed using the Slide-A-Lyser® Dialysis cassette with 20,000 molecular weight cut-off (MWCO) and 0.5–3.0 mL capacity (Thermo Scientific, Waltham, MA, USA) following the manufacturer's protocol. SDS-PAGE confirmed the purity of the recombinant λ1 protein. Protein concentrations for both μNS and λ1 were determined using the DC Protein Assay Reagent Package (Bio-Rad), with bovine serum albumin (BSA; Sigma-Aldrich, St. Louis, MO, USA) as protein standard.

2.4. Immunization of Rabbits

The purified recombinant proteins were used for immunization of rabbits and generation of antisera named anti-μNS #R320684 and anti-λ1 #K273. In the first injection, Freund's complete adjuvant was added, thereafter the rabbits were boosted three times with Freund's incomplete adjuvant weekly. The amount of µNS and λ1 antigen used per immunization was in the range of 45–500 µg. The rabbit sera produced were tested by Western blotting (WB) and fluorescent microscopy after transfection of epithelioma papulosum cyprini (EPC; ATCC CRL-2872) cells with pcDNA3.1 µNS N-FLAG [31] or pcDNA3.1 λ1 N-HA [31] (see description below). Antisera controls were collected prior to immunization. WB and immunofluorescent microscopy confirmed that the rabbit µNS and λ1 antisera recognized the µNS and λ1 proteins in transfected EPC cells (Figure S1). No staining was detected using the pre-immunization sera (data not shown).

2.5. Specificity of Antisera

EPC cells were cultivated in Leibovitz-15 medium (L15; Life Technologies) supplemented with 10% heat inactivated FBS, 2 mM L-glutamine, 0.04 mM mercaptoethanol and 0.05 mg/mL gentamycin-sulphate (Life Technologies), and seeded at a density of 1.5×10^4 cells/well in a 24-well plate 24 h prior to transfection. Plasmids pcDNA3.1-µNS N-FLAG and pcDNA3.1-λ1 N-HA were transfected using Lipofectamine LTX reagent (Life Technologies) according to the manufacturer's instructions. The cells were fixed and stained 48 h post-transfection with an Intracellular Fixation and Permeabilization Buffer Set (eBioscience, San Diego, CA, USA) following the manufacturer's protocol. Antisera against µNS (1:1000) and λ1 (1:500); secondary antibody against rabbit IgG conjugated with Alexa Fluor 488 (Life Technologies) and Hoechst trihydrochloride trihydrate (Life Technologies) were used for staining. Images were captured on an inverted fluorescence microscope (Olympus IX81). Transfected EPC cells were also used to further verify anti-µNS and anti-λ1 in WB. A total of 3×5 million EPC cells were pelleted by centrifugation, resuspended in 100 µL Ingenio Electroporation Solution (Mirus, Madison, WI, USA) and transfected with 4 µg pcDNA3.1 µNS N-FLAG or pcDNA3.1 λ1 N-HA. The transfected cells were transferred to 75 cm^2 culture flasks containing 20 mL pre-equilibrated L-15 growth medium (described above) and collected 72 h post-transfection. The cell pellets were lysed in Nonidet-P40 lysis buffer (1% NP-40, 50 mM Tris–HCl pH 8.0, 150 mM NaCl, 2 mM EDTA) containing Complete ultra mini protease inhibitor cocktail (Roche, Mannheim, Germany). The mix was incubated on ice for 30 min, and then centrifuged at $5000 \times g$ for 5 min at 4 °C. The supernatant was mixed with Sample Buffer (Bio-Rad) and Reducing Agent (Bio-Rad), denatured for 5 min at 95 °C and run in SDS-PAGE, using 4%–12% Bis–Tris Criterion XT gel (Bio-Rad). Magic Mark^{TM} XP Standard (Invitrogen) was used as a molecular size marker. Following SDS-PAGE, the proteins were blotted onto a polyvinylidene fluoride (PVDF) membrane (Bio-Rad) and anti-µNS and anti-λ1 were used as primary antibodies and anti-Rabbit IgG-HRP (GE Healthcare, Buchinghamshire, UK) as secondary antibody. Protein bands were detected by chemiluminescence (Amersham ECL Plus, GE Healthcare).

2.6. Experimental Challenge of Salmon

A cohabitation challenge experiment was performed at VESO Vikan aquatic research facility, (Vikan, Norway). The fish had an average weight of 30 grams at the onset of the experiment with a maximum stocking density of 80 kg/m^3, and were kept in 0.4 m^3 tanks supplied with filtered and UV-radiated fresh water, 12 °C ± 1 °C with a 12 h light/12 h dark regime. Water discharge of the tanks was provided by a tube overflow system with 7.2 L/min flow rate. The fish were acclimatized for two weeks prior to challenge, fed according to standard procedures and anesthetized by bath immersion (2–5 min) in benzocaine chloride (0.5 g/10 L water, Apotekproduksjon AS, Oslo, Norway) before handling. Briefly, the experimental study included one group of shedder fish (50%) marked at the time of PRV-injection by cutting off the adipose fin and one naïve cohabitant group (50%). The PRV inoculum was prepared from a batch of pooled heparinized blood samples from a previous PRV challenge experiment [19].

On day 0 of the challenge, the heparinized blood was diluted 1:2 in PBS and 0.1 mL of the inoculum was intraperitoneal (i.p.) injected into the shedders. The inoculum was confirmed negative for salmon viruses such as infectious pancreatic necrosis virus (IPNV), infectious salmon anemia virus

(ISAV), salmonid alphavirus (SAV) and piscine myocarditis virus (PMCV) by reverse transcription quantitative PCR (RT-qPCR). Samples from six fish were collected before initiation of the experiment to provide time-0 uninfected control material for protein assays. Heparinized blood was collected from six cohabitant fish at each sampling point; 3, 4, 5, 6, 7 and 8 weeks post challenge (wpc). In addition, a second cohabitation challenge experiment lasting 10 weeks was performed at the same facility following a similar experimental design. In this study, six fish sampled prior to PRV challenge were used to provide uninfected control material for protein and RT-qPCR assays, and heparinized blood was collected from six cohabitant fish at 4, 6, 8 and 10 wpc. The second challenge experiment was otherwise performed under the same conditions as the first experiment. Both experiments were approved by the Norwegian Animal Research Authority and followed the European Union Directive 2010/63/EU for animal experiments.

2.7. RNA Isolation and Reverse Transcription Quantiative Polymerase Chain Reaction (RT-qPCR)

Total RNA was isolated from 20 μL heparinized blood homogenized in 650 μL QIAzol Lysis Reagent (Qiagen, Hilden, Germany) using 5 mm steel beads, TissueLyser II (Qiagen) and RNeasy Mini spin column (Qiagen) as recommended by the manufacturer. RNA was quantified using a NanoDrop, ND-1000 spectrophotometer (Thermo Fisher Scientific, Wilmington, DE, USA). The Qiagen OneStep kit (Qiagen) was used for RT-qPCR with a standard input of 100 ng (5 μL of 20 ng/μL) of the isolated total RNA per reaction in a total reaction volume of 12.5 μL. The template RNA was denatured at 95 °C for 5 min prior to RT-qPCR targeting PRV gene segments S1, M2 and M3. The following conditions were used for S1: 400 nM primer, 300 nM probe, 400 nM dNTPs, 1.26 mM $MgCl_2$, 1:100 RNase Out (Invitrogen) and 1 × ROX reference dye with the following cycle parameters: 30 min at 50 °C, 15 min at 94 °C, 40 cycles of 94 °C/15 s, 54 °C/30 s and 72 °C/15 s in an AriaMx (Agilent, Santa Clara, CA, USA). Similar conditions and cycle parameters were also used targeting M2 and M3, although primer concentration was adjusted to 600 nM and annealing temperature to 58 °C. All samples were run in duplicates, and a sample was defined as positive if both parallel samples had a Ct <35. The fluorescence threshold for S1, M2 and M3 was set at ΔRn 0.261, 0.028 and 0.021, respectively. The primers and probes are listed in Table S1. For analysis of antiviral gene expression, cDNA was prepared from 500 ng RNA using the QuantiTect reverse transcription kit with gDNA elimination (Qiagen) following the instructions from the manufacturer. Quantitative PCR was performed in triplets on 384-well plates using cDNA corresponding to 5 ng RNA in a total volume of 10 μL per parallel, SsoAdvanced™ Universal SYBR® Green Supermix, and 500 nM forward and reverse primers (Table S2). The qPCRs were run for 40 cycles of 94 °C/15 s and 60 °C/30 s. All samples in the sample set were analyzed on the same plate using the same fluorescence threshold, and the cut-off value was set to Ct 37. The specificity of the SYBR green assays was confirmed by melting point analysis. Levels of Elongation factor (EF1α) mRNA were used for normalization of all assays by the ΔΔCt method.

2.8. Flow Cytometry

Samples consisting of 1.25 μL heparinized blood (diluted 1:20 in PBS) from each of the cohabitant fish in the first challenge experiment were plated into 96-well plates for intracellular staining as previously described [19] using anti-μNS and anti-σ1 [4]. The corresponding zero serum, anti-μNS Zero and anti-σ1 Zero [4] were used as negative controls for background staining. Samples originating from 5 and 8 wpc were fixed, stained and analyzed immediately, while samples from 4 and 7 wpc were fixed and stored for one week and samples from 0, 3 and 6 wpc were fixed and stored for two weeks in flowbuffer (PBS, 1% BSA, 0.05% azide) before analysis. The cells were analyzed on a Gallios Flow Cytometer (Beckman Coulter, Miami, FL, USA), counting 50,000 cells per sample, and the data were analyzed using the Kaluza software (Becton Dickinson). Cells were gated according to size and granularity to include only intact cells and samples from 0 wpc were used as negative controls. Due to slight variation in background staining, the flow charts were gated individually to discriminate between negative and positive peaks.

2.9. Immunofluorescence Microscopy

Following flow cytometry analysis, the cells were prepared for immunofluorescence microscopy. The nuclei were stained with Hoechst trihydrochloride trihydrate (Life Technologies) and the cells were mounted to glass slides using Fluoroshield (Sigma-Aldrich, St. Louis, MO, USA) and cover slips. Images were captured on an inverted fluorescence microscope (Olympus IX81).

2.10. Transmission Electron Microscopy (TEM)

Samples consisting of 20 µL heparinized blood from each cohabitant fish in the first experimental challenge were diluted in 1 mL PBS, centrifuged at $1000 \times g$ for 5 min at 4 °C, washed twice in PBS and fixed in 3% glutaraldehyde overnight at 4 °C. All samples were further washed twice in PBS and prepared for transmission electron microscopy (TEM) as described earlier [19]. The sections were examined in a FEI MORGAGNI 268, and photographs were recorded using a VELETA camera.

2.11. Western Blotting (WB)

Heparinized blood from each cohabitant fish in the first challenge experiment was analyzed separately and as pooled samples from the different time-points. The samples were centrifuged at $5000 \times g$ and the blood pellets was lysed in Nonidet-P40 lysis buffer containing Complete ultra mini protease inhibitor cocktail and prepared for WB as described above. Anti-µNS (1:1000), anti-µ1C (1:500) [4], anti-σ1 (1:1000) [4], anti-σ3 (1:500) [2] and anti-λ1 (1:500) were used as primary antisera, Rabbit Anti-Actin (Sigma-Aldrich, St. Louis, MO, USA) was used to standardize the blots and Anti-Rabbit IgG-HRP (GE Healthcare) was used as secondary antibody. Blood collected at 0 wpc was used as negative control. In addition, heparinized blood from six of the cohabitant fish sampled at 0, 4, 6, 8 and 10 wpc in the second challenge experiment were prepared and analyzed in the same manner.

2.12. Immunoprecipitation (IP)

Blood from six cohabitants in the first challenge experiment sampled at 4, 5 and 8 wpc were pooled and lysed in Nonidet-P40 lysis buffer containing Complete ultra mini protease inhibitor cocktail as described above. The supernatants were transferred to new tubes and added anti-µNS or anti-µ1C (1:50) and incubated at 4 °C overnight with rotation. The Immunoprecipitation Kit Dynabeads Protein G (Novex, Life Technologies) was used for protein extraction and the beads were prepared according to the manufacturer's protocol. The cell–lysate–antibody mixtures were mixed with the protein G-coated beads and incubated 2 h at 4 °C. The beads–antibody–protein complexes were washed according to the manufacturer's protocol and run in SDS-PAGE. The SDS-gel was blotted onto PVDF membranes (Bio-Rad) and the proteins were detected using anti-µNS, anti-µ1C [4], anti-σ1 [4], anti-σ3 [2] and anti-λ1.

2.13. Liquid Chromatography–Mass Spectrometry (LC–MS)

Five and three fragments immunoprecipitated with anti-µNS (4 and 5 wpc) and anti-µ1C (5 wpc), respectively, that were not observed at 0 wpc, were excised and in-gel digested with 0.1 µg of trypsin in 20 µL of 50 mM ammonium bicarbonate, pH 7.8 for 16 h at 37 °C (Promega, Madison, WI, USA). The peptides were purified with µ-C18 ZipTips (Millipore, Billerica, MA, USA), and analyzed using an Ultimate 3000 nano-UHPLC system (Dionex, Sunnyvale, CA, USA) connected to a Q Exactive mass spectrometer (ThermoElectron, Bremen, Germany). Liquid chromatography and mass spectrometry was performed as previously described [37]. Data were acquired using Xcalibur v2.5.5 and raw files were processed to generate peak list in Mascot generic format (*.mgf) using ProteoWizard release (Version 3.0.331). Database searches were performed using Mascot (Version 2.4.0) against the protein sequences of λ1, λ2, λ3, µNS, µ1, µ2, σNS, σ1, σ2 and σ3 assuming the digestion enzyme trypsin and semi-trypsin, at a maximum of one missed cleavage site, fragment ion mass tolerance of 0.05 Da, parent ion tolerance of 10 ppm and oxidation of methionines, propionamidylation of cysteines, acetylation of the protein N-terminus as variable modifications. Scaffold 4.4.8 (Proteome Software Inc., Portland, OR, USA) was used to validate MS/MS based peptide and protein identifications.

2.14. Computational Analysis

Theoretical molecular weights for proteins were calculated using the Compute pI/Mw tool [38]. PSI-blast based secondary structure PREDiction (PSIPRED; Version 3.3) was used to predict protein secondary structure [39].

2.15. Statistical Analysis

Differences in gene expression levels of innate antiviral genes was analyzed using one-way Anova with Tukey's multiple comparison test. Correlation analysis between PRV S1/M3 RNA levels and antiviral and immune gene expression were performed using nonparametric Spearman correlation.

3. Results

3.1. Viral RNA Load in Blood Cells

RT-qPCR targeting PRV genomic segments S1, M2 and M3 revealed high viral RNA loads in blood cells from 3 to 8 wpc (Figure 1). RNA from segments S1, M2 and M3 were first detected at 3 wpc and peaked at 5 wpc with mean Ct-values of 17.2 ($\pm$0.4), 14.5 ($\pm$0.3) and 14.6 ($\pm$0.4). From 5 wpc, the S1 RNA load decreased, and by 8 wpc the mean Ct-value was 26.4 ($\pm$0.6). However, a similar decrease was not observed for the M2 and M3 RNAs, and by 8 wpc mean Ct-values for these genomic segments were 17.4 ($\pm$0.5) and 17.7 ($\pm$0.4), respectively. RT-qPCR targeting genomic segment S1 in blood from six fish sampled at 0, 4, 6, 8 and 10 wpc in the second challenge experiment was also performed and gave similar results (Figure S2).

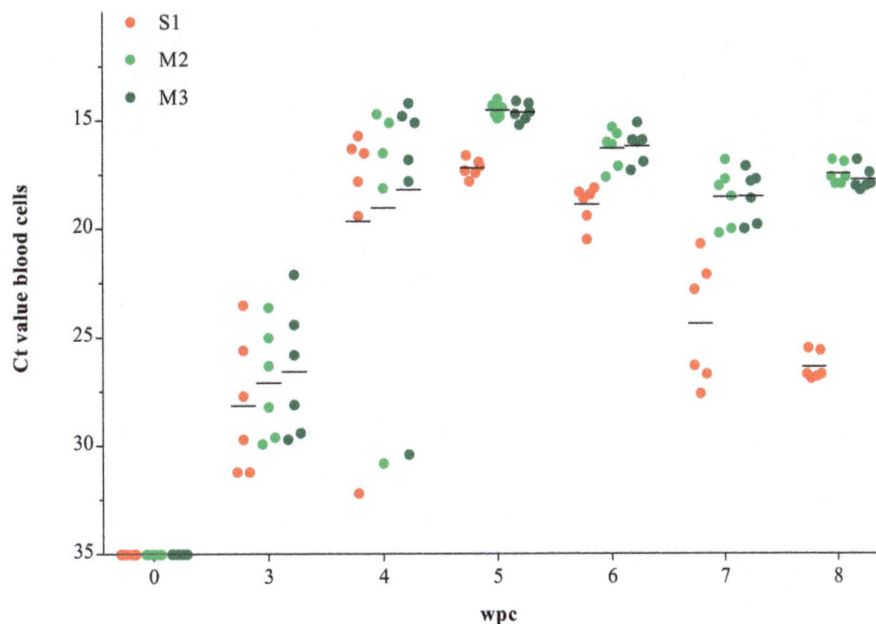

Figure 1. *Piscine orthoreovirus* (PRV) RNA load in blood cells. Reverse transcription quantitative polymerase chain reaction (RT-qPCR) of PRV gene segments S1, M2 and M3 in blood cells from cohabitant fish. Individual (**dots**) and mean (**line**) Ct-values, *n* = 6 per time-point. wpc = weeks post challenge.

3.2. Expression of Innate Antiviral Genes in PRV Infected Blood Cells

The innate antiviral immune response in blood following PRV infection was studied by RT-qPCR targeting Atlantic salmon type I interferon (IFNab), viperin, interferon-stimulated gene 15 (ISG15), dsRNA-activated protein kinase (PKR) and IFNγ. All innate antiviral genes analyzed were statistically significantly upregulated during the peak phase of PRV infection from 4 to 6 wpc, increasing 5- to 20-fold compared to the level at 3 wpc (Figure 2a, Figure S3). The Ct values for S1 and M3 RNA correlated with the relative levels of gene expression for all innate antiviral genes, but not for the T-cell marker genes CD4 and CD8 (Figure 2b). When comparing the early phase up to the peak of infection (3–5 wpc) with the later phase (6–8 wpc), S1 RNA was correlated with the innate antiviral response in both phases, whereas M3 only showed significant correlation in the early phase (Figure 2b). EF1α were stably expressed during PRV infection and were used for normalization of all other assays by the $\Delta\Delta$Ct method (Figure S4) [15,40].

(a)

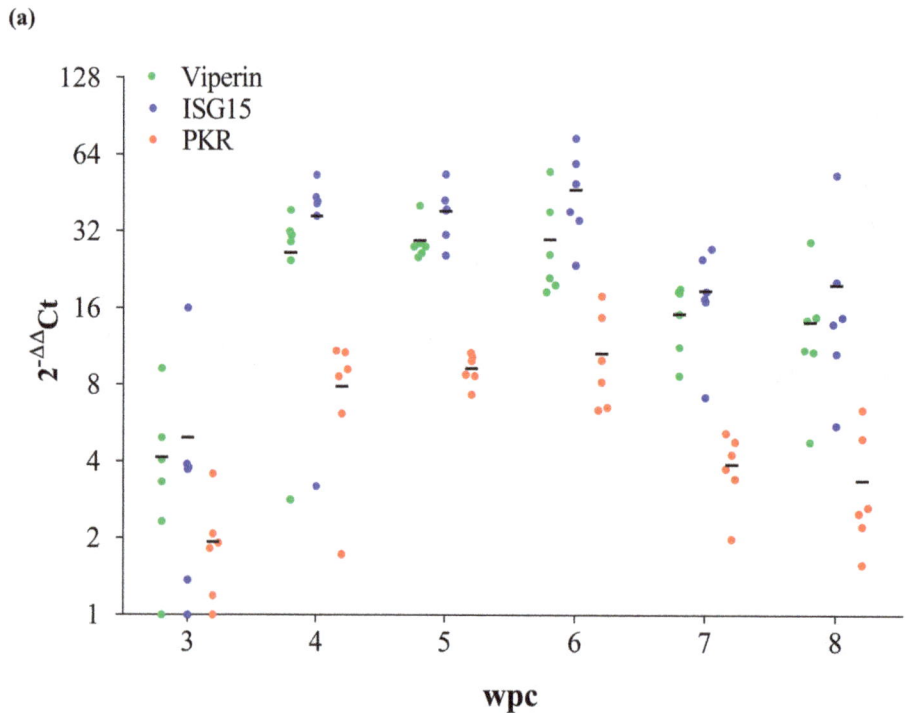

(b)

	n	IFN1ab	Viperin	ISG15	PKR	IFNγ	CD4	CD8α
PRV S1 total	36	****	****	****	****	****	ns	ns
S1 3-5 wpc	18	**	**	**	***	**	ns	ns
S1 6-8 wpc	18	****	***	**	***	**	ns	ns
PRV M3 total	36	****	****	****	****	****	ns	ns
M3 3-5 wpc	18	**	*	*	**	***	ns	ns
M3 6-8wpc	18	*	ns	ns	ns	*	ns	ns

Correlation *$P<0,05$, **$P<0,01$, ***$P<0,001$, ****$P<0,0001$, ns: not significant

Figure 2. Expression of immune genes in blood cells. (**a**) Immune genes were assayed at 3–8 wpc by RT-qPCR in blood cells from cohabitant fish (n = 6 per time point). Data are normalized against EF1α and the lowest ΔCt level at 3 wpc (n = 6), and 2-$\Delta\Delta$Ct values are calculated. Mean relative expression is indicated. ISG = interferon-stimulated gene, PKR = double-stranded RNA (dsRNA)-activated protein kinase; (**b**) Correlation between Ct values for S1/M3 RNA and relative levels of antiviral gene expression for a set of immune genes.

3.3. Flow Cytometry Indicates a Transient Peak in Blood Cells

Blood cells stained intracellularly with anti-μNS and anti-σ1 were analyzed by flow cytometry (Figure 3a, Figure S5). A PRV positive population of blood cells was observed from 4 wpc as a marked shift in the histograms compared to negative samples. Five out of six fish were positive for μNS by flow cytometry at 4 wpc, consistent with the RT-qPCR data where the positive fish had lower Ct-values (18.2 ± 5.6) compared to the negative fish (30.4). At 5 wpc, the PRV positive blood cell population decreased, but was still visible for all individuals. From 6 wpc and onwards, no PRV-positive cell populations were observed. The pattern for σ1 positive cells was similar to that described for μNS.

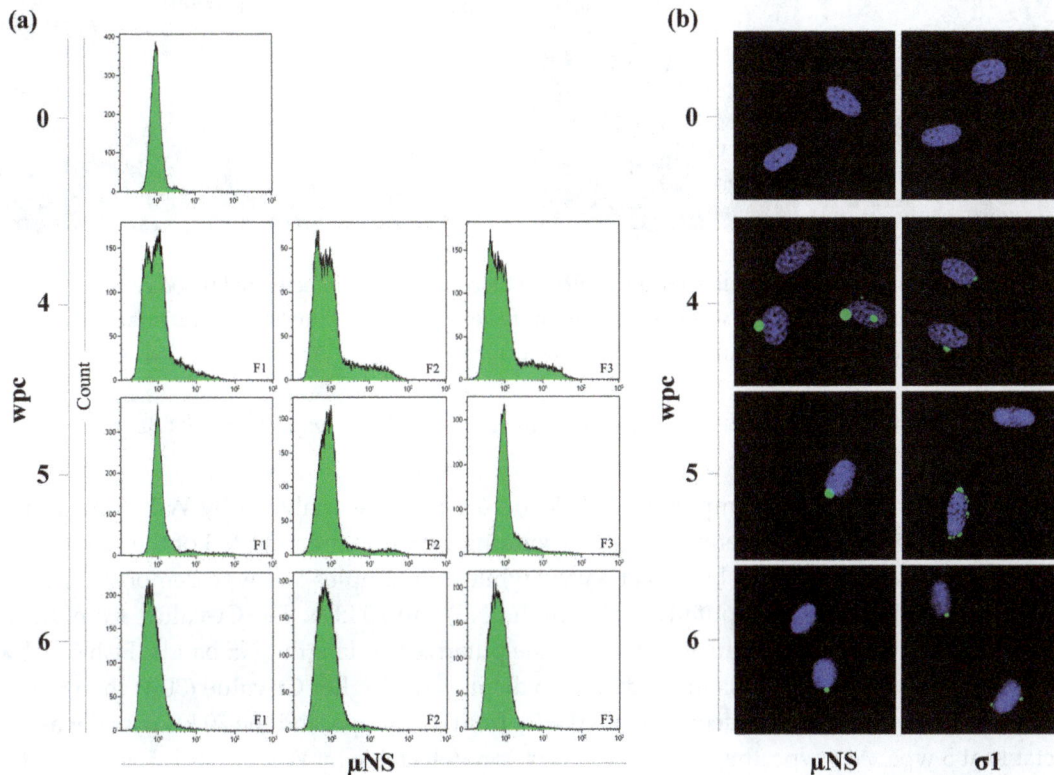

Figure 3. Presence of PRV μNS and σ1 in blood cells. (**a**) Intracellular staining of μNS in blood cells analyzed by flow cytometry from three cohabitant fish sampled at 4, 5 and 6 wpc. The negative control staining is one fish sampled at 0 wpc. A total of 50,000 cells were counted per sample and 30,000 were gated for analysis; (**b**) Fluorescent labeling of μNS (left) and σ1 (right) displaying viral factory-like inclusions (green) in infected red blood cells sampled 0 (negative control), 4, 5 and 6 wpc. The nuclei were stained with Hoechst (blue).

3.4. Viral Factories Observed in Blood Cells

Both μNS and σ1 were detected by immunofluorescence as cytoplasmic globular inclusions in erythrocytes at 4, 5 and 6 wpc (Figure 3b). The inclusions varied in both size and number. At 4 and 5 wpc, they were predominantly large and perinuclear. Inclusions stained with anti-σ1 were generally smaller and more variable in size than those stained with anti-μNS. At 6 wpc, the number and size of the inclusions were considerably reduced and at 7 wpc and onward no inclusions were detected. These findings correlated with the results obtained from flow cytometry.

3.5. TEM of PRV Infected Blood Cells

TEM of PRV infected blood cells sampled at 0, 4, 5 and 6 wpc are shown in Figure 4. The control cells (0 wpc) contained circular cytoplasmic vesicles (200–500 nm) that were apparently devoid of specific content. In addition, a few control cells contained lamellar structures up to 300 nm in size. At 4 wpc, lamellar structures were frequent and a few large cytoplasmic inclusions (~800 nm) containing particles with reovirus-like morphology were observed. The viral particles were naked with an electron dense core that resembled previous TEM descriptions of PRV [19]. At 5 wpc, several small (200–500 nm) and large (~800 nm) cytoplasmic inclusions containing reovirus-like particles were detected. The larger inclusions contained a mixture of reovirus-like particles and lamellar structures, some enclosed within membrane-like structures. At 6 wpc, large inclusions were frequent, but only a few contained viral particles.

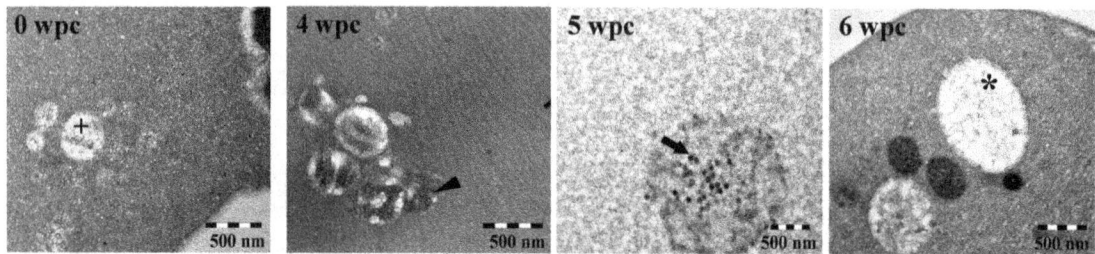

Figure 4. Transmission electron microscopy (TEM) of blood cells. PRV-infected red blood cells sampled at 0 (negative control), 4, 5 and 6 wpc show small empty vesicles (cross), lamellar structures (arrowhead), reovirus-like particles (arrow) and large empty inclusions (star).

3.6. µNS Protein Expression in Individual Fish Correlate with viral RNA only during the Acute Phase of Infection

Blood cells from six fish sampled at 3, 4, 5 and 6 wpc were analyzed by WB using anti-µNS and compared to Ct-values targeting the corresponding genomic segment M3 of the same samples (Figure 5). No fish were positive by WB at 3 wpc, while five samples at 4 wpc demonstrated bands at molecular weight (MW) 83.5 (putative full-length µNS) and 70 kDa. The Ct-values from the same samples corresponded to the positive staining of the putative full-length µNS bands. Fish 6 at 4 wpc, was negative for µNS by WB; this individual also displayed a higher Ct-value (30.4) than the other cohabitants. The amount of µNS decreased markedly from 4 to 5 wpc, and the 70 kDa band was barely detectable at 5 wpc. At 6 wpc, the µNS protein was non-detectable by WB in fish 1, 5 and 6, and only barely detectable in the remaining fish. Although the µNS protein level decreased below the detection limit for WB, the corresponding viral RNA levels (genomic segment M3) remained high throughout the challenge. Thus, µNS protein and M3 RNA levels only correlated at 4 wpc.

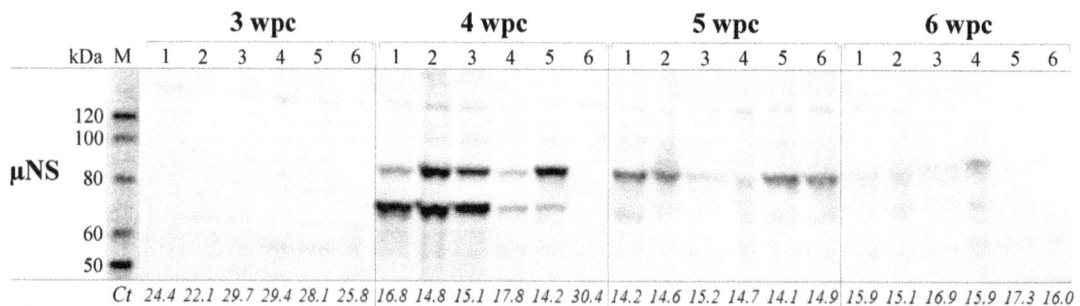

Figure 5. Detection of PRV uNS protein in blood cells compared to viral RNA load. Blood cells from 3, 4, 5 and 6 wpc (*n* = 6) analyzed for µNS by Western blotting. Ct-values for gene segment M3 (µNS) from the same samples are shown below each lane. M = molecular weight standard; Lane 1–6 refers to individual fish (1–6) per time point.

3.7. PRV Protein Levels Display a Transient Peak in Blood Cells

The load of structural proteins λ1, μ1, σ1 and σ3, and the non-structural protein μNS, displayed a similar transient peak at 4–6 wpc in blood cells (Figure 6). All five proteins appeared at 4 wpc and were non-detectable at 7 wpc. In addition to the putative full-length μNS, a band with the MW of about 70 kDa was observed at 4 wpc, consistent with findings from individual fish (Figure 5). The putative full-length μ1 protein (74.2 kDa) was detected at 4 wpc. However, at 5 wpc, this band was not present but replaced by three bands of approximately 70 kDa, 37 kDa and 32 kDa in size. At 7 and 8 wpc, only one band of approximately 35 kDa was detected. The same staining patterns for the λ1, μNS, μ1, σ1 and σ3 proteins were observed when blood from the second challenge experiment was analyzed (Figure S6).

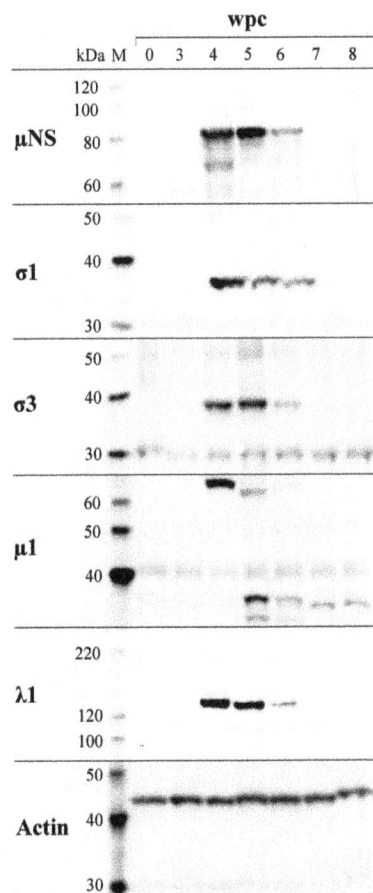

Figure 6. Presence of PRV proteins in blood cells. Pooled blood cell samples (*n* = 6) from each week were analyzed by Western blotting, targeting μNS, σ1, σ3, μ1 and λ1. M = molecular weight standard. Actin was used as control for protein load.

3.8. PRV Proteins Interact with μNS

Interaction between μNS and other PRV proteins was studied by IP and WB (Figure 7). At 4 wpc, μNS was detected as a 70 kDa protein and at the same time-point the structural proteins λ1, μ1, σ1 and σ3 were co-immunoprecipitated. At 5 wpc, μNS was detected in three different sizes ranging from 70 kDa to 83.5 kDa (putative full-length μNS). However, the only structural proteins co-immunoprecipitating with μNS at 5 wpc were σ3 and the 35 kDa fragment of μ1 (see above). Interactions between μNS and other viral proteins were also investigated by liquid chromatography–mass spectrometry (LC–MS; Table 1) and peptides corresponding to λ1, λ2, λ3, μNS, μ1, σNS and σ1 were identified.

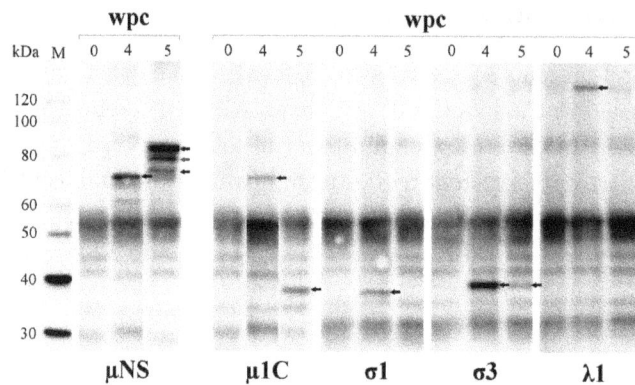

Figure 7. μNS interacts with multiple PRV proteins. Pooled blood cell lysate ($n = 6$) immunoprecipitated with μNS-antiserum, followed by Western blotting with primary antibodies detecting μNS, μ1C, σ1, σ3 and λ1 (arrows). M = molecular weight standard.

Table 1. Identified *piscine orthoreovirus* (PRV) peptides following immunoprecipitation with anti-μNS and mass spectrometry (MS).

* Band Excised from SDS-PAGE (kDa)	Identified PRV Proteins	Unique Peptides	Theoretical PRV Protein Size (kDa)
140 (5 wpc)	μNS	1	83.5
130 (5 wpc)	λ3	2	144.5
	λ2	7	143.7
	λ1	14	141.5
	μNS	9	83.5
	σ1	1	34.6
80 (5 wpc)	λ1	11	141.5
	μNS	24	83.5
70 (4 wpc)	μNS	16	83.5
35 (5 wpc)	μNS	4	83.5
	δ †	3	37.7
	σNS	1	39.1
	σ1	2	34.6

* Approximate size of proteins excised from bands following IP with anti-μNS antisera at four and five weeks post challenge (wpc). † Proteolytic fragment of μ1 proposed in the present work.

3.9. μNS Exists in Two Forms

WB of infected blood cells consistently produced two μNS bands of approximately 83.5 and 70 kDa (Figures 5 and 6). Due to the presence of two translation initiation sites in MRV segment M3, the LC–MS data were analyzed to identify putative shorter variants of the PRV μNS. The peptide distribution along the full-length μNS sequence and their spectrum matches are shown in Figure S7a. The μNS peptides and total spectrum matches obtained from the two bands are shown in Figure S7b. Several N-terminal μNS peptides were identified from the 83.5 kDa band that were not observed in the 70 kDa band. Furthermore, the peptide spectrum matches from the 83.5 kDa and 70 kDa bands in the 200 amino acid N-terminus were 10 to 1, respectively. In contrast, for the remaining C-terminal μNS sequence, the 83.5 kDa and 70 kDa bands produced similar or identical peptide spectrum matches, with a ratio of 66 to 63 (Figure S7). These results point to the presence of a second translation initiation site in the 5'- region of the μNS ORF. Start sites at M_{85}, M_{94}, M_{115} or M_{169} would provide proteins with predicted sizes of 74.5, 73.6, 71.1 and 65.5 kDa, respectively. M_{115} is the most likely candidate due to its size and presence in all PRV strains.

3.10. μ1 Has Two Putative Proteolytic Cleavage Sites

WB targeting the μ1 protein showed that the protein is present in different forms during infection. The putative full-length μ1 (74.1 kDa) was detected at 4 wpc (Figures 6 and 7). In contrast, smaller versions, with estimated sizes of 70, 37 and 32 kDa, replaced the full-length variant at 5 wpc (Figure 6). The three size variants from 5 wpc were subjected to LC–MS analysis (Figure S8). The 70 kDa band most likely represents μ1C following pre-cleavage at $N_{42}P_{43}$ (MW 69.8 kDa). Of the fourteen peptide spectrum matches identified from the 70 kDa band, two were found to overlap $N_{42}P_{43}$ (Figure S8a). This is most likely due to carryover of slightly larger full-length μ1 (74.1 kDa) following gel excision. No peptides stretching N-terminal to $N_{42}P_{43}$ were identified from the 37 and 32 kDa bands. Additional semi-tryptic peptides, i.e., peptides generated by trypsin cleavage at one end but not the other, were identified from both the 37 and 32 kDa bands (Figure S8a). Among these is a peptide identified from the 32 kDa band harboring an N-terminal S_{388}. Cleavage of μ1C at $F_{387}S_{388}$ would yield N- and C-terminal fragments of 37.7 kDa and 32.1 kDa, respectively. The distribution of peptide sequences and peptide spectrum matches provides support for proteolytic cleavage at or close to $F_{387}S_{388}$ (Figure S8). The results suggest that the 37 kDa and 32 kDa bands represent the PRV homologues of MRV μ1 fragments δ and φ, respectively. Besides the μ1 peptide sequences, peptides originating from other PRV proteins with sizes close to the sizes of the three excised fragments, were also identified. Peptide sequences matching λ1 and μNS (one peptide spectrum match each) were identified from the 70 kDa band, sequences matching σ1, σ3 and σNS were identified from the 37 kDa band (2, 2 and 11 peptide spectrum matches, respectively) and σ2 sequences were identified from the 32 kDa band (four peptide spectrum matches).

4. Discussion

Screening of farmed Atlantic salmon has indicated that PRV is ubiquitous in seawater and causes a persistent infection [9,11,41,42]. The study of PRV pathogenesis has been hampered by the lack of susceptible cell lines, and is currently dependent upon in vivo experiments. The fish in this experiment were challenged by cohabitation, i.e., through a natural transmission route. To ensure coordinated onset of infection, a high ratio of shedder fish was used. We found that PRV infection of salmon blood cells is acute and transient, with a peak lasting for 1–2 weeks under these experimental conditions.

Erythrocytes are major target cells for PRV [19]. Piscine erythrocytes are nucleated and contain the transcriptional and translational machinery enabling virus replication both in vivo and ex vivo [16,19]. We detected various PRV proteins in blood cells from 4 wpc, and the amount of protein was reduced at 6 wpc. Innate antiviral gene expression also peaked at 4–6 wpc and all selected genes were significantly induced during the peak period, in line with PRV protein production. In contrast to the transient peak displayed by PRV proteins, the viral RNA levels in blood cells persisted. The viral RNA level though, varied for the targeted genomic segments; the level of M2 (μ1) and M3 (μNS) remained high throughout the trial, while S1 (σ3) transcripts decreased from 6 wpc. TEM analysis corresponded well with viral protein production, i.e., the lamellar structures observed at 4 wpc developed into inclusions containing reovirus-like particles at 5 wpc, while no virus particles could be observed at 7 wpc. The findings support PRV, causing an acute infection in blood cells where high PRV protein and particle production are sustained 1–2 weeks before the infection becomes persistent. Our study shows that, after the acute phase, the PRV RNA level as determined by RT-qPCR does not reflect the virus load in blood.

The salmon does not appear to be able to eliminate PRV. Challenge experiments have shown that PRV RNA can be detected at a steady level in heart and liver until 36 wpc (end of experiment) [41], and in blood for more than a year after challenge [9,19]. In an experiment where the infectious potential of persistently PRV infected Atlantic salmon was studied, sentinel fish were added at 59 wpc, but no transmission to the sentinel fish was observed [9]. This indicates that fish persistently infected with PRV do not continuously shed the virus. Viral persistence is common in fish and has been demonstrated for several RNA viruses [9,43–47]. The only PRV protein that could be detected after the peak of virus protein production was a fragment of μ1, suggesting a possible role for this protein in persistent infection. In farmed salmon, where the size of the population in a net pen may exceed a hundred thousand individuals, and in the whole farm be more than a million fish, viral persistence in the population, but not necessarily in the individual, is also a critical parameter.

PRV infection in erythrocytes has previously been shown to induce expression of type I interferon and interferon-regulated genes [16,18]. In this study, the level of viral RNA correlated with the innate antiviral response in individual fish, with the exception of M3 expression after the virus peak (6–8 wpc). The continuous production of M3 RNA indicates that the innate antiviral immune response primarily inhibits virus replication post transcriptionally, which is in line with the functions of PKR and ISG15 on translation and protein modification, respectively [48,49].

Orthoreoviruses generate viral factories in the cytoplasm of infected cells [21,27,50–52], and PRV forms cytoplasmic globular viral factories resembling the structures produced by MRV T3D [16,19,31]. Viral factories are structures where virus replication and assembly occur, and thus where the viral proteins co-localize. The secluded nature of the viral factories modulates the level of the innate antiviral immune response. The orthoreoviral protein μNS is orchestrating the construction of the factories and in this study and earlier studies we have found that λ1, λ2, λ3, μ1, σNS, σ1, σ2 and σ3 interact with μNS [31]. The σ3 protein co-precipitated with μNS but was not identified by MS, however WB can be more sensitive than LC–MS [52]. This suggests that μNS interacts directly or indirectly with all three λ-proteins, the μ1 protein, and possibly all four σ-proteins. The μNS protein was detected in different molecular sizes at specific time points. Further investigations led to the finding of four possible internal translation initiation sites in the μNS gene. The M_{115} residue was determined to be the best candidate as M_{94} is not conserved among all PRV isolates, and M_{85} and M_{169} are unlikely due to the sizes of the proteins generated. Post-translational cleavage to generate μNSC as shown for ARV μNS cannot be excluded, although the specific proteolytic cleavage site in the ARV protein is not conserved in PRV [2,33,53]. The different μNS size variants, i.e., full-length μNS and the 70 kDa variant with putative translation initiation at M_{115}, may differ in their interactions with other PRV proteins. At 4 wpc, when only the 70 kDa variant of μNS was detected following IP, all targeted structural proteins co-precipitated. However, at 5 wpc, when full-length μNS was dominant, only the σ3 protein and the assumed μ1 fragment δ co-precipitated. Studies previously performed on aquareoviruses and ARV indicate that recruitment of viral proteins into viral factories occurs in a predefined order through direct or indirect association with μNS [50,54].

Four different molecular sizes of the μ1 protein were observed in the infected blood cells. Previous multiple sequence alignments of the μ1 amino acid sequence showed absolute conservation of the G2-myristoylation site and the autolytic $N_{42}P_{43}$ cleavage site, both regarded as crucial for reovirus μ1-mediated membrane penetration [2]. The band observed at 4 wpc represents the full-length μ1 protein while the 70 kDa band at 5 wpc most likely represents μ1C.

Although peptides containing amino acid sequences overlapping the $N_{42}P_{43}$ site were observed from the 70 kDa band following LC–MS, peptides ending in P_{43} were present in equal amount. We conclude that the presence of the $N_{42}P_{43}$ overlapping peptides originate from carryover of the slightly larger full-length μ1 following gel excision. In addition, proteins can exhibit different abilities to separate in SDS-PAGE. This explains the presence of a minor fraction of peptides from the δ fragment (i.e., 37 kDa band) in the φ fragment (i.e., 32 kDa band) and vice versa. No peptide sequences overlapping $N_{42}P_{43}$ were identified from the 37 and 32 kDa bands. Rather, a higher number of peptides with an N-terminal P_{43} generated by non-tryptic cleavage were identified, providing additional support for cleavage at $N_{42}P_{43}$.

MRV μ1 contains a second cleavage site in its C-terminal region which, upon cleavage by exogenous proteases, generates the additional fragments δ and φ [55]. In the present study, we propose that the 37 kDa and 32 kDa bands represent the PRV homologues of the MRV δ and φ proteins. Hence, PRV φ contains a larger N-terminal portion of μ1 compared to MRV φ. Although there is only 28% identity at the amino acid level [2], the secondary structure of the PRV μ1 monomer predicted by

PSIPRED [39] (not shown) is very similar to that of MRV μl [56]. This includes the helix-rich region in the C-terminal end [57], which for MRV largely constitutes the φ fragment shown to be crucial for membrane penetration, apoptosis induction and intracellular localization [57,58]. An interesting observation is that the three PRV μ1 peptide sequences detected in the 35 kDa band following IP with anti-μNS were all N-terminal to the proposed φ region, suggesting that μNS-interacting sites on μ1 may be located in the proposed δ region, between P_{43} and F_{387}. From 7 wpc and onwards, the only PRV protein detected was a ~35 kDa protein which could represent the δ proteolytic fragment.

Production-related diseases are often multifactorial and the outcome of a PRV infection is influenced by viral strain, age of the fish, production and environmental factors. Recently, PRV was demonstrated to be the etiologic agent of EIBS, causing anemia and mass mortality in juvenile Coho salmon [7]. The level of anemia in EIBS corresponded well with the level of viral replication in blood and it is therefore tempting to suggest that EIBS is a consequence of acute PRV infection, i.e., the direct effect of virus PRV replication in erythrocytes. PRV is also the causative agent of HSMI [1,4], which appears 2–3 weeks after virus replication peaks in blood cells. The dominance of CD8 positive inflammatory cells found in the HSMI specific heart lesions indicates that immune mediated mechanisms are a major cause of the myocarditis.

In this study, we show that PRV infection has an acute phase in blood cells with high virus production before the infection subsides to a low persistent level. The continued transcription of viral RNA in the persistent phase suggests that the innate antiviral immune response may act to inhibit the virus infection post transcriptionally.

Supplementary Materials
Figure S1: Specificity of μNS and λ1 antisera; Figure S2: PRV RNA load in blood cells (second challenge experiment); Figure S3: Expression of immune genes in blood cells; Figure S4: Expression of EF1α during PRV infection; Figure S5: PRV σ1 positive blood cells detected by flow cytometry; Figure S6: Presence of PRV proteins in blood cells (second challenge experiment); Figure S7: LC–MS analyses of PRV μNS; Figure S8: LC–MS analyses of PRV μ1; Table S1: Primers and probes used for construction of plasmids and expression of viral RNA levels; Table S2: Primers used for analysis of antiviral gene expression.

Acknowledgments: The Research Council of Norway supported the research with grant #237315/E40 and #235788. We would also like to thank Elisabeth Furuseth Hansen and Ida Aksnes from the University of Life Sciences, and Elisabeth Dahl Nybø at the Norwegian Veterinary Institute (NVI) for technical and scientific assistance.

Author Contributions: H.M.H.: Study design (parts), experiments, analysis, interpretation of data, drafting, revising and approving the manuscript. Ø.W.: Flow cytometry experiment (parts), analysis and interpretation of data, revising and approving the manuscript. T.M.: Data analysis and interpretation, writing, revising and approving the manuscript. M.L.: Challenge experiments, revising and approving the manuscript. B.T.: LC–MS experiments and data analysis, revising and approving the manuscript. I.B.N.: Antibody production, revising and approving the manuscript. S.B.: Flow cytometry experiment (parts), interpretation of data, revising and approving the manuscript. M.K.D.: Analysis, interpretation of data, drafting (parts), revising and approving the manuscript. E.R.: Study design, analysis, interpretation of data, drafting, revising and approving the manuscript.

References

1.   Palacios, G.; Lovoll, M.; Tengs, T.; Hornig, M.; Hutchison, S.; Hui, J.; Kongtorp, R.T.; Savji, N.; Bussetti, A.V.; Solovyov, A.; et al. Heart and skeletal muscle inflammation of farmed salmon is associated with infection with a novel reovirus. *PLoS ONE* **2010**, *5*, e11487. [CrossRef] [PubMed]

2.   Markussen, T.; Dahle, M.K.; Tengs, T.; Lovoll, M.; Finstad, O.W.; Wiik-Nielsen, C.R.; Grove, S.; Lauksund, S.; Robertsen, B.; Rimstad, E. Sequence analysis of the genome of piscine orthoreovirus (PRV) associated with heart and skeletal muscle inflammation (HSMI) in Atlantic salmon (*Salmo salar*). *PLoS ONE* **2013**, *8*, e70075. [CrossRef]

3.   Day, J.M. The diversity of the orthoreoviruses: Molecular taxonomy and phylogentic divides. *Infect. Genet. Evol.* **2009**, *9*, 390–400. [CrossRef] [PubMed]

4. Finstad, O.W.; Falk, K.; Lovoll, M.; Evensen, O.; Rimstad, E. Immunohistochemical detection of piscine reovirus (PRV) in hearts of Atlantic salmon coincide with the course of heart and skeletal muscle inflammation (HSMI). *Vet. Res* **2012**, *43*, 27. [CrossRef] [PubMed]

5. Olsen, A.B.; Hjortaas, M.; Tengs, T.; Hellberg, H.; Johansen, R. First description of a new disease in Rainbow Trout (*Oncorhynchus mykiss* (Walbaum)) similar to Heart and skeletal muscle inflammation (HSMI) and detection of a gene sequence related to Piscine Orthoreovirus (PRV). *PLoS ONE* **2015**, *10*, e0131638. [CrossRef] [PubMed]

6. Sibley, S.D.; Finley, M.A.; Baker, B.B.; Puzach, C.; Armien, A.G.; Giehtbrock, D.; Goldberg, T.L. Novel reovirus associated with epidemic mortality in wild Largemouth Bass (*Micropterus salmoides*). *J. Gen. Virol.* **2016**, *97*, 2482–2487. [PubMed]

7. Takano, T.; Nawata, A.; Sakai, T.; Matsuyama, T.; Ito, T.; Kurita, J.; Terashima, S.; Yasuike, M.; Nakamura, Y.; Fujiwara, A.; et al. Full-Genome sequencing and confirmation of the causative agent of Erythrocytic inclusion body syndrome in Coho Salmon identifies a new type of Piscine Orthoreovirus. *PLoS ONE* **2016**, *11*, e0165424. [CrossRef] [PubMed]

8. Kibenge, M.J.; Iwamoto, T.; Wang, Y.; Morton, A.; Godoy, M.G.; Kibenge, F.S. Whole-genome analysis of piscine reovirus (PRV) shows PRV represents a new genus in family Reoviridae and its genome segment S1 sequences group it into two separate sub-genotypes. *Virol. J.* **2013**, *10*, 230. [CrossRef] [PubMed]

9. Garver, K.A.; Johnson, S.C.; Polinski, M.P.; Bradshaw, J.C.; Marty, G.D.; Snyman, H.N.; Morrison, D.B.; Richard, J. Piscine Orthoreovirus from Western North America is transmissible to Atlantic Salmon and Sockeye Salmon but fails to cause Heart and skeletal muscle inflammation. *PLoS ONE* **2016**, *11*, e0146229. [CrossRef] [PubMed]

10. Ferguson, H.W.; Kongtorp, R.T.; Taksdal, T.; Graham, D.; Falk, K. An outbreak of disease resembling heart and skeletal muscle inflammation in Scottish farmed salmon, *Salmo salar* L., with observations on myocardial regeneration. *J. Fish Dis.* **2005**, *28*, 119–123. [CrossRef] [PubMed]

11. Lovoll, M.; Alarcon, M.; Bang Jensen, B.; Taksdal, T.; Kristoffersen, A.B.; Tengs, T. Quantification of Piscine reovirus (PRV) at different stages of Atlantic salmon *Salmo salar* production. *Dis. Aquat. Organ.* **2012**, *99*, 7–12. [CrossRef] [PubMed]

12. Bjorgen, H.; Wessel, O.; Fjelldal, P.G.; Hansen, T.; Sveier, H.; Saebo, H.R.; Enger, K.B.; Monsen, E.; Kvellestad, A.; Rimstad, E.; et al. *Piscine orthoreovirus* (PRV) in red and melanised foci in white muscle of Atlantic salmon (*Salmo salar*). *Vet. Res.* **2015**, *46*, 89. [CrossRef] [PubMed]

13. Kongtorp, R.T.; Halse, M.; Taksdal, T.; Falk, K. Longitudinal study of a natural outbreak of heart and skeletal muscle inflammation in Atlantic salmon, *Salmo salar* L. *J. Fish Dis.* **2006**, *29*, 233–244. [CrossRef] [PubMed]

14. Wiik-Nielsen, C.R.; Ski, P.M.; Aunsmo, A.; Lovoll, M. Prevalence of viral RNA from piscine reovirus and piscine myocarditis virus in Atlantic salmon, *Salmo salar* L., broodfish and progeny. *J. Fish Dis.* **2012**, *35*, 169–171. [CrossRef] [PubMed]

15. Johansen, L.H.; Dahle, M.K.; Wessel, O.; Timmerhaus, G.; Lovoll, M.; Rosaeg, M.; Jorgensen, S.M.; Rimstad, E.; Krasnov, A. Differences in gene expression in Atlantic salmon parr and smolt after challenge with *Piscine orthoreovirus* (PRV). *Mol. Immunol.* **2016**, *73*, 138–150. [CrossRef] [PubMed]

16. Wessel, O.; Olsen, C.M.; Rimstad, E.; Dahle, M.K. *Piscine orthoreovirus* (PRV) replicates in Atlantic salmon (*Salmo salar* L.) erythrocytes ex vivo. *Vet. Res.* **2015**, *46*, 26. [CrossRef] [PubMed]

17. Morera, D.; Roher, N.; Ribas, L.; Balasch, J.C.; Donate, C.; Callol, A.; Boltana, S.; Roberts, S.; Goetz, G.; Goetz, F.W.; et al. RNA-Seq reveals an integrated immune response in nucleated erythrocytes. *PLoS ONE* **2011**, *6*, e26998. [CrossRef] [PubMed]

18. Dahle, M.K.; Wessel, O.; Timmerhaus, G.; Nyman, I.B.; Jorgensen, S.M.; Rimstad, E.; Krasnov, A. Transcriptome analyses of Atlantic salmon (*Salmo salar* L.) erythrocytes infected with piscine orthoreovirus (PRV). *Fish Shellfish Immunol.* **2015**, *45*, 780–790. [CrossRef] [PubMed]

19. Finstad, O.W.; Dahle, M.K.; Lindholm, T.H.; Nyman, I.B.; Lovoll, M.; Wallace, C.; Olsen, C.M.; Storset, A.K.; Rimstad, E. *Piscine orthoreovirus* (PRV) infects Atlantic salmon erythrocytes. *Vet. Res.* **2014**, *45*, 35. [CrossRef] [PubMed]

20. Kongtorp, R.T.; Taksdal, T.; Lyngoy, A. Pathology of heart and skeletal muscle inflammation (HSMI) in farmed Atlantic salmon *Salmo salar*. *Dis. Aquat. Organ.* **2004**, *59*, 217–224. [CrossRef] [PubMed]

21. Netherton, C.; Moffat, K.; Brooks, E.; Wileman, T. A guide to viral inclusions, membrane rearrangements, factories, and viroplasm produced during virus replication. *Adv. Virus Res.* **2007**, *70*, 101–182. [PubMed]

22. Knipe, D.M.; Howley, P.M. *Fields Virology*, 5th ed.; Wolters Kluwer/Lippincott Williams & Wilkins Health: Philadelphia, PA, USA, 2007; pp. 1854–1858.

23. Lee, P.W.; Hayes, E.C.; Joklik, W.K. Protein sigma 1 is the reovirus cell attachment protein. *Virology* **1981**, *108*, 156–163. [CrossRef]

24. Nibert, M.L.; Fields, B.N. A carboxy-terminal fragment of protein mu 1/mu 1C is present in infectious subvirion particles of mammalian reoviruses and is proposed to have a role in penetration. *J. Virol.* **1992**, *66*, 6408–6418. [PubMed]

25. Thete, D.; Snyder, A.J.; Mainou, B.A.; Danthi, P. Reovirus mu1 protein affects infectivity by altering virus-receptor interactions. *J. Virol.* **2016**, *90*, 10951–10962. [CrossRef] [PubMed]

26. Becker, M.M.; Peters, T.R.; Dermody, T.S. Reovirus σNS and μNS proteins form cytoplasmic inclusion structures in the absence of viral infection. *J. Virol.* **2003**, *77*, 5948–5963. [CrossRef] [PubMed]

27. Schiff, L.A.; Nibert, M.L.; Tyler, K.L. Orthoreoviruses and their replication. In *Fields virology*, 5th ed.; Knipe, D.M., Howley, P.M., Fields, B.N., Eds.; Wolters Kluwer/Lippincott Williams & Wilkins: Philadelphia, PA, USA, 2007; Volume 2, pp. 1853–1915.

28. Jayasuriya, A.K.; Nibert, M.L.; Fields, B.N. Complete nucleotide sequence of the M2 gene segment of reovirus type 3 dearing and analysis of its protein product mu 1. *Virology* **1988**, *163*, 591–602. [CrossRef]

29. Duncan, R. The low pH-dependent entry of avian reovirus is accompanied by two specific cleavages of the major outer capsid protein mu 2C. *Virology* **1996**, *219*, 179–189. [CrossRef] [PubMed]

30. Wiener, J.R.; Joklik, W.K. Evolution of reovirus genes: A comparison of serotype 1, 2, and 3 M2 genome segments, which encode the major structural capsid protein mu 1C. *Virology* **1988**, *163*, 603–613. [CrossRef]

31. Haatveit, H.M.; Nyman, I.B.; Markussen, T.; Wessel, O.; Dahle, M.K.; Rimstad, E. The non-structural protein μNS of piscine orthoreovirus (PRV) forms viral factory-like structures. *Vet. Res.* **2016**, *47*, 5. [CrossRef] [PubMed]

32. Parker, J.S.; Broering, T.J.; Kim, J.; Higgins, D.E.; Nibert, M.L. Reovirus core protein mu2 determines the filamentous morphology of viral inclusion bodies by interacting with and stabilizing microtubules. *J. Virol.* **2002**, *76*, 4483–4496. [CrossRef] [PubMed]

33. Busch, L.K.; Rodriguez-Grille, J.; Casal, J.I.; Martinez-Costas, J.; Benavente, J. Avian and mammalian reoviruses use different molecular mechanisms to synthesize their microNS isoforms. *J. Gen. Virol.* **2011**, *92*, 2566–2574. [CrossRef] [PubMed]

34. Touris-Otero, F.; Martinez-Costas, J.; Vakharia, V.N.; Benavente, J. Avian reovirus nonstructural protein microNS forms viroplasm-like inclusions and recruits protein sigmaNS to these structures. *Virology* **2004**, *319*, 94–106. [CrossRef] [PubMed]

35. Wiener, J.R.; Bartlett, J.A.; Joklik, W.K. The sequences of reovirus serotype 3 genome segments M1 and M3 encoding the minor protein mu 2 and the major nonstructural protein mu NS, respectively. *Virology* **1989**, *169*, 293–304. [CrossRef]

36. McCutcheon, A.M.; Broering, T.J.; Nibert, M.L. Mammalian reovirus M3 gene sequences and conservation of coiled-coil motifs near the carboxyl terminus of the microNS protein. *Virology* **1999**, *264*, 16–24. [CrossRef] [PubMed]

37. Koehler, C.J.; Bollineni, R.C.; Thiede, B. Application of the half decimal place rule to increase the peptide identification rate. *Rapid Commun. Mass Spectrom.* **2016**, *31*, 227–233. [CrossRef] [PubMed]

38. ExPASy Bioinformatics Resource Portal. Available online: http://web.expasy.org/compute_pi/ (accessed on 1 April 2016).

39. UCL Department Of Computer Science. Available online: http://bioinf.cs.ucl.ac.uk/psipred/ (accessed on 1 April 2016).

40. Su, J.; Zhang, R.; Dong, J.; Yang, C. Evaluation of internal control genes for qRT-PCR normalization in tissues and cell culture for antiviral studies of grass carp (*Ctenopharyngodon idella*). *Fish Shellfish Immunol.* **2011**, *30*, 830–835. [CrossRef] [PubMed]

41. Lovoll, M.; Wiik-Nielsen, J.; Grove, S.; Wiik-Nielsen, C.R.; Kristoffersen, A.B.; Faller, R.; Poppe, T.; Jung, J.; Pedamallu, C.S.; Nederbragt, A.J.; et al. A novel totivirus and piscine reovirus (PRV) in Atlantic salmon (*Salmo salar*) with cardiomyopathy syndrome (CMS). *Virol. J.* **2010**, *7*, 309. [CrossRef] [PubMed]

42. Marty, G.D.; Morrison, D.B.; Bidulka, J.; Joseph, T.; Siah, A. Piscine reovirus in wild and farmed salmonids in British Columbia, Canada: 1974–2013. *J. Fish Dis.* **2015**, *38*, 713–728. [CrossRef] [PubMed]

43. Julin, K.; Johansen, L.H.; Sommer, A.I.; Jorgensen, J.B. Persistent infections with infectious pancreatic necrosis virus (IPNV) of different virulence in Atlantic salmon, *Salmo salar* L. *J. Fish Dis.* **2015**, *38*, 1005–1019. [CrossRef] [PubMed]

44. Gjessing, M.C.; Kvellestad, A.; Ottesen, K.; Falk, K. Nodavirus provokes subclinical encephalitis and retinochoroiditis in adult farmed Atlantic cod, *Gadus morhua* L. *J. Fish Dis.* **2009**, *32*, 421–431. [CrossRef] [PubMed]

45. Amend, D.F. Detection and transmission of infectious hematopoietic necrosis virus in rainbow trout. *J. Wildl. Dis.* **1975**, *11*, 471–478. [CrossRef] [PubMed]

46. Neukirch, M. Demonstration of persistent viral haemorrhagic septicaemia (VHS) virus in rainbow trout after experimental waterborne infection. *Zentralbl Veterinarmed B* **1986**, *33*, 471–476. [CrossRef] [PubMed]

47. Hershberger, P.K.; Gregg, J.L.; Grady, C.A.; Taylor, L.; Winton, J.R. Chronic and persistent viral hemorrhagic septicemia virus infections in Pacific herring. *Dis. Aquat. Organ.* **2010**, *93*, 43–49. [CrossRef] [PubMed]

48. Dalet, A.; Gatti, E.; Pierre, P. Integration of PKR-dependent translation inhibition with innate immunity is required for a coordinated anti-viral response. *FEBS Lett.* **2015**, *589*, 1539–1545. [CrossRef] [PubMed]

49. Durfee, L.A.; Lyon, N.; Seo, K.; Huibregtse, J.M. The ISG15 conjugation system broadly targets newly synthesized proteins: Implications for the antiviral function of ISG15. *Mol. Cell* **2010**, *38*, 722–732. [CrossRef] [PubMed]

50. Touris-Otero, F.; Cortez-San Martin, M.; Martinez-Costas, J.; Benavente, J. Avian reovirus morphogenesis occurs within viral factories and begins with the selective recruitment of sigmaNS and lambdaA to microNS inclusions. *J. Mol. Biol.* **2004**, *341*, 361–374. [CrossRef] [PubMed]

51. Carroll, K.; Hastings, C.; Miller, C.L. Amino acids 78 and 79 of Mammalian Orthoreovirus protein microNS are necessary for stress granule localization, core protein lambda 2 interaction, and de novo virus replication. *Virology* **2014**, *448*, 133–145. [CrossRef] [PubMed]

52. Aebersold, R.; Burlingame, A.L.; Bradshaw, R.A. Western blots versus selected reaction monitoring assays: time to turn the tables? *Mol. Cell. Proteomics* **2013**, *12*, 2381–2382. [CrossRef] [PubMed]

53. Rodriguez-Grille, J.; Busch, L.K.; Martinez-Costas, J.; Benavente, J. Avian reovirus-triggered apoptosis enhances both virus spread and the processing of the viral nonstructural muNS protein. *Virology* **2014**, *462–463*, 49–59. [CrossRef] [PubMed]

54. Yan, L.; Zhang, J.; Guo, H.; Yan, S.; Chen, Q.; Zhang, F.; Fang, Q. Aquareovirus NS80 initiates efficient viral replication by retaining core proteins within replication-associated viral inclusion bodies. *PLoS ONE* **2015**, *10*, e0126127. [CrossRef] [PubMed]

55. Nibert, M.L.; Odegard, A.L.; Agosto, M.A.; Chandran, K.; Schiff, L.A. Putative autocleavage of reovirus mu1 protein in concert with outer-capsid disassembly and activation for membrane permeabilization. *J. Mol. Biol.* **2005**, *345*, 461–474. [CrossRef] [PubMed]

56. Liemann, S.; Chandran, K.; Baker, T.S.; Nibert, M.L.; Harrison, S.C. Structure of the reovirus membrane-penetration protein, Mu1, in a complex with is protector protein, Sigma3. *Cell* **2002**, *108*, 283–295. [CrossRef]

57. Coffey, C.M.; Sheh, A.; Kim, I.S.; Chandran, K.; Nibert, M.L.; Parker, J.S. Reovirus outer capsid protein micro1 induces apoptosis and associates with lipid droplets, endoplasmic reticulum, and mitochondria. *J. Virol.* **2006**, *80*, 8422–8438. [CrossRef] [PubMed]

58. Danthi, P.; Coffey, C.M.; Parker, J.S.; Abel, T.W.; Dermody, T.S. Independent regulation of reovirus membrane penetration and apoptosis by the mu1 phi domain. *PLoS Pathog.* **2008**, *4*, e1000248. [CrossRef] [PubMed]

2

A Chinese Variant Marek's Disease Virus Strain with Divergence between Virulence and Vaccine Resistance

Guo-rong Sun [†], Yan-ping Zhang [†], Hong-chao Lv, Lin-yi Zhou, Hong-yu Cui, Yu-long Gao, Xiao-le Qi, Yong-qiang Wang, Kai Li, Li Gao, Qing Pan, Xiao-mei Wang * and Chang-jun Liu *

Division of Avian Immunosuppressive Diseases, State Key Laboratory of Veterinary Biotechnology, Harbin Veterinary Research Institute, Chinese Academy of Agricultural Sciences, Harbin 150069, China; sgrshenhua@hotmail.com (G.-r.S.); zhyp_77@hvri.ac.cn (Y.-p.Z.); lv6739533@163.com (H.-c.L.); zlyi123321@126.com (L.-y.Z.); cuihy@hvri.ac.cn (H.-y.C.); ylg@hvri.ac.cn (Y.-l.G.); qxl@hvri.ac.cn (X.-l.Q.); yqw@hvri.ac.cn (Y.-q.W.); likaihvri@163.com (K.L.); gaoli0820@163.com (L.G.); panqing20050101@126.com (Q.P.)
* Correspondence: xmw@hvri.ac.cn (X.-m.W.); liucj93711@hvri.ac.cn (C.-j.L.)

† These authors contributed equally to this study.

Academic Editor: Joanna Parish

Abstract: Marek's disease (MD) virus (MDV) has been evolving continuously, leading to increasing vaccination failure. Here, the MDV field strain BS/15 was isolated from a severely diseased Chinese chicken flock previously vaccinated with CVI988. To explore the causes of vaccination failure, specific-pathogen free (SPF) chickens vaccinated with CVI988 or 814 and unvaccinated controls were challenged with either BS/15 or the reference strain Md5. Both strains induced MD lesions in unvaccinated chickens with similar mortality rates of 85.7% and 80.0% during the experimental period, respectively. However, unvaccinated chickens inoculated with BS/15 exhibited a higher tumor development rate (64.3% vs. 40.0%), but prolonged survival and diminished immune defects compared to Md5-challenged counterparts. These results suggest that BS/15 and Md5 show a similar virulence but manifest with different pathogenic characteristics. Moreover, the protective indices of CVI988 and 814 were 33.3 and 66.7 for BS/15, and 92.9 and 100 for Md5, respectively, indicating that neither vaccine could provide efficient protection against BS/15. Taken together, these data suggest that MD vaccination failure is probably due to the existence of variant MDV strains with known virulence and unexpected vaccine resistance. Our findings should be helpful for understanding the pathogenicity and evolution of MDV strains prevalent in China.

Keywords: Marek's disease virus; vaccination failure; virulence; pathogenicity; vaccine efficacy; evolution

1. Introduction

Marek's disease (MD) is a lymphoproliferative disease in chickens that has caused considerable economic losses to the commercial poultry industry worldwide. The MD virus (MDV) is the etiological agent of MD. It is an oncogenic alpha-herpesvirus that belongs to the genus *Mardivirus* and *Gallid herpesviruses* species, which includes *Gallid herpesvirus* type 2 (GaHV-2; represented as MDV in this manuscript), GaHV-3, and *Meleagrid herpesvirus* type 1 (MeHV-1; represented as HVT in this manuscript) [1]. However, GaHV-3 and MeHV-1 are non-oncogenic, with only MDV causing the disease in susceptible hosts. Moreover, MDV can be further classified into four pathotypes, including mild (m), virulent (v), very virulent (vv), and very virulent plus (vv+) MDV strains, based on the induction of MD lesions in unvaccinated and vaccinated chickens, in line with the

MDV pathotyping assay developed at the Avian Disease and Oncology Laboratory (ADOL) of the United States Department of Agriculture [2,3]. The virulence and vaccine resistance of the MDV strains are generally associated, i.e., the more virulent MDV strains are more likely to induce more serious MD lesions in susceptible chickens, with the associated vaccines less likely to provide protection.

Vaccinations are the primary approach used for controlling MD in chickens, and three types of vaccines have been developed against MD, including HVT, nonpathogenic GaHV-3, and attenuated GaHV-2 [4–6]. MD has been successfully controlled by vaccination, significantly reducing economic losses to the domestic poultry industry [7,8]. However, the vaccines used for MD control cannot induce sterile immunity and allow for replication and transmission of virulent MDV strains in a live host, leading to complicated interactions between the pathogens, vaccines, and hosts [9–12]. Thus, MDV has continued to evolve and obtain enhanced virulence over the last few decades owing to the widespread use of these vaccines. The relationship between increased MDV virulence and the introduction of different vaccines was summarized as a step-wise evolution, in which MDV strains have shown continuous evolution to maintain virulence, acquiring the ability to overcome immune responses induced by vaccines [13]. Successive generations of MD vaccines have been introduced to protect birds from increasingly virulent MDV strains; however, the virus has countered each new vaccine.

In China, MDV was first reported in the 1970s. At present, laying and breeding chickens are vaccinated with vaccine CVI988 (an effective attenuated GaHV-2 vaccine used commercially worldwide), 814 (an effective attenuated GaHV-2 vaccine used widely in China), or a bivalent vaccine (CVI988 plus HVT) via subcutaneous or intramuscular injection at the time of hatching in the vast majority of Chinese poultry enterprises. The economical HVT vaccine is normally used to vaccinate commercial meat-producing chickens, which are typically raised for only 7–8 weeks and have a lower risk for MD. Although the use of vaccines has effectively controlled MDV, MD vaccination failure cases have occasionally occurred, although little is known about the causes of vaccination failure.

In this study, we isolated a field MDV strain BS/15 from severely diseased chickens that were vaccinated with the commercial vaccine CVI988. Furthermore, the pathogenicity of BS/15 to specific pathogen-free (SPF) chickens was analyzed, and the protective efficacy of vaccine CVI988 and vaccine 814 against BS/15 was evaluated. Our study was created in order to increase understanding of the pathogenic characteristics of the newly isolated MDV strain, the causes of MD vaccination failure, and the evolution of MDV, as well as providing guidance for MD control.

2. Materials and Methods

2.1. Collection of Clinical Samples

A layer flock in Baishan City of the Jilin Province in China developed serious MD in 2015. Chickens had been vaccinated with the commercial vaccine CVI988 at 1 day of age, and clinical symptoms of MD were observed beginning at about 55 days of age. The incidence of MD in the chicken flocks quickly reached about 36% by 120 days of age. Visible tumors were frequently found in the visceral organs of dead or diseased chickens by postmortem examination. For molecular diagnosis and viral isolation, feather pulps were collected from the chickens with suspected MD.

2.2. Viral Isolation and Identification

Feather pulps collected from the diseased chickens described above were used for viral isolation, as previously described [14]. Briefly, SPF duck or chicken embryos were obtained from the Harbin Veterinary Research Institute (HVRI), Chinese Academy of Agricultural Sciences (CAAS) for the preparation of duck embryo fibroblasts (DEFs) or chicken embryo fibroblasts (CEFs). Feather pulp-inoculated primary DEFs were incubated at 37 °C in an atmosphere containing 5% CO_2, and blind passages were performed until cytopathogenic effects (CPEs) were observed. Viruses from infected cells with CPEs were plaque- purified in DEFs, and then the viruses were propagated and stored in liquid nitrogen.

Viruses were identified by a polymerase chain reaction (PCR)-based method targeting the *meq* gene and genomic 132-base pair repeat sequence (132bpr) of MDV, which we used to clearly distinguish between the vaccine and pathogenic wild-type MDV strains in a previous study [15]. DNA samples from inoculated CEFs were used as PCR templates, and the PCR primers used for MDV strain identification are listed in Table 1. Furthermore, an indirect immunofluorescence assay (IFA) was carried out to identify the viral plaques in CEFs using monoclonal antibodies, which were produced by our laboratory and are specific for the gI protein of MDV.

Table 1. Polymerase chain reaction (PCR) primers for Marek's disease virus (MDV) strain identification.

Target	Primer Sequence	Product Size (bp)
meq	F: 5′-GGGAAATGACAGGTGAATTGTG-3′ R: 5′-TAAGGAAAATTTGTTACCCCAG-3′	1403/1580 [a]
132bpr	F: 5′-TGCGATGAAAGTGCTATGGAGG-3′ R: 5′-GAGAATCCCTATGAGAAAGCGC-3′	316–844 [b]

[a] Exact size is strain-dependent, based on inclusion of a 177-bp insertion; [b] Exact size depends on 132-base pair repeat sequence (132bpr) copy number.

2.3. Screening for Causative Agents

Due to the multiple causes of oncosis in Chinese chicken flocks, it is necessary to detect and then cull chickens infected with one of several causative agents of oncogenicity. To detect avian leucosis virus (ALV), PCR and enzyme-linked immunosorbent assays (ELISAs) were performed, as previously described [16,17]. Additionally, PCR and IFA were used to detect reticuloendotheliosis virus (REV), as previously described [16,18]. Finally, a previously described PCR method was used to detect chicken infectious anemia virus (CIAV) [19].

2.4. Animal Experiments

The MDV strain BS/15 was used as a challenge virus, and the standard vv MDV strain Md5 was used as a reference strain. The CVI988 vaccine and 814 vaccine purchased from commercial companies were used in animal experiments to assess vaccine efficacy.

In total, 105 one-day-old SPF White Leghorn chickens were obtained from the Experimental Animal Center (EAC) of HVRI. The birds were housed in negative-pressure-filtered air isolators and were randomly divided into seven groups (n = 15 birds each). Vaccination was performed on day 1 and both vaccines were used at a dose of 2000 plaque forming units (PFU) in 200 μL dedicated diluent provided by the manufacturers, being administered via the subcutaneous route (groups 3 and 6: CVI988; groups 4 and 7: 814). In addition, groups 1, 2, 3, 5, and 6 were inoculated with 200 μL diluent in the same manner. On day 7 post-vaccination, MDV challenge was performed via the intra-abdominal route with 1000 PFU of MDV in 200-μL diluent (groups 2, 3, and 4: Md5; groups 5, 6, and 7: BS/15). Chickens of group 1 received the same amount of diluent in the same manner and served as controls.

In this study, the birds were observed daily for clinical signs of MD, and feather pulps were randomly plucked from five birds at 4, 7, 14, 21, 28, 35, 42, and 90 days post-challenge (dpc; n = 2 in the BS/15-challenged group, and n = 3 in the Md5-challenged control group at 90 dpc). In addition, the body weights and immune organ (bursa, thymus, and spleen) weights of the surviving chickens at the end-point of the experiment (90 dpc) were recorded for further analysis. The MD status of the experimental animals was estimated mainly by monitoring for early mortality syndrome, immune organ damage, and tumor formation, as described previously [14]. The days that various chickens died during the experimental period in each group were recorded for survival analysis.

The birds were randomly assigned to each group and numbered by workers at the EAC, HVRI. To ensure a blinded study, the group number was only known by the workers who dealt directly with the researchers. All animal experiments were approved by the ethical review board of HVRI, CAAS, and performed in accordance with approved animal care guidelines and protocols (approval number: SYXK (Heilongjiang) 2011022). These standard procedures included using as few animals as possible (n = 15 in each group); ensuring that the workers of the EAC of HVRI were trained strictly, in accordance with animal policy, and complied with all standard operating procedures; the feeding environment was clean, spacious, and comfortable, allowing the animals to move and feed freely; the poultry feed was nutritionally balanced, and the drinking water was fresh and clean; during the experimental operation, the treatments were applied using gentle movements, to avoid frightening the animals; and at the end of the experiment, the animals were euthanized immediately.

2.5. DNA Extraction and TaqMan Real-Time Polymerase Chain Teaction (qPCR)

The collected feather pulps were homogenized in phosphate buffer solution (PBS), and DNA was extracted using the AxyPrep Body Fluid Viral DNA/RNA Miniprep Kit (Corning Life Sciences Co., Ltd., Suzhou, China) according to the manufacturer's instructions. The MDV meq gene and the chicken ovotransferrin gene were used as a real-time PCR (qPCR) target gene in the MDV genome and an internal reference gene in the host cell genome, respectively, and qPCR detection was performed as previously described [20,21].

2.6. Statistical Analysis

The vaccine protective index (PI) was calculated as previously described [22], using the following formula: PI = ((%MD in unvaccinated chickens; %MD in vaccinated chickens))/%MD in unvaccinated chickens) × 100. The absolute numbers of the MDV genome per million cells from the collected feather pulps were calculated based on the standard curves generated and were normalized to the viral load. The normalized viral load was calculated using the formula: normalized viral load = $\log_{10}$ ((MDV genome copy number/chicken genome copy number) × 10^6). Viral load, body weight, immune organ index, and survival analysis data were analyzed using GraphPad Prism (Version 7.02; GraphPad Software, Inc., San Diego, CA, USA). Comparisons of the viral load between each group at each time point were determined using Multiple t tests (Holm–Sidak method, with alpha = 0.05, by GraphPad Prism), and the statistical significance of body weights and immune organ indexes between each group was evaluated using the same method. Survival patterns between each two groups were compared by Log-rank (Mantel–Cox) test. Differences were considered to be statistically significant at $p < 0.05$.

3. Results

3.1. Marek's Disease Virus (MDV) Strain BS/15 Isolated in China

MDV strain BS/15 was isolated from diseased chickens from a layer farm located in the Jilin Province of China in 2015. PCR detection of MDV meq gene (Figure 1A) and 132bpr (Figure 1B) showed that MDV strain BS/15 has the unique molecular characteristics of pathogenic MDV strains. CEF cultures inoculated with BS/15 exhibited typical CPEs, consistent with those induced by MDV infection observed by light microscopy (Figure 2A). Then, the CPEs were confirmed in IFAs using MDV gI-specific monoclonal antibodies, and no CPEs were observed in uninfected CEF cultures using a fluorescence microscope (Figure 2B). In addition, no amplification products were obtained using PCR assays designed to detect ALV, REV, and CIAV, and the ALV ELISA test was negative, as was the IFA for REV; thus, the MDV isolate BS/15 was confirmed to be free of ALV, REV, and CIAV.

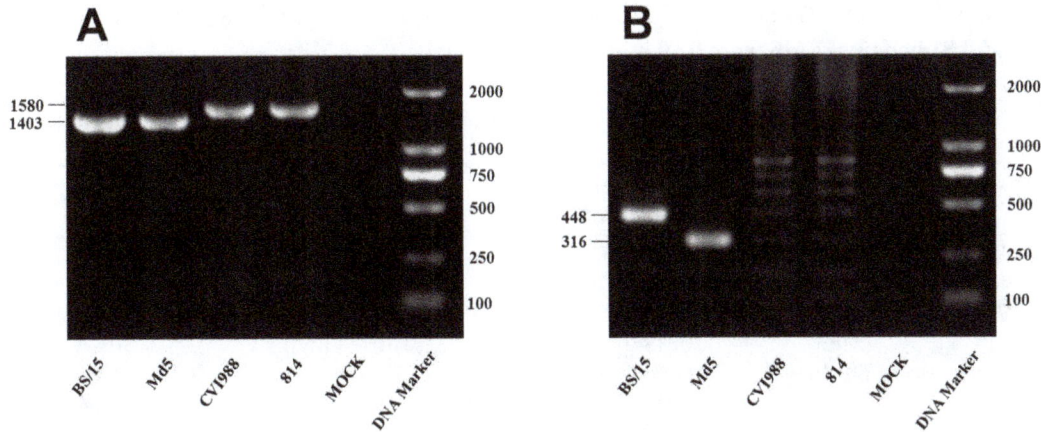

Figure 1. Detection of Marek's disease virus (MDV) by polymerase chain reaction (PCR). DNA of MDV-infected chicken embryo fibroblasts (CEFs) was used as the template for PCR amplification. (**A**) PCR amplification of the *meq* gene of MDV. The PCR products of BS/15 and Md5 were 1403 base pair (bp) long, while those of CVI988 and 814 were 1580 bp, as they have a 177-bp insertion in the *meq* gene; (**B**) PCR amplification of the 132bpr of MDV. The PCR product of BS/15 132bpr was 448 bp long with a copy number of 3, while the PCR product length for Md5 was 316 bp with a copy number of 2, CVI988 and 814 showed PCR products 316–844 bp long and contained multiple copies of the 132bpr.

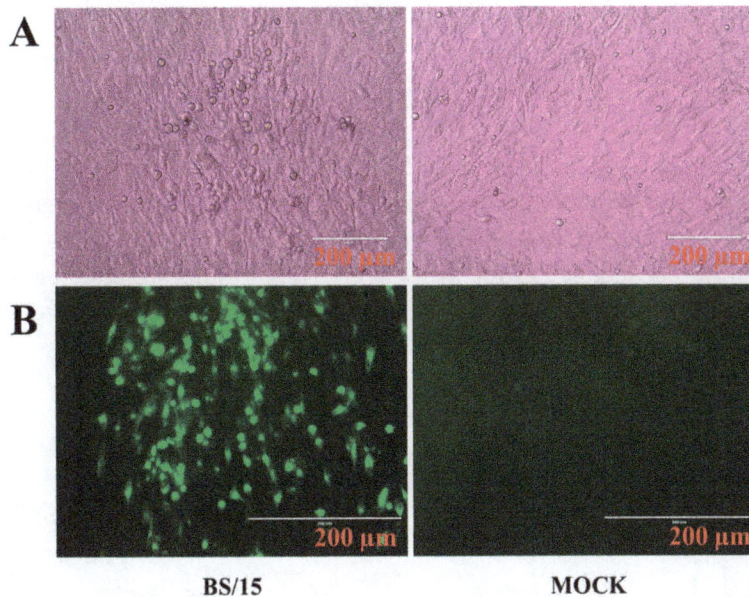

Figure 2. Viral plaques of CEF cultures caused by MDV infection, and immunofluorescence assays (IFAs). (**A**) Viral plaques of CEF cultures caused by MDV BS/15 were evident by light microscopy at 120 h post-inoculation; (**B**) Specific staining of the viral plaques with an MDV gI-specific monoclonal antibody was observed by fluorescence microscopy. Scale bar: 200 μm.

3.2. Virulence Studies of MDV BS/15 and Protective Potency Evaluation of Vaccines

3.2.1. MD Incidence, Mortality, and Tumor Rates per Group

MD incidence, mortality, and tumor rates in each group were analyzed and are summarized in Table 2. No diseased chicken was observed in the control group and all of the unvaccinated BS/15- or Md5-challenged chickens developed MD during the experimental period. When vaccinated with vaccines CVI988 or 814, the MD incidences induced by BS/15 were 66.7% and 33.3%, respectively, while those of the vaccinated Md5-challenged groups were 7.1% and 0, respectively. Thus, the PIs of vaccines CVI988 and 814 against BS/15 were 33.3 and 66.7, respectively, while the PIs of vaccines CVI988 and 814 against Md5 were 92.9 and 100, respectively.

Table 2. Marek's disease (MD) incidence, mortality, and tumor rates in each group at 90 dpc.

Vaccine	Challenge	MD Incidence Diseased/Total (%)	PI	Mortality Deaths/Total (%)	Tumor Incidence [a]
None	None	0/14 (0%)	-	0/14 (0%)	0%
None	Md5	15/15 (100%)	-	12/15 (80.0%)	40.0%
CVI988	Md5	1/14 (7.1%)	92.9	0/14 (0%)	0%
814	Md5	0/13 (0%)	100	0/14 (0%)	0%
None	BS/15	14/14 (100%)	-	12/14 (85.7%)	64.3%
CVI988	BS/15	10/15 (66.7%)	33.3	8/15 (53.3%)	13.3%
814	BS/15	5/15 (33.3%)	66.7	5/15 (33.3%)	20.0%

PI, protective index; [a] Percent of birds that developed tumors; dpc, days post-challenge.

BS/15 and Md5 induced similar mortality in unvaccinated SPF chickens (85.7% and 80.0%, respectively). In addition, 53.3% and 33.3% of chickens died in the CVI988- or 814-vaccinated BS/15-challenged groups, respectively, while none of the vaccinated Md5-challenged chickens died. Furthermore, BS/15 infection caused many unvaccinated chickens (64.3%) to develop tumors in their visceral organs, whereas the tumor rate in Md5-challenged unvaccinated chickens was only 40.0%, showing that BS/15 has stronger oncogenicity. In contrast, only 13.3% and 20.0% of chickens in the CVI988- or 814-vaccinated, BS/15-challenged groups developed tumors, while vaccination completely prevented tumors induced by Md5.

3.2.2. Survival Analysis

The earliest deaths appeared at 22 dpc in the BS/15-challenged unvaccinated group, while the first death in the Md5-challenged reference group occurred at 19 dpc. Furthermore, deaths in unvaccinated BS/15-challenged chickens peaked at 9–12 weeks post-challenge (wpc), which was much later than that in the unvaccinated Md5-challenged group at 4–7 wpc. These results suggested that BS/15 infection induced death slower than Md5 infection in SPF chickens. Survival analysis revealed that the death patterns of chickens challenged with BS/15 and Md5 were different ($p < 0.05$), as shown in Figure 3.

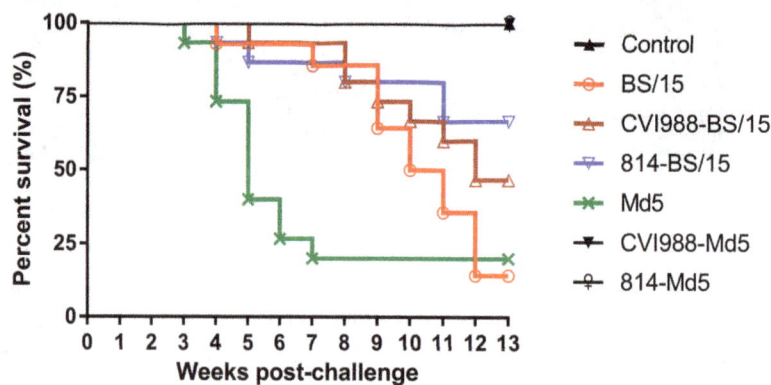

Figure 3. Survival curves for each group. The survival patterns between the BS/15-challenged group and Md5-challenged control group showed significant differences ($p < 0.05$) by Log-rank (Mantel-Cox) test.

3.2.3. Developmental Disorders and Immune Organ Damages

At the end of the animal experiment at 90 dpc, the body weights and immune organ index of the surviving chickens in each group were calculated and analyzed (Figure 4). The body weights of

BS/15- and Md5-challenged chickens (n = 2 and 3, respectively, whereas n = 5 in the other groups) were much lower than that of the control group ($p < 0.05$ for each comparison). Furthermore, the body weights of BS/15-challenged chickens were higher than those of Md5-challenged chickens ($p < 0.05$). These results suggested that both MDV strains affect the growth of chicken, and BS/15 inhibited chicken development less than Md5.

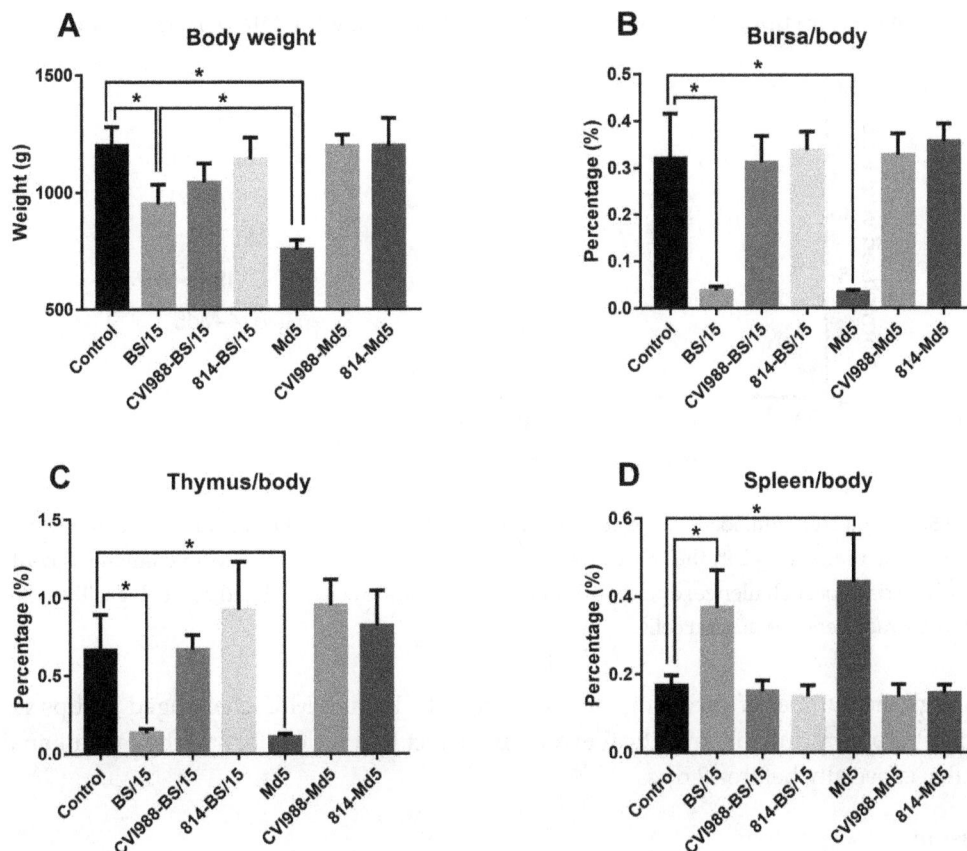

Figure 4. Body weights and ratios of immune organ weight/body weight in the surviving chickens of each group. The data are shown as mean with *SD* (n = 2 in the BS/15-challenged group, and n = 3 in the Md5-challenged control group, whereas n = 5 in the other groups), and differences were considered to be statistically significant at $p < 0.05$ (*). (**A**) Body weights of the surviving chickens in each group; (**B**) Ratios of bursa weight/body weight in the surviving chickens of each group; (**C**) Ratios of thymus weight/body weight in the surviving chickens of each group; (**D**) Ratios of spleen weight/body weight in the surviving chickens of each group.

Immune organ damages are another typical symptom of MD. In this study, both BS/15 and Md5 infection could cause severe bursa and thymus atrophy (Figure 4B,C) and splenic enlargement (Figure 4D) compared to the controls ($p < 0.05$ for each comparison). Immune organ damages caused by BS/15 and Md5 showed no difference ($p > 0.05$ for each comparison).

Although the low survival number used for analysis in the BS/15- and Md5-challenged groups might allow for some statistical error, these data showed a trend towards the differential effects of MDV BS/15 and Md5 on chicken growth and immune suppression.

3.2.4. MDV Genome Load in Feather Pulps

qPCR detection was performed with DNA from the feather pulps obtained from the BS/15- and Md5-challenged (n = 2 and 3, respectively, in the two groups at 90 dpc), vaccinated or unvaccinated chickens. The viral load of each sample was calculated and analyzed (Figure 5). The results showed that the MDV genome load of BS/15-challenged unvaccinated chickens was significantly lower than that of Md5-challenged unvaccinated chickens at 14 dpc, but became significantly higher than that of Md5-challenged unvaccinated chickens at 35 and 42 dpc ($p < 0.05$ for each comparison). In addition, the viral loads of BS/15-challenged vaccinated chickens were significantly higher than those of Md5-challenged vaccinated chickens between 35 and 90 dpc ($p < 0.05$ for each comparison). These results suggested that the vaccines could inhibit the replication of BS/15 and Md5 to some extent, and the vaccines better suppressed the replication of Md5 during the late period of viral infection.

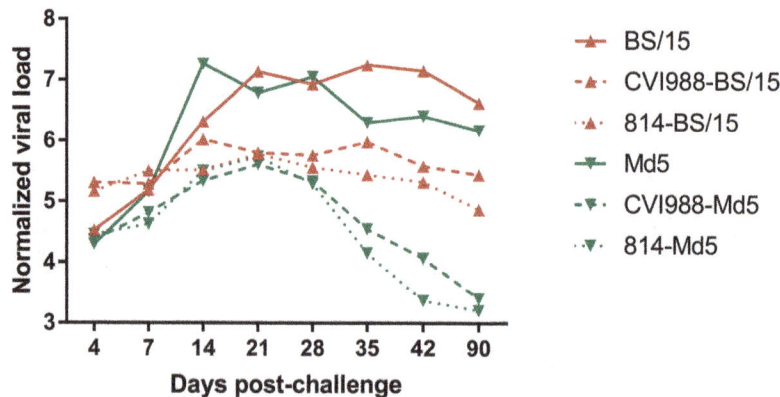

Figure 5. Normalized viral loads in the feather pulps of five birds from the various treatment groups at different time points ($n = 2$ in the BS/15-challenged group, and $n = 3$ in the Md5-challenged control group at 90 days post challenge; dpc). Normalized viral loads were calculated as the logarithm of the MDV copy numbers per million cells.

Although the number of surviving animals in the BS/15- and Md5-challenged groups were less, which is likely to introduce an analytical error, this defect had little effect on the overall analysis of these results, especially before 42 dpc.

4. Discussion

Although MDV has been continually evolving and has obtained enhanced virulence over the last few decades, no report has described MDV stains with more virulence than vv+, and no trend of a large-scale MD outbreak has been shown. However, vaccination failure of MD has often occurred in China in recent years [14,23–26]. Here, a field MDV strain BS/15 was isolated from a Chinese chicken flock in 2015 that had been vaccinated with the commercial vaccine CVI988 and had developed severe MD. Determination of MDV pathotypes is useful for investigating the cause of excessive MD losses in vaccinated flocks [27]. The ADOL method of MDV pathotyping is widely recognized throughout the world. However, because of the limitations of ADOL (requirements for specific chicken types, large numbers of birds, and objective statistical methods to measure lesion responses), it is hardly used in other laboratories. Thus, researchers at the ADOL proposed that the pathotypes of MDV could be evaluated by comparing the pathogenicity and vaccine efficacy between the MDV isolates and the reference MDV strains in local SPF chickens [27]. In this study, to investigate the causes of vaccine immune failure, unvaccinated and CVI988- or 814-vaccinated SPF White Leghorn chickens were challenged with BS/15 and the standard vv MDV strain Md5 at day 7 post-vaccination, and the birds were observed for 90 dpc.

Animal experimental results showed that BS/15 and Md5 challenge both induced 100% MD incidence with similar mortality (85.7% and 80.0%, respectively) in unvaccinated SPF chickens. Unexpectedly, BS/15 induced much higher tumor rates than did Md5 in unvaccinated chickens (64.3% and 40.0%, respectively). In addition, although the number of surviving animals in the BS/15- and Md5-challenged groups at 90 dpc ($n = 2$ and 3, respectively) was less than that in the other groups ($n = 5$), which might allow for some statistical error, our data, to some extent, showed that BS/15 infection caused fewer serious effects on chicken growth and similar immune organ damages in unvaccinated chickens compared to Md5 (Figure 4). In vaccine protection tests, the vaccines CVI988 and 814 only provided PIs of 33.3 and 66.7 to SPF chickens against BS/15, while the PIs of vaccine CVI988 and 814 against Md5 were 92.9 and 100, respectively. Moreover, these results suggested that the virulence of BS/15 was similar to that of the vv MDV strain Md5 with different pathogenic characteristics, but vaccines could not provide effective protection against BS/15, especially the vaccine CVI988. These results confirmed that the virulence and vaccine resistance of MDV are different traits, as previously discussed [14,28]. Thus, we could not clearly classify the pathotype of MDV BS/15 by the ADOL pathotyping method. However, this did not affect our understanding of the pathogenicity and the potential threat of MDV BS/15. Data generated in this study demonstrated the existence of variant Chinese strains with known virulence with altered pathogenic characteristics, which were highly resistant to existing vaccines. Thus, the presence of such strains in the field may be a cause of MD vaccination failure in China.

The MDV genome load in the feather pulps of chickens plays an important role in monitoring the MD status and vaccination performance [21]. In this study, we quantified the viral load in chicken feather pulps (Figure 5). The results showed that the viral loads of BS/15-challenged unvaccinated chickens were lower than those of Md5-challenged unvaccinated chickens at 14 dpc ($p < 0.05$), but were higher than those of Md5-challenged unvaccinated chickens at 35 and 42 dpc ($p < 0.05$ for each comparison). The lower viral loads observed in BS/15-challenged unvaccinated chickens during an early experimental period (about 7–21 dpc) may be associated with the delayed pathogenesis of BS/15. As shown in Figure 3, increasing deaths occurred in the BS/15-challenged group beginning at 9 wpc, while none of the chickens died in the Md5-challenged control group after 7 wpc. These data indicated that chickens in the BS/15-challenged group were suffering from MD, while some chickens in the Md5-challenged control group were already resistant to MD at 35 and 42 dpc, as consistently revealed by measuring the viral load in the two groups at the same time point. Furthermore, the vaccines could not effectively inhibit the replication of BS/15 compared to Md5 after 28 dpc ($p < 0.05$ for each comparison), which may explain the different protective effects of vaccines against BS/15 and Md5 to a certain extent.

Previous epidemiological studies have demonstrated the lack of efficacy of vaccines against Chinese MDV isolates with virulence of no obvious enhancement. SD2012-1 has a long latent period and causes subsequent death with a mortality rate of 70.3% in unvaccinated SPF chickens; however, SD2012-1 can break through the protection provided by vaccine HVT (FC126 strain) or bivalent vaccine HVT plus SB-1, causing mortality rates of over 60% in vaccinated SPF chickens [23]. In another study, the CVI988 vaccine could only protect 83.0% of SPF chickens from SX1301 infection, despite the low mortality rate of 57.0% caused by SX1301 in unvaccinated SPF chickens [24]. In our previous study, the field MDV strain LTS induced a very low mortality (23.1% by 60 dpc) in unvaccinated chickens, but the vaccine CVI988 could not provide efficient protection (PI: 66.7) against LTS [14]. In addition, MDV isolates ZY/1203 and WC/1110 induced very low MD mortality in unvaccinated chickens (20.5% and 21.1% by 60 dpc, respectively) and were similar to that of the standard v MDV strain GA (23.5% by 60 dpc). Although vaccine CVI988 could provide good protection against ZY/1203 and WC/1110 (PI as 82.4 and 83.3, respectively), vaccine CVI988 could not provide complete protection against them compared to that of vaccine CVI988 against GA (PI: 100) [25].

These studies suggested that the virulence of field MDV strains prevalent in China may not have increased in recent years, but several newly isolated MDV strains had obtained the ability to counter the vaccines, such as strain LTS isolated previously and strain BS/15 isolated in this study. We further inferred that the evolution patterns of MDV strains prevalent in China might have changed and no longer synchronously enhanced virulence and vaccine resistance. Thus, the newly isolated Chinese MDV strains often did not show stronger pathogenicity, but their survivability in vaccinated hosts was stronger, leading to increased vaccination failure in recent years. Therefore, the emergence of variant strains and their potential harm to the poultry industry need to be constantly monitored. Given the delayed pathogenesis of these variant MDV strains, their popularity in the field may be a greater threat to laying and breeding chickens than to meat-producing chickens. Thus, we are considering the necessity and feasibility of classifying these variant MDV strains as a new pathotype (such as "late virulence"), to facilitate the analysis and study of these variant strains in the future.

5. Conclusions

To conclude, a field MDV strain BS/15 was isolated from vaccinated chicken flocks in China. Animal experiments showed that the virulence of BS/15 was similar to the reference vv MDV strain Md5 with different pathogenic characteristics. The commercial vaccines could not provide effective protection against BS/15; BS/15, in particular, BS/15 could badly counter the CVI988 vaccine (PI: 33.3). Our study suggested the existence of variant MDV strains, which did not have particularly strong virulence but showed that powerful vaccine resistance exists in China, and these MDV strains may be the etiological cause of MD vaccination failure. Thus, improved vaccines are needed for MDV control. Furthermore, our study provided evidence that the MDV evolution does not always synchronously enhance virulence and vaccine resistance.

Acknowledgments: This study was funded by grants from the Application Technology Research and Development Fund of Harbin (No. 2016AB3BN034), the National Natural Science Foundation of China (No. 31670155), and the National "Twelfth Five-Year" Plan for Science & Technology Support (2015BAD12B03).

Author Contributions: C.L. and X.W. conceived and designed the experiments; G.S., Y.Z., H.L. and L.Z. performed the experiments; C.L., Y.Z. and G.S. analyzed the data; H.C., Y.G., X.Q., Y.W., K.L., L.G. and Q.P. contributed reagents/materials/analysis tools; and G.S. wrote the paper. All authors reviewed and improved the manuscript.

References

1.   Osterrieder, N.; Kamil, J.P.; Schumacher, D.; Tischer, B.K.; Trapp, S. Marek's disease virus: from miasma to model. *Nature Rev. Microbiol.* **2006**, *4*, 283–294. [CrossRef] [PubMed]

2.   Witter, R.L. Characteristics of Marek's disease viruses isolated from vaccinated commercial chicken flocks: association of viral pathotype with lymphoma frequency. *Avian Dis.* **1983**, *27*, 113–132. [CrossRef] [PubMed]

3.   Witter, R.L. Increased virulence of Marek's disease virus field isolates. *Avian Dis.* **1997**, *41*, 149–163. [CrossRef] [PubMed]

4.   Okazaki, W.; Purchase, H.G.; Burmester, B.R. Protection against Marek's disease by vaccination with a herpesvirus of turkeys. *Avian Dis.* **1970**, *14*, 413–429. [CrossRef] [PubMed]

5.   Rispens, B.H.; Van, V.H.; Mastenbroek, N.; Maas, H.J.; Schat, K.A. Control of Marek's disease in the Netherlands. I. Isolation of an avirulent Marek's disease virus (strain CVI988) and its use in laboratory vaccination trials. *Avian Dis.* **1972**, *16*, 108–125. [CrossRef] [PubMed]

6.   Witter, R.L.; Lee, L.F. Polyvalent Marek's disease vaccines: Safety, efficacy and protective synergism in chickens with maternal antibodies. *Avian Pathol.* **1984**, *13*, 75–92. [CrossRef] [PubMed]

7.   Witter, R.L. Protective Efficacy of Marek's Disease Vaccines. *Avian Dis.* **1993**, *37*, 57–90.

8.   Nair, V.; Schudel, A.; Lombard, M. Successful control of Marek's disease by vaccination. *Dev. Biol.* **2004**, *119*, 147–154.

9.   Atkins, K.E.; Woolhouse, M.E.J. Vaccination and reduced cohort duration can drive virulence evolution: Marek's disease virus and industrialized agriculture. *Evolution* **2013**, *67*, 851–860. [CrossRef] [PubMed]

10.   Gandon, S.; Mackinnon, M.; Nee, S.; Read, A. Imperfect vaccination: some epidemiological and evolutionary consequences. *Proc. R. Soc. Lond. [Biol]* **2003**, *270*, 1129–1136. [CrossRef] [PubMed]

11.   Read, A.F.; Baigent, S.J.; Powers, C.; Kgosana, L.B.; Blackwell, L.; Smith, L.P.; Kennedy, D.A.; Walkdenbrown, S.W.; Nair, V.K. Imperfect Vaccination Can Enhance the Transmission of Highly Virulent Pathogens. *PLoS Biol.* **2015**, *13*, 95–109. [CrossRef] [PubMed]

12.   Ewald, P.W. Imperfect Vaccines and the Evolution of Pathogen Virulence. *Nature* **2001**, *414*, 751–756.

13.   Nair, V. Evolution of Marek's disease – A paradigm for incessant race between the pathogen and the host. *Vet. J.* **2005**, *170*, 175–183. [CrossRef] [PubMed]

14. Zhang, Y.P.; Li, Z.J.; Bao, K.Y.; Lv, H.C.; Gao, Y.L.; Gao, H.L.; Qi, X.L.; Cui, H.Y.; Wang, Y.Q.; Ren, X.G. Pathogenic characteristics of Marek's disease virus field strains prevalent in China and the effectiveness of existing vaccines against them. *Vet. Microbiol.* **2015**, *177*, 62–68. [CrossRef] [PubMed]

15. Lv, H.C.; Zhang, Y.P.; Sun, G.R.; Gao, Y.L.; Li, Z.J.; Zheng, H.W.; Bao, K.Y.; Wang, X.M.; Liu, C.J. Assessments of a PCR method for detection and identification of virulent Marek's disease virus and vaccine strain for Marek's disease diagnosis. *Chin. J. Prev. Vet. Med.* **2016**, *38*, 567–571.

16. Gopal, S.; Manoharan, P.; Kathaperumal, K.; Chidambaram, B.; Divya, K.C. Differential detection of avian oncogenic viruses in poultry layer farms and Turkeys by use of multiplex PCR. *J. Clin. Microbiol.* **2012**, *50*, 2668–2673. [CrossRef] [PubMed]

17. Yun, B.; Li, D.; Zhu, H.; Liu, W.; Qin, L.; Liu, Z.; Wu, G.; Wang, Y.; Qi, X.; Gao, H. Development of an antigen-capture ELISA for the detection of avian leukosis virus p27 antigen. *J. Virol. Methods* **2012**, *187*, 278–283. [CrossRef] [PubMed]

18. Xue, M.; Shi, X.; Zhang, J.; Zhao, Y.; Cui, H.; Hu, S.; Gao, H.; Cui, X.; Wang, Y.F. Identification of a conserved B-cell epitope on reticuloendotheliosis virus envelope protein by screening a phage-displayed random peptide library. *PLoS ONE* **2012**, *7*, e49842. [CrossRef] [PubMed]

19. Qin, L.T.; Gao, Y.L.; Pan, W.; Deng, X.Y.; Sun, F.F.; Kai, L.I.; Xiao-Le, Q.I.; Gao, H.L.; Liu, C.N.; Wang, X.M. Investigation of co-infection of ALV-J with REV, MDV, CAV in layer chicken flocks in some regions of China. *Chin. J. Prev. Vet. Med.* **2010**, *32*, 90–93.

20. Zhang, Y.; Liu, C.J.; Qin, Y.A.; Zhang, Y.P.; Zhang, X.W.; Hao, Y.Q. Application of duplex fluorescent quantitative polymerase-chain-reaction for detecting Marek's disease virus serotype 1. *Chin. J. Prev. Vet. Med.* **2007**, *29*, 46–51.

21. Baigent, S.J.; Petherbridge, L.J.; Howes, K.; Smith, L.P.; Currie, R.J.; Nair, V.K. Absolute quantitation of Marek's disease virus genome copy number in chicken feather and lymphocyte samples using real-time PCR. *J. Virol. Methods* **2005**, *123*, 53–64. [CrossRef] [PubMed]

22. Sharma, J.M.; Burmester, B.R. Resistance to Marek's disease at hatching in chickens vaccinated as embryos with the turkey herpesvirus. *Avian Dis.* **1982**, *26*, 134–149. [CrossRef] [PubMed]

23. Gong, Z.; Zhang, L.; Wang, J.; Chen, L.; Hu, S.; Wang, Z.; Ma, H. Isolation and analysis of a very virulent Marek's disease virus strain in China. *Virol. J.* **2013**, *10*, 1–8. [CrossRef] [PubMed]

24. Ning, C.; Shuai, S.; Peng, S.; Zhang, Y.; Ni, H.; Cui, Z. Isolation and pathogenic analysis of virulent Marek's disease virus field strain in China. *Poult. Sci.* **2016**, *95*, 1521–1528.

25. Zhang, Y.P.; Lv, H.C.; Bao, K.Y.; Gao, Y.L.; Gao, H.L.; Qi, X.L.; Cui, H.Y.; Wang, Y.Q.; Li, K.; Gao, L. Molecular and pathogenicity characterization of Gallid herpesvirus 2 newly isolated in China from 2009 to 2013. *Virus Genes* **2015**, *52*, 51–60. [CrossRef] [PubMed]

26. Zhuang, X.; Zou, H.; Shi, H.; Shao, H.; Ye, J.; Ji, M.; Wu, G.; Qin, A. Outbreak of Marek's disease in a vaccinated broiler breeding flock during its peak egg-laying period in China. *BMC Vet. Res.* **2015**, *11*, 1–6. [CrossRef] [PubMed]

27. Witter, R.L.; Calnek, B.W.; Buscaglia, C.; Gimeno, I.M.; Schat, K.A. Classification of Marek's disease viruses according to pathotype: philosophy and methodology. *Avian Pathol.* **2005**, *34*, 75–90. [CrossRef] [PubMed]

28. Walkden-Brown, S.W.; Islam, A.; Islam, A.; Burgess, S.K.; Groves, P.J.; Cooke, J. Pathotyping of Australian isolates of Marek's disease virus in commercial broiler chickens vaccinated with herpesvirus of turkeys (HVT) or bivalent (HVT/SB1) vaccine and association with viral load in the spleen and feather dander. *Aust. Vet. J.* **2013**, *91*, 341–350. [CrossRef] [PubMed]

3

Comparative Study on the Antivirus Activity of Shuang–Huang–Lian Injectable Powder and Its Bioactive Compound Mixture against Human Adenovirus III In Vitro

Qinhai Ma [1,2,†], Dedong Liang [1,2,†], Shuai Song [1,2,†], Qintian Yu [1,2], Chunyu Shi [1,2], Xuefeng Xing [1,2,*] and Jia-Bo Luo [1,2,*]

[1]   School of Traditional Chinese Medical Science, Southern Medical University, Guangzhou 510515, China; 13268268214@163.com (Q.M.); famouscool@163.com (D.L.); ss89112@126.com (S.S.); dkdklm@163.com (Q.Y.); 15626452676@163.com (C.S.)

[2]   Guangdong Provincial Key Laboratory of Chinese Medicine Pharmaceutics, Southern Medical University, Guangzhou 510515, China

*   Correspondence: xiaoxing0610@163.com (X.X.); ljb@smu.edu.cn (J.-B.L.)

†   These authors contributed equally to this work.

Academic Editor: Curt Hagedorn

Abstract: Shuang–Huang–Lian injectable powder (SHL)—a classical purified herbal preparation extracted from *Scutellaria baicalensis*, *Lonicera japonica*, and *Forsythia suspense*—has been used against human adenovirus III (HAdV$_3$) for many years. The combination herb and its major bioactive compounds, including chlorogenic acid, baicalin, and forsythia glycosides A, are effective inhibitors of the virus. However, no comprehensive studies are available on the antiviral effects of SHL against HAdV$_3$. Moreover, it remains unclear whether the mixture of chlorogenic acid, baicalin, and forsythia glycosides A (CBF) has enhanced antiviral activity compared with SHL. Therefore, a comparative study was performed to investigate the combination which is promising for further antiviral drug development. To evaluate their antivirus activity in parallel, the combination ratio and dose of CBF were controlled and consistent with SHL. First, the fingerprint and the ratio of CBF in SHL were determined by high performance liquid chromatography. Then, a plaque reduction assay, reverse transcription polymerase chain reaction (PCR), real-time polymerase chain reaction (qPCR), and enzyme-linked immunosorbent assay (ELISA) were used to explore its therapeutic effects on viral infection and replication, respectively. The results showed that SHL and CBF inhibited dose- and time-dependently HAdV$_3$-induced plaque formation in A549 and HEp-2 cells. SHL was more effective than CBF when supplemented prior to and after viral inoculation. SHL prevented viral attachment, internalization, and replication at high concentration and decreased viral levels within and out of cells at non-toxic concentrations in both cell types. Moreover, the expression of tumor necrosis factor alpha (TNF)-α, interleukin (IL)-1ß, and IL-6 was lower and the expression of interferon (IFN)-γ was higher in both cell types treated with SHL than with CBF. In conclusion, SHL is much more effective and slightly less toxic than CBF.

Keywords: human adenovirus III; anti-viral; Shuang–Huang–Lian injectable powder; effect comparison

1. Introduction

Human adenovirus (HAdV), a nonenveloped DNA virus, is a common causative pathogen of acute respiratory infection. HAdV infection is more common in childcare and overcrowded conditions, and at least one strain was detected in most infected children during the first 5 years after birth. Nowadays, seven species, including many genotypes, have been characterized by genomics and bioinformatics [1,2]. HAdV species B (serotypes 3, 7, 14, and 55), species C (serotypes 1, 2, 5, and 6), and species E (serotype 4) are the most commonly found in patients with respiratory infection. Among these, HAdV$_3$ strains of subspecies B1 are the major epidemic strains responsible for severe respiratory disease epidemics and outbreaks worldwide [3–10]. Currently, there is no effective treatment or vaccine against HAdV$_3$ infection and new anti-HAdV$_3$ drugs urgently need to be developed.

Shuang–Huang–Lian injectable powder (SHL), consisting of *Scutellaria baicalensis*, *Lonicera japonica*, and *Forsythia suspense*, is a classical prescription in traditional Chinese medicine (TCM). Its efficacy against many infectious diseases caused by bacteria and viruses in respiratory traction has been demonstrated [11]. In 1992, SHL was approved in China as a new Chinese patent drug for emergency treatment. Chlorogenic acid, baicalin, and forsythia glycosides, the main effective compounds isolated from the herb extract, are representative markers of its quality recorded in the Chinese Pharmacopoeia (2015). Chlorogenic acid is the major ingredient of *Lonicera japonica* that has been reported to have multi-anti-viral activities against human immunodeficiency virus (HIV), adenovirus, influenza virus (H1N1, H5N1), and EV71 [12–16]. Baicalin, which is derived from the dried root of *Scutellaria baicalensis*, has been demonstrated to inhibit influenza A (H1N1) infection, Dengue virus, and respiratory syncytial virus infection [17–19]. Forsythoside A, the active ingredient of *Forsythia suspense*, has been shown to inhibit syncytial virus and coxsackievirus in vitro [20]. In addition, forsythoside A has the potential to prevent infectious bronchitis virus (IBV) infection in vitro [21].

Multidrug combination is an important strategy in modern antivirus therapy because of increasing drug resistance. Contrary to single-component drugs, the benefits of TCM drugs are often due to the synergistic interactions of multiple ingredients. Therefore, we hypothesized that the antivirus ability of SHL may occur though the activity of multiple ingredients with multiple targets. Accordingly, a comparative antivirus study of SHL and its major bioactive ingredients mixture was investigated. Antivirus effects, interference link, and regulation of inflammatory cytokines were explored in both HEp-2 and A549 cell lines.

2. Materials and Methods

2.1. Reagents

SHL (lot#: 1409422) was provided by Second Chinese Medicine Factory of Harbin Pharm Group Co., Ltd. (Harbin, China). Forsythoside A (94.1%, lot#: 111810–201405), baicalin (93.3%, lot#: 110715–201318), and chlorogenic acid (96.2%, lot#: 110753–201415) were provided by the National Institutes for Food and Drug Control (Beijing, China). Mouse tumor necrosis factor alpha (TNF-α), interleukin (IL)-1ß, IL-6, and interferon gamma (IFN-γ) enzyme-linked immunosorbent assay (ELISA) kits were obtained from Huamei (Wuhang, China). Premix Taq^{TM} (TaKaRa Taq^{TM} Version 2.0) and SYBR Premix Ex Taq II (Tli RNaseH Plus) were purchased from Takara (Tokyo, Japan). The virus DNA extraction kit was obtained from Guangzhou Institute of Respiratory Disease (Guangzhou developed by Pharmaceutical Technology Co., Ltd., Guangzhou, China). Cell counting kit-8 (CCK-8) was from Dojindo (Dojindo, Japan).

2.2. Chromatographic Conditions and Test Samples Preparation

The high performance liquid chromatography (HPLC)-diode array detector (DAD) fingerprint of SHL was performed on an Agilent 1200 series liquid chromatography system (Agilent Technologies, Santa Clara, CA, USA), consisting of a binary pump (Agilent G1312B), an auto-sampler (Agilent G1329B), and a DAD-vis detector (Agilent G1316A). The data were recorded and analyzed using Agilent Chemstation software for the LC-3D system (Rev. B.04.03-SP1). A Cosmosil 5C18-AR-II column (5 μm, 4.6 × 250 mm; Nacalai Tesque Co. Inc., Tokyo, Japan) was used at 30 °C. The mobile phase was composed of (A) phosphoric acid aqueous solution (0.2%, *v/v*) and (B) methanol using a gradient elution of 80–70% (A) at 0–5 min; 70–55% (A) at 5–35 min; and 55–30% (A) at 35–50 min. The sample injection volume was 10 μL with a flow rate of 1.0 mL/min. The detection wavelength was set at 327 nm. Three standard references, which were chlorogenic acid, forsythoside A, and baicalin, were used. The stock solutions of standard references were prepared by dissolving accurately the weighed standards in methanol and storing them in a 10 mL volumetric flask. Otherwise, 20.0 mg SHL was supplemented with 10 mL methanol and then extracted under sonication for 20 min. All the test samples were filtered through a 0.45 μm membrane filter before chromatographic analysis. The HPLC-DAD fingerprint standard of SHL was recorded and compared to these three standard references. HPLC-DAD revealed that SHL contained 1.3% chlorogenic acid, 0.8% forsythoside A, and 21.6% baicalein (Figure 1). Based on the ratios of the three main components of SHL measured by HPLC-DAD, we prepared a mixture of the three main components. Then, the antiviral effect of the mixed standards was compared with that of SHL. The initial concentration of the mixture was 593.3 μg/mL (the rate of the mixture of chlorogenic acid, forsythoside A, and baicalin was 1.3:0.8:21.6). The initial concentration of SHL was 20 mg/mL. The standard mix (including baicalin, chlorogenic acid, and forsythoside A) was prepared according to the composition ratio of SHL. Three chemical standards were weighed accurately and dissolved in Dulbecco's Modified Eagle's Medium (DMEM, Gibco).

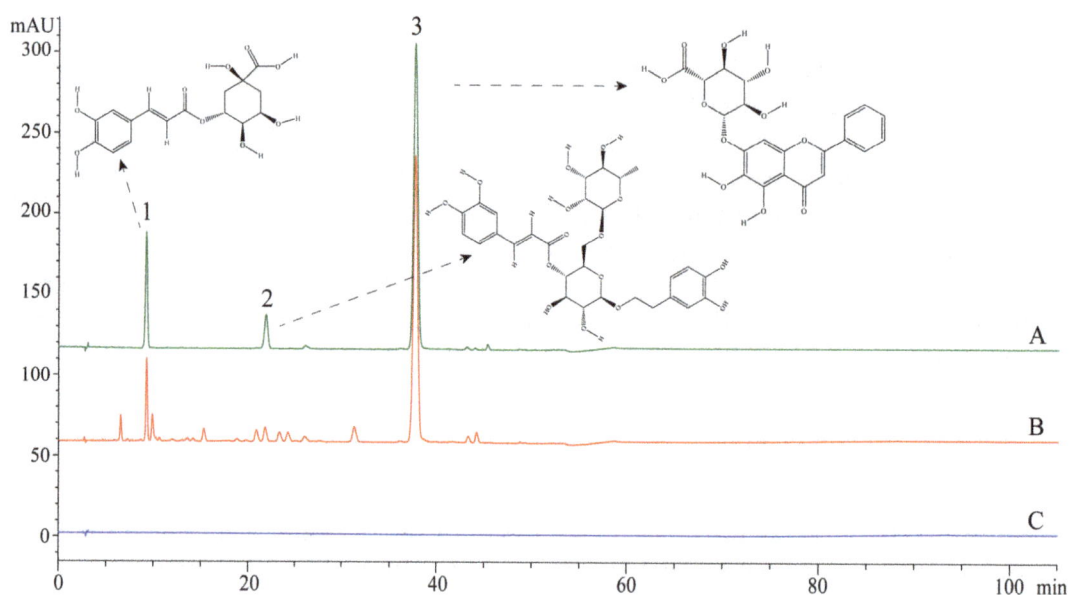

Figure 1. HPLC chromatograms of three effective constituents in SHL at 327 nm. 1. chlorogenic acid; 2. forsythoside A; 3. Baicalin.

2.3. Cells and Virus

HEp-2 cells (ATCC, Manassas, VA, USA) and A549 cells (ATCC, Manassas, VA, USA) were inoculated with HAdV3 (HAdV$_3$, the Chinese Academy of Sciences Wuhan Institute of Virology, Wuhan, China). Cells were propagated at 37 °C under 5% CO_2 in DMEM (Gibco, Carlsbad, CA, USA) cultured with 10% fetal bovine serum (FBS, Gibco) and 1% antibiotics (Gibco). Two percent FBS, instead of 10%, was used to propagate virus-infected cell monolayers. The virus was stored at −80 °C, and its titer was determined by 50% tissue culture infection dose (TCID$_{50}$).

2.4. Cytotoxicity Assay

To determine whether SHL and CBF were toxic to HEp-2 and A549 cells, a cytotoxicity assay was measured using the cell counting kit-8 (CCK-8) assay [22]. Their 50% cytotoxic concentrations (CC$_{50}$) were determined by plotting the percentage of cell growth inhibition against the concentration of the compound drug.

2.5. Antiviral Effect Assay

The antiviral activity of SHL and CBF was examined by plaque reduction assay, as previously described [23]. Briefly, 2×10^4 cells/well were plated in 24-well culture plates at 37 °C under 5% CO_2 for 24 h and inoculated with a mixture of 100 $TCID_{50}$/well virus and various concentrations of SHL and CBF in triplicate at room temperature for 2 h. After supplementation with overlay medium (DMEM plus 2% FBS in 1% methylcellulose), they were cultured at 37 °C under 5% CO_2 for 3 days. The monolayer was then fixed with 10% formalin and stained with 1% crystal violet, and then the plaques were counted. The minimal concentration of drugs required to inhibit 50% of the cytopathic effect (50% inhibitory concentration [IC_{50}]) was calculated by the regression analysis of the dose–response curve generated from the data.

2.6. Time of Addition Assay

The antiviral activity of SHL and CBF was examined at different time points prior to and after viral inoculation by the plaque reduction assay [23]. Cells were seeded and incubated for 24 h as previously described. Various concentrations of SHL and the mixture were supplemented at 2 h (−2 h) or 1 h (−1 h) prior to viral inoculation, or 1 h (+1 h) or 2 h (+2 h) after viral inoculation. Supernatants were removed prior to the supplementation of overlay medium. Cells were incubated for an additional 72 h and examined by plaque assay.

2.7. Attachment Assay

Plaque reduction assay was performed to evaluate the effect of SHL and CBF on viral attachment [23]. Briefly, cells were seeded and incubated for 24 h. Cells were pre-chilled at 4 °C for 1 h, and the medium was replaced by a mixture of 100 $TCID_{50}$/well virus and various concentrations of SHL and the CBF mixture. After incubation at 4 °C for another 3 h, the free virus was removed. The cell monolayer was washed with ice-cold phosphate-buffered saline (PBS) three times, covered with overlay medium, incubated at 37 °C under 5% CO_2 for an additional 72 h, and examined by plaque assay.

2.8. Internalization Assay

The effect of SHL and CBF on viral internalization was also evaluated [23]. Briefly, cells were seeded and incubated for 24 h and pre-chilled at 4 °C for 1 h. Cells were infected with 100 $TCID_{50}$/well virus and incubated at 4 °C for another 3 h. The virus-containing medium was replaced by fresh medium containing various concentrations of SHL and CBF in triplicate. They were shifted to culture at 37 °C. At 20 min, 40 min, and 60 min intervals following the 37 °C shift, un-internalized virus was inactivated by supplementation with acidic PBS (pH 3) for 1 min, followed by alkaline PBS (pH 11) for neutralization. Then, PBS was replaced by fresh overlay medium. After incubation at 37 °C for an additional 72 h, the cell monolayer was examined by plaque assay.

2.9. Polymerase Chain Reaction (PCR) and Quantitative PCR (qPCR)

The antiviral activity of SHL and CBF against viral replication was further examined by PCR semi-quantitatively and by qPCR quantitatively. Briefly, 4×10^5 cells/well were plated into 6-well culture plates for 24 h. A mixture of 100 $TCID_{50}$/well virus and various concentrations of SHL and CBF were supplemented. They were cultured for a further 48 h. Viral DNA was extracted with the virus DNA extraction kit for virus-infected A549 cells, HEp-2 cells, and culture supernatant, according to the manufacturer's instruction. They were then placed on ice or at 4 °C for PCR and qPCR.

The PCR and qPCR sample systems and conditions were determined according to the instruction of Premix TaqTM (TaKaRa TaqTM Version 2.0) and SYBR Premix Ex Taq II (Tli RNaseH Plus), respectively. Amplification products were analyzed semi-quantitatively by 2% agarose gel electrophoresis, and the qPCR was detected by the Step One Real-Time PCR System (Mx3005P,

Stratagene, La, Jolla, CA, USA). The forward primer of Adv3 was 5'-ATCGATGATGCCCCAATGG-3', and the reverse primer was 5'-GGACTCAGGTACTCCGAAGCA-3'. Taking the CT value (cycle threshold) for the vertical axis and the copy number of the logarithm of concentration as abscissa, the standard curve was automatically generated by the qPCR instrument control software. The amount of virus of the experimental groups was calculated from the differences between the CT of the viral control and those of the experimental groups.

2.10. Enzyme-Linked Immunosorbent Assay (ELISA)

After the antiviral effect assays were performed, the culture medium was collected and assayed using the TNF-a, IFN-γ, IL-1ß, and IL-6 ELISA kit according to the manufacturer's instructions. The A450 nm was determined by the ELISA reader (Thermo Scientific, Boston, MA, USA).

2.11. Statistical Analysis

Results are expressed as mean $\pm$ standard deviation (S.D.). Percentage of control (infection rate; %) was calculated from the plaque counts of the experimental groups divided by that of the viral control. Data were analyzed with analysis of variance (ANOVA) by SPSS ver. 19.0 (Armonk, NY, USA). Tukey honestly significant difference (HSD) test was used for post-hoc ANOVA comparisons. $p < 0.05$ was considered statistically significant.

3. Results

3.1. Cytotoxicity of SHL and CBF in A549 and HEp-2 Cells

To determine whether SHL and CBF were toxic to cells, a cytotoxicity assay was performed using the CCK-8 assay. The estimated CC_{50} of SHL and CBF were 297.7 and 148.2 μg/mL on A549 cells and 211.7 and 94.1 μg/mL on HEp-2 cells, respectively (Figure 2). Neither SHL treatment nor CBF treatment were significantly cytotoxic from concentrations of 37.1 μg/mL in the two cell types. Therefore, SHL and CBF at the concentrations of 37.1–1.2 μg/mL were selected in the subsequent experiments.

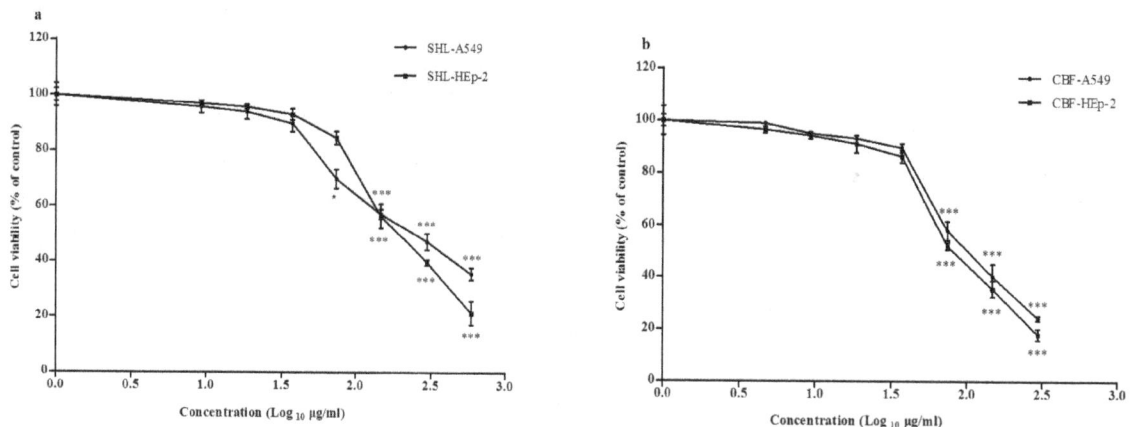

Figure 2. Shuang–Huang–Lian injectable powder (SHL) showed its cytotoxicity against host cells above the concentrations of 74.2 μg/mL on the A549 cells and above the concentrations of 148.3 μg/mL on the HEp-2 cells ($p < 0.05$) (**a**); chlorogenic acid, baicalin, and forsythia glycosides A (CBF) showed cytotoxicity above the concentrations of 74.2 μg/mL on both cells ($p < 0.001$) (**b**). Data are represented as mean $\pm$ S.D. of nine tests. * $p < 0.05$; ** $p < 0.01$; *** $p < 0.001$ were compared to the cell control.

3.2. SHL Attenuated Virus Proliferation More Significantly Than CBF

SHL and CBF dose-dependently decreased virus proliferation in HEp-2 and A549 cells. The effect of SHL, however, showed better suppression than CBF in both HEp-2 cells and A549 cells ($p < 0.05$) (Figure 3). This effect was significantly different at all concentrations except for 2.3 and 1.2 µg/mL on Hep-2 cells ($p < 0.05$). It indicated that SHL was more effective at inhibiting the proliferation of the virus than that by CBF.

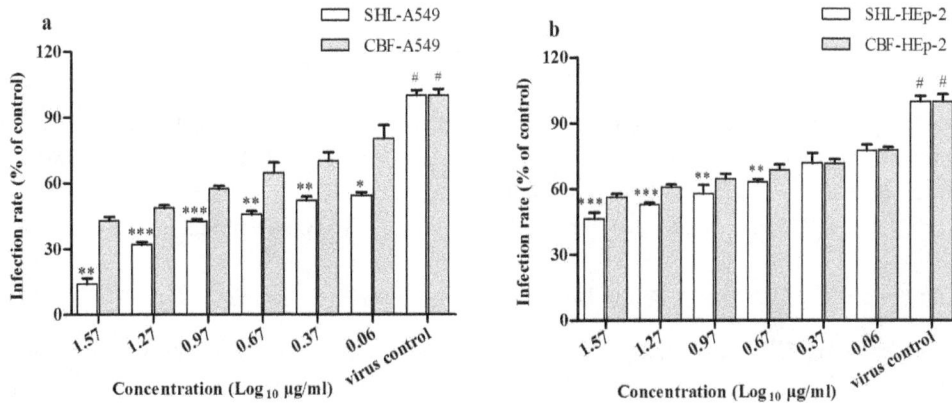

Figure 3. SHL and CBF were dose-dependently effective against human adenovirus III ($HAdV_3$) in both cell types as determined by plaque reduction assay ($p < 0.05$); SHL decreased more plaque formation than CBF at all the concentrations ($p < 0.05$) in A549 cells (**a**) and at the higher concentrations than 4.6 µg/mL ($p < 0.01$) in HEp-2 cells (**b**). Data are represented as mean ± S.D. of nine tests. * $p < 0.05$; ** $p < 0.01$; *** $p < 0.001$ were compared to CBF. # $p < 0.05$ was compared to the virus control.

3.3. SHL Decreased Plaque Formation More Than CBF When Viral Inoculation Was Given in Different Working Points

To better understand the therapeutic intervention during virus invasion, time of addition assay in A549 and HEp-2 cells was employed to explore its working points. SHL and CBF time-dependently and dose-dependently decreased plaque formation in A549 and HEp-2 cells. SHL decreased plaque formation more than CBF when viral inoculation was given in different working points ($p < 0.05$) (Figure 4). It showed that both SHL and CBF were better at inhibiting virus activity when given before viral inoculation than after in the two cells types. As the exposure duration of cells to SHL and CBF before viral inoculation increased, so did the significance of the antiviral activity.

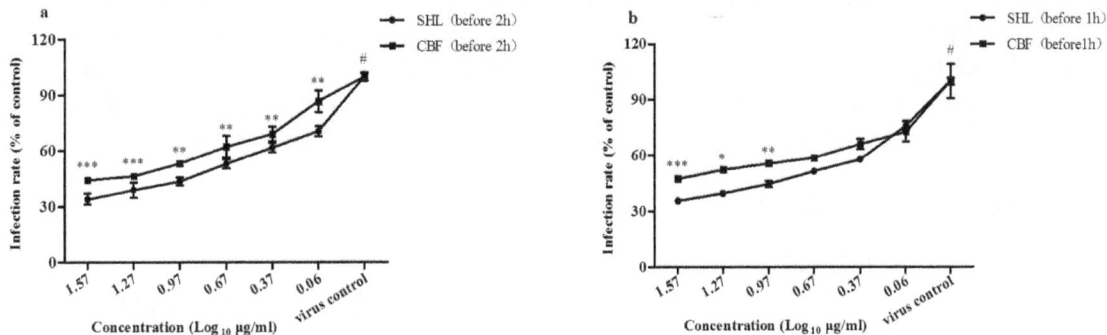

Figure 4. Cont.

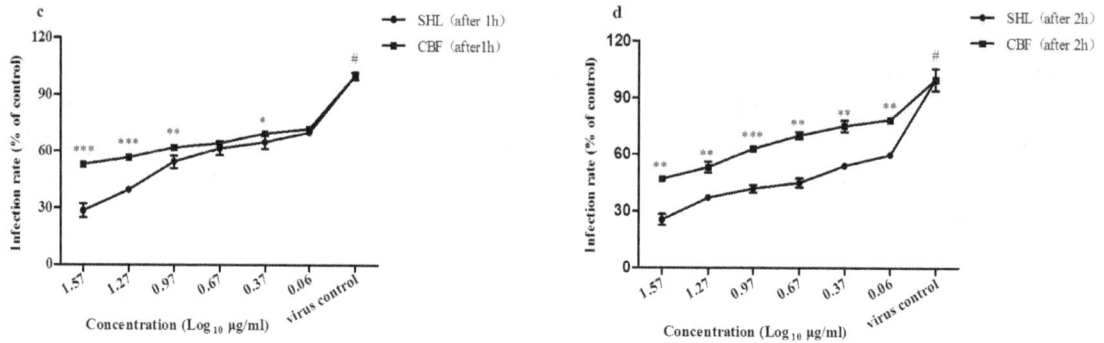

Figure 4. SHL and CBF were time-dependently and dose-dependently effective against $HAdV_3$ when given viral inoculation in different administrations ($p < 0.05$), and SHL decreased more plaque formation than CBF in both cell types ($p < 0.05$). Data are represented as mean $\pm$ S.D. of nine tests. * $p < 0.05$; ** $p < 0.01$; *** $p < 0.001$ were compared to CBF. # $p < 0.05$ was compared to the virus control.

3.4. SHL Inhibited Viral Attachment Better Than CBF in A549 and HEp-2 Cells

Because SHL and CBF anti-virus activity was mainly effective by supplementation before viral inoculation, we predicted that they worked by disrupting viral attachment and that the anti-viral effect of SHL was superior to that of CBF. Results from the attachment assay confirmed this hypothesis, as both SHL and CBF dose-dependently inhibited viral attachment. SHL decreased plaque formation more than CBF at concentrations higher than 4.6 µg/mL in A549 cells ($p < 0.01$) (Figure 5a), and SHL decreased plaque formation more than CBF at all concentrations in HEp-2 cells ($p < 0.01$) (Figure 5b). These results were consistent with those of the anti-viral effect assay (Figure 3) and the time course assay (Figure 4). It demonstrated that viral attachment was inhibited more with SHL than with CBF, and the effect was not significantly different between A549 and HEp-2 cells.

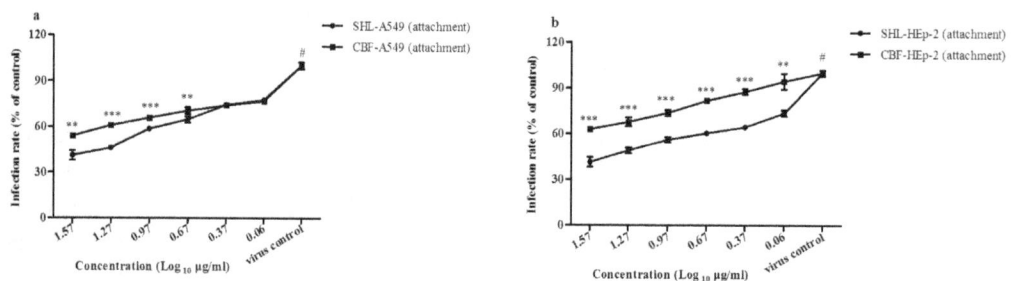

Figure 5. SHL and CBF were dose-dependently effective against viral attachment in both cell types ($p < 0.05$). SHL decreased more plaque formation than CBF at the higher concentration than 4.6 µg/mL in A549 cells ($p < 0.05$) (**a**), and at all the concentrations in HEp-2 cells ($p < 0.05$) (**b**). Data are represented as mean $\pm$ S.D. of nine tests. * $p < 0.05$; ** $p < 0.01$; *** $p < 0.001$ were compared to CBF. # $p < 0.05$ was compared to the virus control.

3.5. SHL Affected Viral Internalization More Than CBF

The results of the internalization assay (Figure 6) were consistent with those of the above assays (Figure 4). SHL decreased plaque formation more than CBF at all concentrations except for 1.2 µg/mL in A549 cells (20 min) ($p < 0.05$) (Figure 6a); at all the concentrations in A549 cells (40 min) ($p < 0.05$) (Figure 6b); and at concentrations higher than 4.6 µg/mL in A549 cells (60 min) ($p < 0.01$) (Figure 6c). In addition, SHL decreased plaque formation more than CBF at concentrations higher than 2.3 µg/mL in HEp-2 cells (20 min) ($p < 0.05$) (Figure 6d); at concentrations higher than 2.3 µg/mL in HEp-2 cells (40 min) ($p < 0.05$) (Figure 6e); and at all concentrations in HEp-2 cells (60 min) ($p < 0.05$) (Figure 6f). It indicated that the antiviral effects enhanced with increase in time.

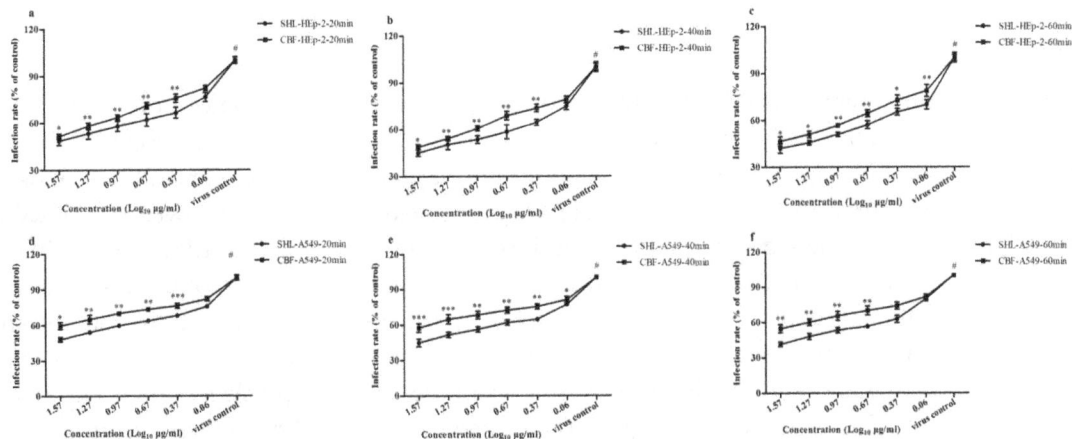

Figure 6. SHL and CBF were time-dependently and dose-dependently effective against viral internalization in both cell types ($p < 0.05$). SHL decreased more plaque formation than CBF at the high concentration ($p < 0.05$). Data are represented as mean ± S.D. of nine tests. * $p < 0.05$; ** $p < 0.01$; *** $p < 0.001$.

3.6. SHL Significantly Suppressed Viral DNA Replication in A549 and HEp-2 Cells Compared to CBF

Data from the semi-quantification of viral DNA by agarose gel electrophoresis were comparable to those of qPCR (Figure 7). The results of the semi-quantification and quantification of viral DNA had different trends in cells and in the suspension. SHL markedly decreased viral amounts compared to CBF at all concentrations ($p < 0.001$) in A549 cells and at concentrations higher than 4.6 μg/mL ($p < 0.01$) in the suspension. In addition, SHL markedly decreased the amount of virus compared to CBF at concentrations higher than 4.6 μg/mL ($p < 0.05$) in HEp-2 cells. The difference, however, was not significant in the suspension.

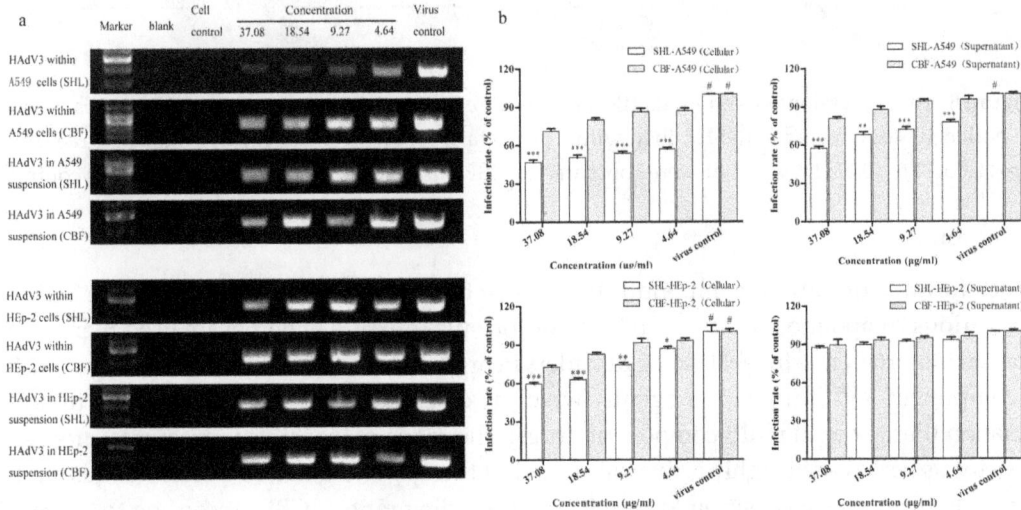

Figure 7. The PCR result of the quantification of viral DNA by the agarose gel electrophoresis (**a**) was comparable to that of the qPCR (**b**). SHL markedly decreased the viral amounts more than CBF at all concentrations ($p < 0.001$) within A549 cells ($p < 0.001$) and in the suspension ($p < 0.05$); and at all the concentrations within HEp-2 cells ($p < 0.05$); and there was no significant difference in the suspension ($p > 0.05$). Data are represented as mean ± S.D. of three tests. * $p < 0.05$; ** $p < 0.01$; *** $p < 0.001$ were compared to CBF. # $p < 0.05$ was compared to the virus control.

3.7. SHL Attenuated Inflammatory Effects Significantly More Than CBF in HAdV$_3$-Stimulated A549 and HEp-2 Cells

Because virus infection can lead to the production of the pro-inflammatory cytokines IFN-γ, TNF-α, IL-1ß, and IL-6, which contribute to inflammation, the expressions of these cytokines were measured. Results showed that SHL and CBF significantly suppressed the secretion of these cytokines in both cell types compared to the virus control group (Figure 8). SHL, however, was more effective than CBF at decreasing the secretion of TNF-α at the concentrations of 37.1 and 18.5 µg/mL in both A549 and HEp-2 cells (Figure 8a,b). In addition, SHL suppressed the secretion of IL-1ß significantly more than CBF at the concentrations of 37.1, 18.5, and 9.3 µg/mL in A549 cells and at the concentration of 37.1 µg/mL in HEp-2 cells (Figure 8c,d). At the concentration of 37.1 µg/mL, the secretion of IL-6 in HEp-2 cells and the secretion of IFN-γ in A549 cells in the SHL group was less than those in the CBF group (Figure 8f,g). There was no significant difference between the treatment groups for IL-6 secretion in A549 cells and IFN-γ secretion in HEp-2 cells (Figure 8e,h). Taken together, these results suggest that SHL can attenuate the inflammatory effects in HAdV$_3$-stimulated A549 and HEp-2 cells more than CBF.

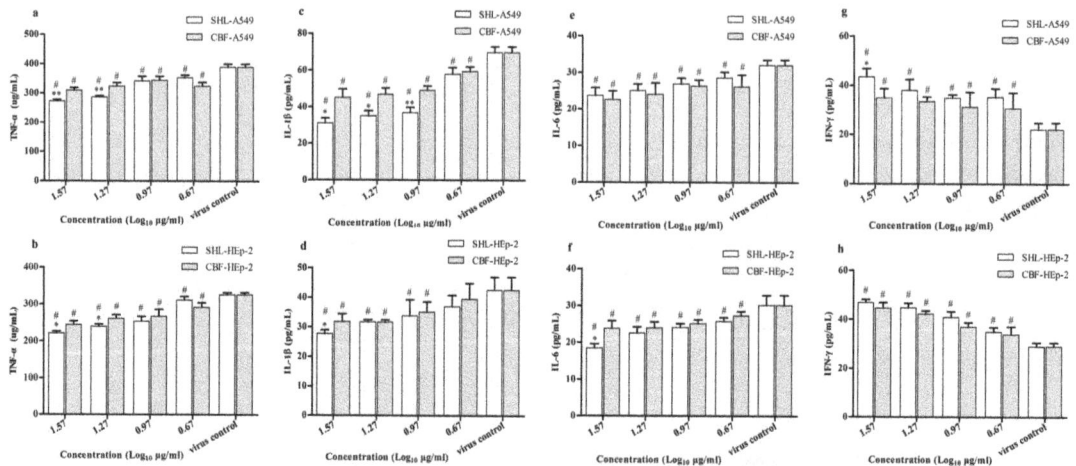

Figure 8. The expressions of TNF-α, IL-1ß, and IL-6 were lower and the expression of IFN-γ was higher in both cell types treated with SHL than with CBF. Data are represented as mean ± S.D. of three tests. * $p < 0.05$; ** $p < 0.01$; *** $p < 0.001$ were compared to CBF. # $p < 0.05$ was compared to the virus control.

4. Discussion

Herb formula or natural compound drugs have been shown to be effective in the TCM clinic. Indeed, various compounds contained in herbs or formula (mixture of herbs) are used therapeutically and play an important role in rebalancing disorders in organisms. TCM formulas are always considered multi-component and multi-target agents, which is essentially the same strategy as that during combination therapy with multi-component drugs. The ratios of bioactive ingredients affect efficacy and toxicity, as occurs with highly active antiretroviral therapy and ginaton. Therefore, this study was performed to determine a consistent dose and compounding ratio of CBF. According to the results of HPLC (Figure 1), chlorogenic acid (1.3%), forsythoside A (0.8%), and baicalein (21.6%) were the major components of SHL. CBF was a combination of the ratio of the results of HPLC.

We used a plaque reduction assay to determine the effects of SHL and CBF in both HEp-2 and A549 cells. Variation of the anti-virus activities of SHL and BCF was correlated with its working point after administration. According to our study, the antiviral ability of SHL was greater than that of CBF when the cells were treated before viral inoculation. This antiviral potency of SHL was further confirmed by its ability to inhibit viral attachment and internalization (Figures 5 and 6), suggesting that the anti-viral ability of SHL is mainly a preventative effect. Therefore, longer pre-incubation times improve the anti-viral effects of the two drugs. Moreover, different anti-virus assays and different respiratory tract cell lines were tested. Compared with CBF, SHL had a high CC_{50} and inhibited virus entry and replication. Therefore, SHL is safer, more effective, and more readily available than CBF for the prevention and management of virus-induced airway injuries.

Although both SHL and CBF can disrupt virus DNA amplification and destroy virus particles, SHL treatment reduced virus DNA load in cells more than CBF. The reduction in virus DNA load was greater with SHL than CBF within or out of the two cell types, A549 and HEp-2 cells. In cell suspensions, the anti-viral effects of intervention with SHL were greater than the effects with CBF in A549 cells but not HEp-2 cells, suggesting that SHL was a more effective drug at improving the lower respiratory tract infections in virus. What is more, the sensitivity of the virus to the two cells was different; the amount of the virus released was also different; and the low levels of virus in the supernatant of the cells may be one reason for the absence of differences between the two cells.

Virus-induced secretion of cytokines, such as INF, TNF, and IL, contribute to innate immunity against viral infection. Therefore, variations in the levels of IFN-γ, TNF-α, IL-1ß, and IL-6 were compared between SHL and CBF treatment groups [23–27]. According to our study, both SHL and CBF attenuated the production of pro-inflammatory cytokines IL-1ß, IL-6, and TNF-α and promoted the release of INF-γ in vitro as compared to the virus control group, suggesting that SHL and CBF can suppress inflammation stimulated by virus. In addition, the regulation of inflammation markers was greater for SHL than that for CBF in both cell types. Consistent with this finding, SHL had a multi-channel anti-virus effect and protected against respiratory injury from $HAdV_3$ infection better than the monomer combination.

SHL has been used as an empirical therapy to treat virus infection in the TCM clinic for many years. This study is the first to evaluate comprehensively the role of SHL as an antiviral agent against $HAdV_3$. Furthermore, the antiviral activity of SHL was compared to CBF—the main compound combination in SHL—to evaluate the potential anti-virus activity of the effective parts and its composition. Parallel doses and combination ratios of these ingredients were determined and controlled by HPLC. We demonstrated that both SHL and CBF effectively inhibited $HAdV_3$-induced injuries by preventing viral penetration; un-coating; mRNA translation; protein synthesis; genome replication; and virus assembly and release to counteract viral infection. SHL appeared to be very promising in terms of efficacy and toxicity. This study also provides a primary explanation for the hypothesis that multiple bioactive ingredients with multiple targets of TCM are involved in the antivirus mechanism of SHL. Although CBF was not an ideal combination of the compound ingredient, our findings are helpful for related drug discovery and merit further exploration.

Acknowledgments: This work was supported by Grinds Gathers Project of Guangdong Province, China (2010A090200076).

Author Contributions: J.-B.L. and X.X. designed the study. Q.M., D.L. and S.S. wrote the paper. Q.M., Q.Y. and C.S. collected the samples. Q.M., X.X., D.L., S.S., Q.Y. and C.S. performed tests and analyzed the data. Q.M. thoroughly revised the manuscript.

Abbreviations

The following abbreviations are used in this manuscript:

ANOVA	analysis of variance
CBF	chlorogenic acid, baicalin, forsythia glycosides A
CC50	50% cytotoxic concentrations
CCK-8	cell counting kit-8
DAD	diode array detector
DMEM	Dulbecco's Modified Eagle's Medium
ELISA	enzyme-linked immunosorbent assay

FBS	fetal bovine serum
HAdV	human adenovirus
HIV	human immunodeficiency virus
HPLC	high performance liquid chromatography
HSD	honestly significant difference
IC50	50% inhibitory concentration
IFN	interferon
IL	interleukin
PCR	polymerase chain reaction
qPCR	quantitative polymerase chain reaction
S.D.	standard deviation
SHL	Shuang–Huang–Lian injectable powder
TCID50	50% tissue culture infection dose
TCM	traditional Chinese medicine
TNF-α	tumor necrosis factor alpha
DNA	deoxyribonucleic acid

References

1. Dehghan, S.; Liu, E.B.; Seto, J.; Torres, S.F.; Hudson, N.R.; Kajon, A.E.; Metzgar, D.; Dyer, D.W.; Chodosh, J.; Jones, M.S.; et al. Five genome sequences of subspecies B1 human adenoviruses associated with acute respiratory disease. *J. Virol.* **2012**, *86*, 635–636. [CrossRef] [PubMed]

2. Robinson, C.M.; Zhou, X.; Rajaiya, J.; Yousuf, M.A.; Singh, G.; DeSerres, J.J.; Walsh, M.P.; Wong, S.; Seto, D.; Dyer, D.W.; et al. Predicting the next eye pathogen: Analysis of a novel adenovirus. *MBio* **2013**, *4*, e00595:1–e00595:12. [CrossRef] [PubMed]

3. Yun, H.C.; Fugate, W.H.; Murray, C.K.; Cropper, T.L.; Lott, L.; McDonald, J.M. Pandemic influenza virus 2009 H1N1 and adenovirus in a high risk population of young adults: Epidemiology, comparison of clinical presentations, and coinfection. *PLoS ONE* **2014**, *9*, e85094. [CrossRef] [PubMed]

4. Lu, Q.B.; Tong, Y.G.; Wo, Y.; Wang, H.Y.; Liu, E.M.; Gray, G.C.; Liu, W.; Cao, W.C. Epidemiology of human adenovirus and molecular characterization of human adenovirus 55 in China, 2009–2012. *Influenza Other Respir. Viruses* **2014**, *8*, 302–308. [CrossRef] [PubMed]

5. Barrero, P.R.; Valinotto, L.E.; Tittarelli, E.; Mistchenko, A.S. Molecular typing of adenoviruses in pediatric respiratory infections in Buenos Aires, Argentina (1999–2010). *J. Clin. Virol.* **2012**, *53*, 145–150. [CrossRef] [PubMed]

6. Alkhalaf, M.A.; Guiver, M.; Cooper, R.J. Genome stability of adenovirus types 3 and 7 during a simultaneous outbreak in Greater Manchester, UK. *J. Med. Virol.* **2015**, *87*, 17–24. [CrossRef] [PubMed]

7. Ampuero, J.S.; Ocana, V.; Gomez, J.; Gamero, M.E.; Garcia, J.; Halsey, E.S.; Laguan-Torres, V.A. Adenovirus respiratory tract infections in Peru. *PLoS ONE* **2012**, *7*, e46898. [CrossRef] [PubMed]

8. Guo, L.; Gonzalez, R.; Zhou, H.; Wu, C.; Vernet, G.; Wang, Z.; Wang, J. Detection of three human adenovirus species in adults with acute respiratory infection in China. *Eur. J. Clin. Microbiol. Infect. Dis.* **2012**, *31*, 1051–1058. [CrossRef] [PubMed]

9. Lai, C.Y.L.C. Adenovirus serotype 3 and 7 infection with acute respiratory failure in children in Taiwan, 2010–2011. *PLoS ONE* **2013**, *8*, e53614.

10. Lee, W.J.; Jung, H.D.; Cheong, H.M.; Kim, K. Molecular epidemiology of a post-influenza pandemic outbreak of acute respiratory infections in Korea caused by human adenovirus type 3. *J. Med. Virol.* **2015**, *87*, 10–17. [CrossRef] [PubMed]

11. Wang, H.S.; Cheng, F.; Shi, Y.Q.; Li, Z.G.; Qin, H.D.; Liu, Z.P. Hypotensive response in rats and toxicological mechanisms induced by Shuanghuanglian, an herbal extract mixture. *Drug Discov. Ther.* **2010**, *4*, 13–18. [PubMed]

12. McDougall, B.; King, P.J.; Wu, B.W.; Hostomsky, Z.; Reinecke, M.G.; Robinson, W.J. Dicaffeoylquinic and dicaffeoyltartaric acids are selective inhibitors of human immunodeficiency virus type 1 integrase. *Antimicrob. Agents Chemother.* **1998**, *42*, 140–146. [PubMed]

13. Chiang, L.C.; Chiang, W.; Chang, M.Y.; Ng, L.T.; Lin, C.C. Antiviral activity of Plantago major extracts and related compounds in vitro. *Antivir. Res.* **2002**, *55*, 53–62. [CrossRef]

14. Liu, Z.; Zhao, J.; Li, W.; Shen, L.; Huang, S.; Tang, J.; Duan, J.; Fang, F.; Huang, Y.; Chang, H.; et al. Computational screen and experimental validation of anti-influenza effects of quercetin and chlorogenic acid from traditional Chinese medicine. *Sci. Rep.* **2016**, *6*. [CrossRef] [PubMed]

15. Gamaleldin, E.K.M.; Matei, M.F.; Jaiswal, R.; Illenberger, S.; Kuhnert, N. Neuraminidase inhibition of Dietary chlorogenic acids and derivatives—Potential antivirals from dietary sources. *Food Funct.* **2016**, *7*, 2052–2059. [CrossRef] [PubMed]

16. Li, X.; Liu, Y.; Hou, X.; Peng, H.; Zhang, L.; Jiang, Q.; Shi, M.; Ji, Y.; Wang, Y.; Shi, W. Chlorogenic acid inhibits the replication and viability of enterovirus 71 in vitro. *PLoS ONE* **2016**, *7*, 2052–2059. [CrossRef] [PubMed]

17. Zhang, C.J.; Gu, L.G.; Yu, H.T. Antagonism of baicalin on cell cyclical distribution and cell apoptosis in A549 cells infected with influenza A (H1N1) virus. *Bing Du Xue Bao* **2011**, *27*, 108–116. (In Chinese). [PubMed]

18. Moghaddam, E.; Teoh, B.T.; Sam, S.S.; Lani, R.; Hassandarvish, P.; Chik, Z.; Yueh, A.; Abubakar, S.; Zandi, K. Baicalin, a metabolite of baicalein with antiviral activity against dengue virus. *Sci. Rep.* **2014**, *4*. [CrossRef] [PubMed]

19. Cheng, K.; Wu, Z.; Gao, B.; Xu, J. Analysis of influence of baicalin joint resveratrol retention enema on the TNF-alpha, SIgA, IL-2, IFN-gamma of rats with respiratory syncytial virus infection. *Cell Biochem. Biophys.* **2014**, *70*, 1305–1309. [CrossRef] [PubMed]

20. Li, H.; Wu, J.; Zhang, Z.; Ma, Y.; Liao, F.; Zhang, Y.; Wu, G. Forsythoside A inhibits the avian infectious bronchitis virus in cell culture. *Phytother. Res.* **2011**, *25*, 338–342. [CrossRef] [PubMed]

21. Chen, L.; Zhang, Y.; Liu, J.; Wei, L.; Song, B.; Shao, L. Exposure of the murine RAW 264.7 macrophage cell line to dicalcium silicate coating: Assessment of cytotoxicity and pro-inflammatory effects. *J. Mater. Sci. Mater. Med.* **2016**, *27*. [CrossRef] [PubMed]

22. Chang, J.S.; Wang, K.C.; Shieh, D.E.; Hsu, F.F.; Chiang, L.C. Ge-Gen-Tang has anti-viral activity against human respiratory syncytial virus in human respiratory tract cell lines. *J. Ethnopharmacol.* **2012**, *139*, 305–310. [CrossRef] [PubMed]

23. Bartee, E.; Mohamed, M.R.; McFadden, G. Tumor necrosis factor and interferon: Cytokines in harmony. *Curr. Opin. Microbiol.* **2008**, *11*, 378–383. [CrossRef] [PubMed]

24. McFadden, G.; Mohamed, M.R.; Rahman, M.M.; Bartee, E. Cytokine determinants of viral tropism. *Nat. Rev. Immunol.* **2009**, *9*, 645–655. [CrossRef] [PubMed]

25. Fang, J.; Cheng, Q. Etiological mechanisms of post-stroke depression: A review. *Neurol. Res.* **2009**, *31*, 904–909. [CrossRef] [PubMed]

26. Trefler, J.; Paradowska-Gorycka, A.; Lacki, J.K. Influence of genetic factors on development and severity of rheumatoid arthritis—Part II. *Pol. Merkur. Lekarski* **2009**, *27*, 161–165. [PubMed]

27. Masters, S.L.; Simon, A.; Aksentijevich, I.; Kastner, D.L. Horror autoinflammaticus: The molecular pathophysiology of autoinflammatory disease. *Annu. Rev. Immunol.* **2009**, *27*, 621–668. [CrossRef] [PubMed]

4

Seasonal Dynamics of Haptophytes and dsDNA Algal Viruses Suggest Complex Virus-Host Relationship

Torill Vik Johannessen [1], Aud Larsen [2], Gunnar Bratbak [3], António Pagarete [3], Bente Edvardsen [4], Elianne D. Egge [4] and Ruth-Anne Sandaa [3],*

[1]   Vaxxinova Norway AS, Kong Christian Frederiks plass 3, 5006 Bergen, Norway; torillvjohannessen@gmail.com

[2]   Uni Research Environment, N-5008 Bergen, Norway; aud.larsen@bio.uib.no

[3]   Department of Biology, University of Bergen, N-5020 Bergen, Norway; gunnar.bratbak@bio.uib.no (G.B.); antonio.pagarete@uib.no (A.P.)

[4]   Department of Biosciences, University of Oslo, 0316 Oslo, Norway; bente.edvardsen@ibv.uio.no (B.E.); elianne.egge@gmail.com (E.D.E.)

*   Correspondence: ruth.sandaa@bio.uib.no

Academic Editors: Mathias Middelboe and Corina P.D. Brussaard

Abstract: Viruses influence the ecology and diversity of phytoplankton in the ocean. Most studies of phytoplankton host–virus interactions have focused on bloom-forming species like *Emiliania huxleyi* or *Phaeocystis* spp. The role of viruses infecting phytoplankton that do not form conspicuous blooms have received less attention. Here we explore the dynamics of phytoplankton and algal viruses over several sequential seasons, with a focus on the ubiquitous and diverse phytoplankton division Haptophyta, and their double-stranded DNA viruses, potentially with the capacity to infect the haptophytes. Viral and phytoplankton abundance and diversity showed recurrent seasonal changes, mainly explained by hydrographic conditions. By 454 tag-sequencing we revealed 93 unique haptophyte operational taxonomic units (OTUs), with seasonal changes in abundance. Sixty-one unique viral OTUs, representing Megaviridae and *Phycodnaviridae*, showed only distant relationship with currently isolated algal viruses. Haptophyte and virus community composition and diversity varied substantially throughout the year, but in an uncoordinated manner. A minority of the viral OTUs were highly abundant at specific time-points, indicating a boom-bust relationship with their host. Most of the viral OTUs were very persistent, which may represent viruses that coexist with their hosts, or able to exploit several host species.

Keywords: Haptophyta; *Phycodnaviridae*; Megaviridae; viral–host interactions; metagenomics; marine viral ecology

---

1. Introduction

Marine phytoplankton account for approximately 50% of global primary production and have a strong impact on global nutrient cycling [1]. As key components within the phytoplankton community in both coastal and open oceans, and at all latitudes [2], haptophytes play important roles both as primary producers but also as mixotrophs, grazing on bacteria and protist [3]. Blooms of haptophytes can have significant ecological and economic impacts both through the amount of organic matter being produced and through production of toxins harmful to marine biota [4]. Most haptophyte species, however, do not usually form blooms, but rather appear at low concentrations at all times [5–7].

Phytoplankton diversity, abundance, and community composition change through the seasons, driven by variations in environmental conditions and biological processes. Viruses can, in theory, significantly condition those dynamics. Viral-based phytoplankton lysis can be at least as significant as grazing [8,9] and have the potential to drastically change host community structure [10]. Viral activity related to bloom forming haptophytes like *Emiliania huxleyi*, *Phaeocystis pouchetii*, and *Phaeocystis globosa* has been well studied [9,11–14]. During such blooms, viruses exhibit a strong regulatory role, and contribute to the termination of the bloom in what may be referred to as a "boom and bust" relationship [11,15,16]. Viruses may also prevent bloom formation by keeping host population at non-blooming levels [16–18]. Such interactions between host and virus have been described as a stable coexistence and explained by viral resistance, immunity and/or strain specificity [17,19–23].

The low diversity and high abundance, which characterize phytoplankton blooms, give species of specific viruses ample possibilities to find susceptible hosts. Most haptophyte species, such as species belonging to the Prymnesiales, however, are part of highly-diverse communities and occur at low concentrations [5–7], which decrease their chance of being infected by viruses with specific host requirements. Nevertheless, viruses infecting both *Prymnesium* and *Haptolina* species (order Prymnesiales) have been isolated, but have several characteristics that distinguish them from viruses infecting bloom-forming haptophytes like *E. huxleyi* [24,25]. Studies describing the seasonal diversity and abundances of these viruses and their potential host communities (haptophytes) in the environment are scarce.

All known haptophyte viruses have double-stranded DNA (dsDNA) genomes and belong to two related viral families, the *Phycodnaviridae* and Megaviridae, within a monophyletic group of nucleocytoplasmic large DNA viruses (NCLDV) [26]. Phycodnaviruses infect prasinophytes, chlorophytes, raphidophytes, phaeophytes, and haptophytes [27]. The Megaviridae family, not yet recognized as a taxon by the International Committee on Taxonomy of Viruses (ICTV), consists of NCLDVs infecting both non-photosynthetic protists such as *Acanthamoeba* and *Cafeteria roenbergensis* [28,29], as well as photosynthetic ones including prasinophytes, pelagophytes and prymnesiophytes (haptophytes) [14,30,31]. Both *Phycodnaviridae* and Megaviridae are abundant in aquatic environments but the majority are uncultured and not yet described [31–37]. The diversity within these two families is high, and available primers only match a fraction of its representatives [32,38,39]. Moreover, only few polymerase chain reaction (PCR) primer-sets that target *Phycodnaviridae* and Megaviridae families are currently available [32,38]. The DNA polymerase B primers (polB) capture a wide diversity within the *Phycodnaviridae* family including the prasinoviruses and chloroviruses [36–39] whereas the major capsid protein (MCP) primers are better suited for capturing the diversity of the Megaviridae family including prymnesioviruses that infect various haptophytes ([32,39], this study). Coccolithoviruses (e.g., *Emiliania huxleyi virus* (EhV)), a diverged group in the *Phycodnaviridae* family, are not targeted by any of these primer-sets.

In previous studies, we have described the microbial community dynamics of the seasonal spring- and fall-blooms in a West Norwegian open fjord system (Raunefjorden) [40,41]. Virus infection seems to be one of the factors that drive the succession in the haptophyte community after the typical diatom spring bloom [40]. In the present study, we follow up on these investigations using methods with higher taxonomic resolution that enable a more specific focus on haptophytes and their potential viruses. By following dynamics and diversity in virus and haptophyte communities over a two-year period, we aimed at revealing the possible regulatory role of viruses, not only during blooms but also during periods with higher community diversity and lower productivity such as late fall and winter.

2. Results

2.1. Microbial Abundance and Abiotic Factors

The phytoplankton spring bloom, identified as elevated chlorophyll *a* (Chl *a*) fluorescence, started in late February before any stratification of the water masses, and lasted longer in 2011 than in 2010 (Figure 1). The water masses in Raunefjorden started to stratify in March–April, and the stratification was more pronounced and deeper in 2011 than in 2010 (Figure 1).

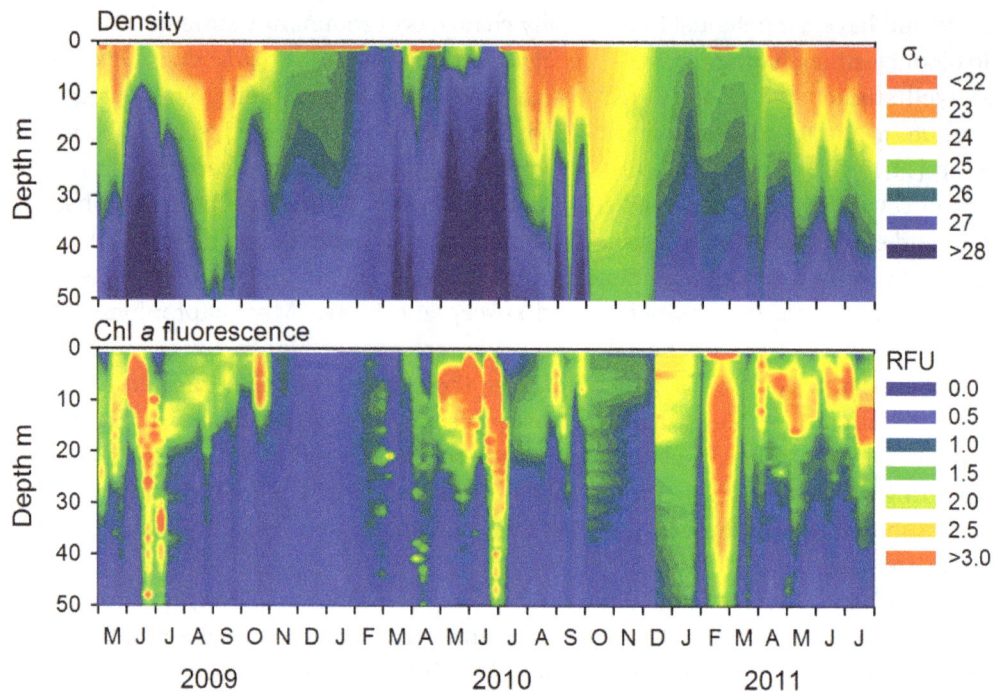

Figure 1. Isopleth diagrams showing seawater density (σt) and chlorophyll *a* (Chl *a*) fluorescence (RFU = relative fluorescence units) at the sampling station in Raunefjorden, respectively.

Several minor upwelling events and exchange of water masses were evident in spring (e.g., in June 2009, April 2010 and June 2010); concurrently May and June were characterized by several successive blooms with high Chl *a* levels (2–8 µg per L). The pycnocline deepened throughout summer and fall before the seasonal inflow and upwelling caused deep mixing in late fall, which corresponded to a temporary, slight increase in Chl *a* concentrations in fall each year (October–November). The water masses were well mixed through fall and winter.

The first increase in pico- and nano-eukaryote abundance, as measured by flow cytometry, was observed in late February (Figure 2A). The cell numbers increased throughout spring and summer with maximum abundance of both groups in August 2010 and May/June 2011. Total bacterial abundance was variable with a decreasing trend in fall and winter and an increasing trend in spring and summer-fall (Figure 2B), while the *Synechococcus* (cyanobacteria) abundance peaked once each year in late summer-fall (Figure 2B).

Viral abundance increased in the spring and summer. The highest values were found during summer and fall. The abundance of all three viral groups varied in synchrony (V2 vs. V1: $r = 0.603$, df = 26, $p < 0.0007$, V3 vs. V1: $r = 0.483$, df = 26, $p < 0.0091$) and the smaller viruses (V1) outnumbered the larger viruses (V2 and V3) by a factor of approximately 5–20 and 50–300, respectively (Figure 2C). Correlation analysis showed that the viral abundance was correlated ($p < 0.01$) with the abundance of bacteria, cryptophytes and *Synechococcus* (Table S1). The abundance of small-sized viruses (V1) also correlated with Chl *a* and abundance of nanoeukaryotes, while the abundance of intermediate-sized viruses (V2) correlated with abundance of picoeukaryotes and nanoeukaryotes.

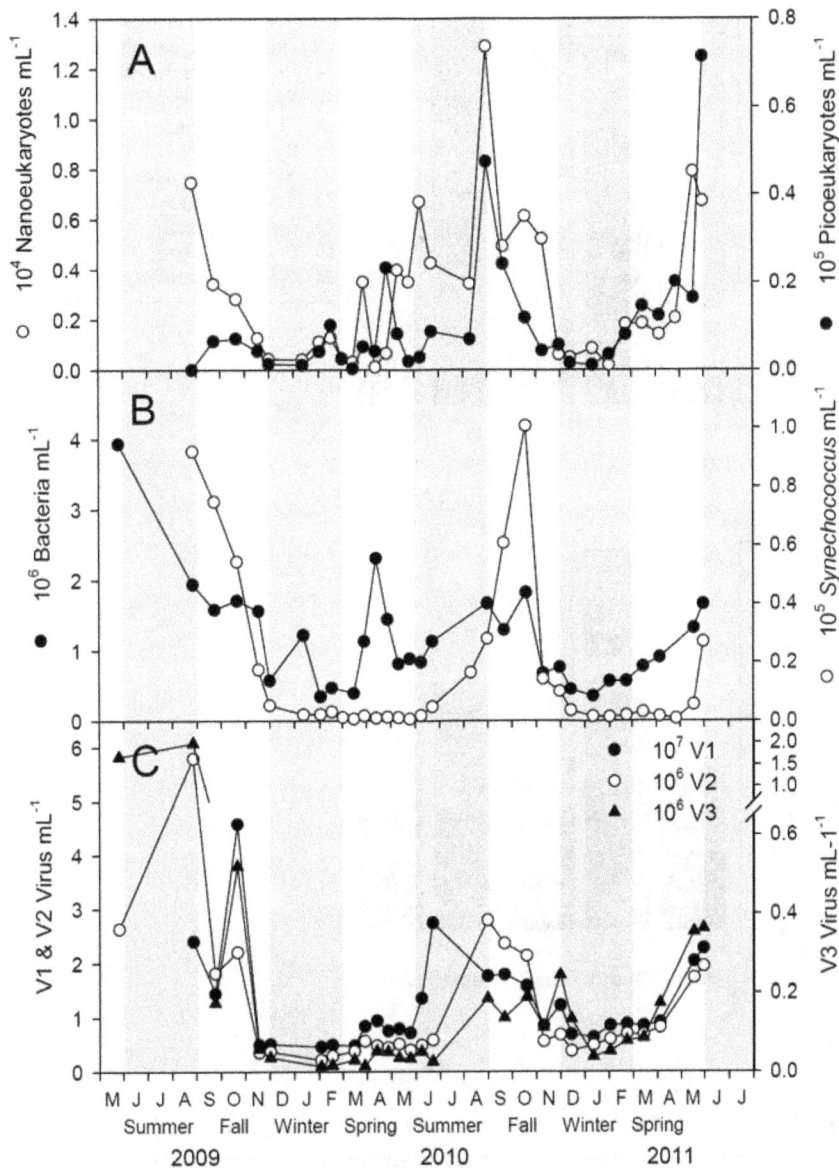

Figure 2. Abundance of microbial plankton at the sampling station in Raunefjorden measured by flow cytometry. (**A**) Phototrophic picoeukaryotes (filled circles) and nanoeukaryotes (open circles); (**B**) *Synechococcus* (open circles) and total bacteria (filled circles); and (**C**) V1 (low fluorescence viruses), V2 (intermediate fluorescence viruses) and V3 (high fluorescence viruses).

2.2. Haptophytes

Haptophyte reads (sequences), clustered based on 98% nucleotide sequence similarity, formed a total of 93 operational taxonomic units (OTUs) (Table S2). OTUs were classified against a curated Haptophyta reference sequence database [42] to the lowest reliable taxonomic level (Table S2). The classified OTUs were placed into one of seven haptophyte orders: Pavlovales, Phaeocystales, Zygodiscales, Syracosphaerales, Isochrysidales, Coccolithales and Prymnesiales (Figure 3). A number of the reads could not be classified to these formally-accepted taxa, and were assigned to defined clades without cultured representatives according to [42], here named Haptophyta unclassified (Clades HAP2, HAP3, HAP4, and HAP5) and Prymnesiophyceae unclassified (Clades B3, B4, D, E and F) (Table S2). Prymnesiales is, in Figure 3, divided into the families Chysochromulinaceae and Prymnesiaceae. OTUs assigned to the order Isochrysidales all belonged to the *E. huxleyi* cluster. A more detailed classification of the 93 haptophyte OTUs to is shown in Table S2.

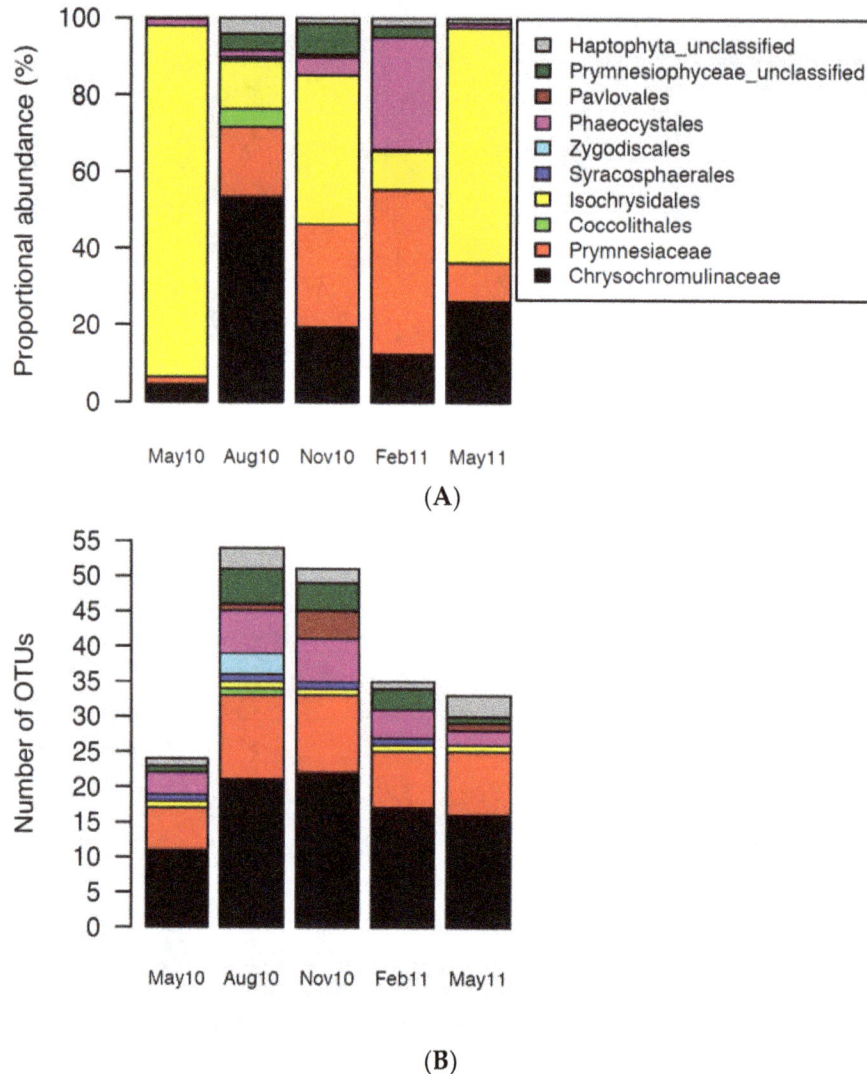

Figure 3. (**A**) Relative abundance of operational taxonomic units (OTUs) in the different haptophyte orders or clades. The OTUs represented seven accepted haptophyte orders: Pavlovales, Phaeocystales, Coccolithales, Isochrysidales, Syracosphaerales, Zygodiscales and Prymnesiales. The latter is represented by the families Chysochromulinaceae and Prymnesiaceae. The reads that did not belong to any of these orders were assigned to class Prymnesiophyceae (unclassified) or Haptophyta (unclassified); (**B**) Number of unique OTUs (richness) in each sample after rarifying to lowest read abundance (i.e., subsampling to obtain equal sample size).

Diversity and community composition varied between the samples, with highest diversity in August and lowest in May (Shannon diversity values of 2.97 and 0.50 respectively, Table S3). Based on the Bray-Curtis dissimilarity analysis we found that the August 2010 sample differed most from the rest (Figure 4).

OTUs assigned to Isochrysidales (only *E. huxleyi*) were present in all samples, with particularly high relative abundance in May. OTUs assigned to Prymnesiales, i.e., the families Prymnesiaceae and Chrysochromulinaceae, dominated the samples from August, November and February, while OTUs belonging to Phaeocystales occurred in high relative abundance only in February. Diversity was highest within Prymnesiales, with 35 and 21 unique OTUs assigned to Chrysochromulinaceae and Prymnesiaceae, respectively (Figure 3B). Ten different haptophyte OTUs were present in all samples, one was classified to *E. huxleyi*, one to Clade F, and eight to Prymnesiales (Table S2).

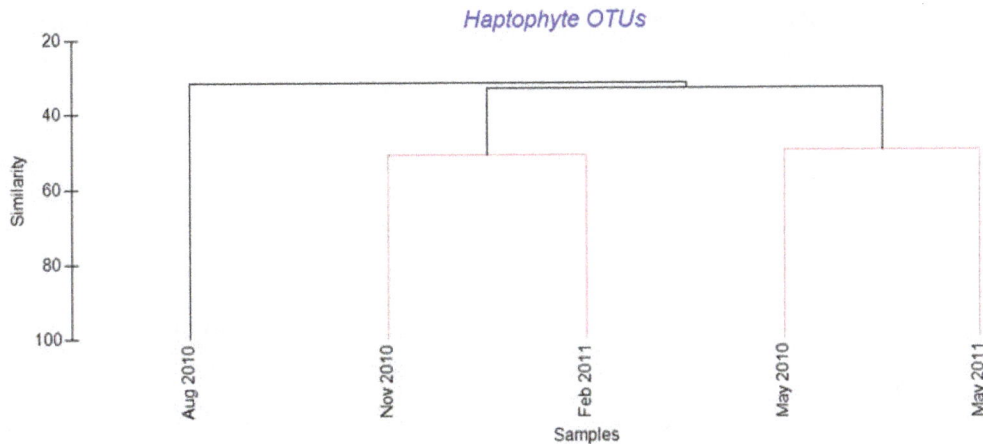

Figure 4. Cluster dendrogram illustrating Bray-Curtis dissimilarity in the haptophyte OTU compositions between the five samples from Raunefjorden. OTUs were defined as reads with $\geq 98\%$ nucleotide similarity. Sequences were normalized to 100 in each sample and log-transformed prior to similarity calculations. Samples connected by red lines were not significantly differentiated (SIMPER permutation test). Black lines indicate significant differentiation ($p < 0.05$, SIMPER permutation test).

2.3. Megaviridae and Phycodnaviridae

All the quality-trimmed viral reads showed similarity to algal viruses in the Megaviridae and *Phycodnaviridae* families, with BLAST scores between 50 and 90% amino acid sequence identity. OTU clustering based on 95% amino acid identity gave a total of 161 OTUs containing 10 or more sequences (Table S4), with 61 being unique (Table S5). Forty-one and 20 of these OTUs showed highest similarity to the Megaviridae and *Phycodnaviridae* families, respectively (Table S5). Half of the OTUs (53%) were rare, each comprising less than 1% of the total reads (Table S5). The diversity was highest in May 2010 and lowest in February 2011 (Shannon diversity values of 2.66 and 1.45, respectively) (Table S4). Based on Bray-Curtis dissimilarity, the May 2010 sample differed significantly from the other 4 samples (Figure 5).

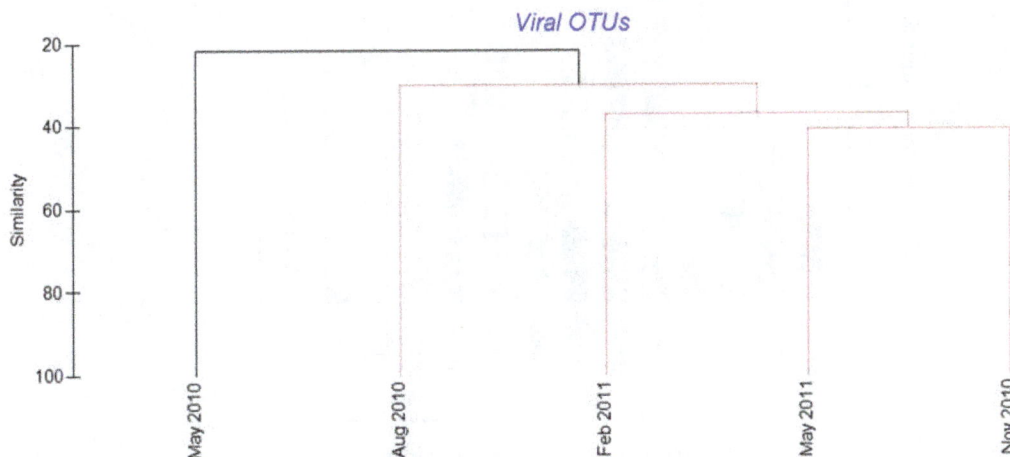

Figure 5. Cluster dendrogram illustrating Bray-Curtis dissimilarity between the virus OTU compositions in five samples from Raunefjorden. OTUs were defined as sequences with >95% amino acid similarity. The sequence data were normalized to 100 in each sample and log-transformed prior to similarity calculations. Samples connected by red lines could not be significantly differentiated (SIMPER permutation test). Black lines indicate significant differentiation ($p < 0.05$, SIMPER permutation test).

Fifty-eight percent of the viral OTUs were found in more than two of the samples (Table S5). Five viral OTUs (OTU009, OTU002, OTU001, OTU003, OTU008) were present in all the samples and dominated the samples from November, February2010 and May 2011. Four of these clustered within the *Phycodnaviridae* family. Others, such as OTU006, OTU373 and OTU010 that dominated the samples taken in May and August 2010, respectively, were almost absent or undetectable the rest of the year (Table S5, Figure 6).

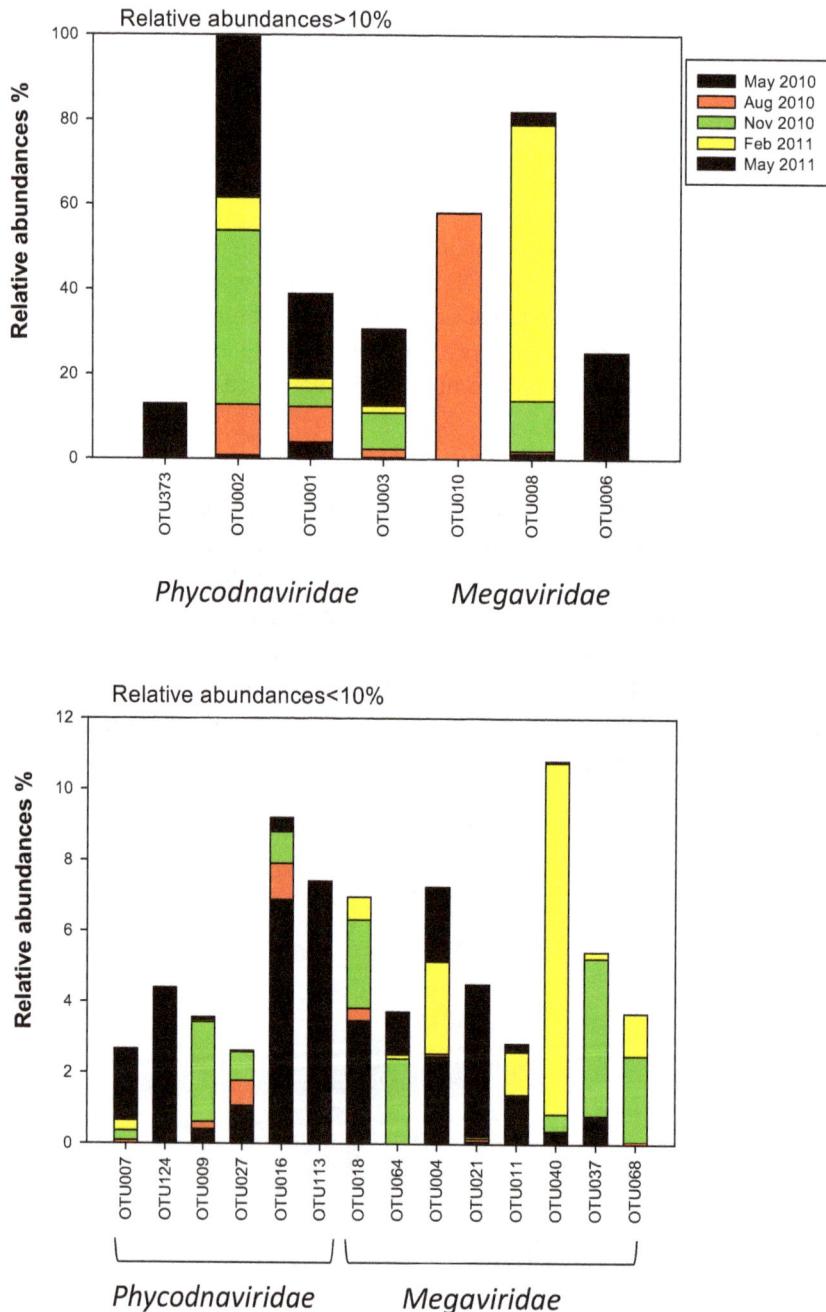

Figure 6. Relative abundances of 21 different OTUs containing more than 50 reads in the different samples.

The Megaviridae-like OTUs dominated the samples from August and February (Figure 6, Table S5). Three OTUs occurred in relative abundances over 10% (OTU010, OTU008 and OTU006). Two of them dominated the samples in August and February, the third dominated in May 2010. Further, eight OTUs occurred in low relative abundances (<10%) and were present in three or more samples. The most abundant algal Megaviridae-like OTU, OTU008, was distantly related to a virus infecting the haptophyte *Prymnesium kappa* (RF01) [25], with low bootstrap support (41%) (Figure 7). OTU008 dominated the sample from February (65%) and was present in all samples (relative abundances between 0.32–65%) (Figure 6, Table S5). The single most abundant OTU in May samples (OTU006) was also grouped within Megaviridae and clustered together with viruses infecting different haptophytes such as *Chrysochromulina, Phaeocystis* and *Prymnesium* (Figures 6 and 7). Three other OTUs (OTU037, OTU068 and OTU040) clustered within this clade as well. OTU037 and OTU040 were present at relatively low abundances (0.04–9.9%) in May and November 2010, and February 2011, while OTU068 were present at relative low abundances (0.09–2.4%) in samples from August and November 2010, and February 2011 (Figures 6 and 7, Table S5). Three new branches were made next to the Megaviridae family consisting of 4 OTUs from this study (Figure 7) where OTU010 and OTU18 clustered together with an environmental sequence from an earlier study at the same site [32].

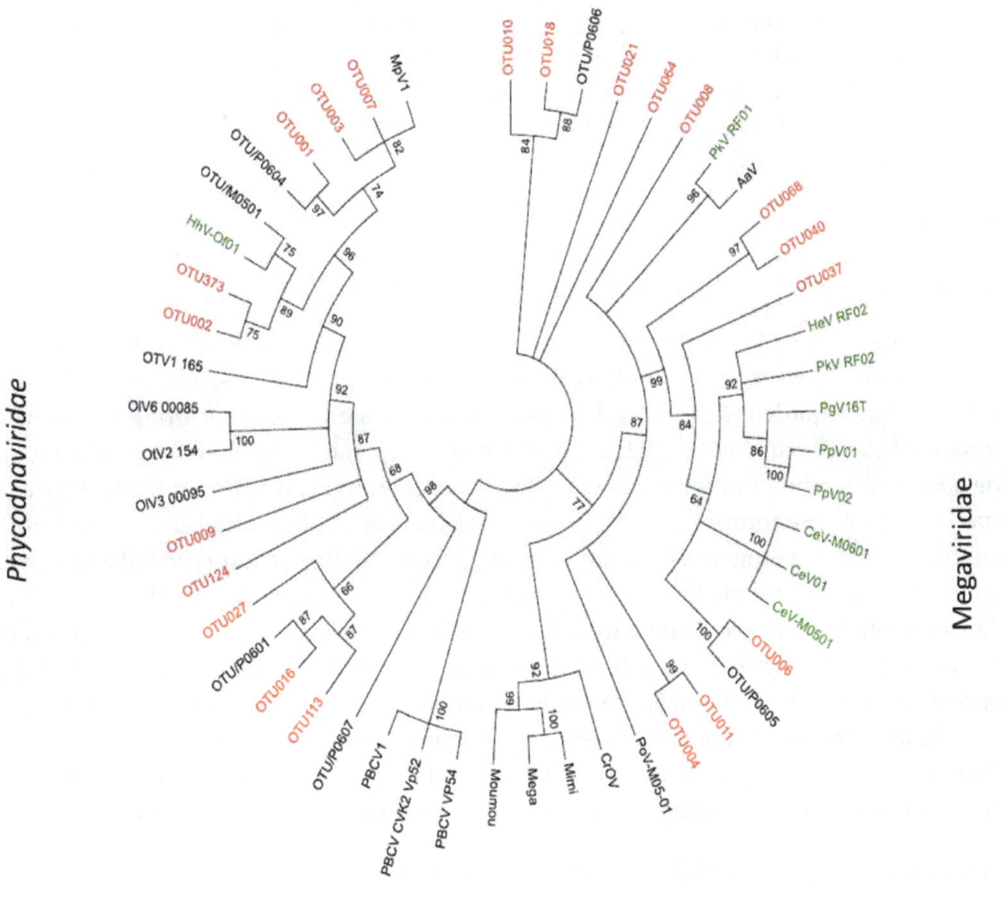

Figure 7. Midpoint rooted phylogenetic tree of the most abundant OTUs (>50 reads, marked in red) with similarities to the Megaviridae and *Phycodnaviridae* families, respectively. The tree was calculated based on the DNA-sequences encoding partial MCP-genes (FastTree v2.1.8 with default parameters). Bootstrap values form 100 replicates and aLRT- likelihood-values >0.5 are shown on nodes. Abbreviations: CroV; *Cafeteria roenbergensis virus*, Moumou; *Moumouvirus goulette*, Mimi; *Mimivirus*, Mega; *Megavirus chiliensis*, AaV; *Aureococcus anophagefferens virus*, PoV; *Pyramimonas orientalis virus*, PkV; *Prymnesium kappa virus*, HeV; *Haptolina ericina virus*, HhV; *Haptolina hirta virus*, CeV; *Chrysochromulina ericina virus*, PgV; *Phaeocystis globosa virus*, PpV; *Phaeocystis pouchetii virus*, PBCV; *Paramecium bursaria Chlorella virus*, MpV; *Micromonas pusilla virus*, OsV; *Ostreococcus sp. virus*, OlV; *Ostreococcus lucimarinus virus*. OTU/P0605, OTU/P0606, OTU/P0604, OTU/P0607, OTU/P0601, OTU/M0501 are all sequences from an earlier study in Raunefjorden [32]. Reference strains marked in green are haptophyte-infecting viruses maintained in culture.

OTU010 dominated in the sample from August (58%) and occurred at low relative abundances (0.04–0.05%) in the samples from November 2010 and May 2011 (Table S5). OTU18 occurred at low abundances in samples from May 2010, August, November and February. The two other collapsed branches comprised OTU064 and OTU021, both occurring at low abundances at three and four sample periods, respectively (Figure 6, Table S5). The two last OTUs in the Megaviridae family (OTU011, OTU004) showed highest similarity to a virus infecting the chlorophyte *Pyromimonas orientalis*. Both occurred at less than 10% in samples from May 2010, February 2011 and May 2011.

The *Phycodnaviridae* OTUs (Figure 6, Table S5) consisted of four OTUs (OTU373, OTU002, OTU001, OTU003) with relative abundances over 10% and six OTUs (OTU007, OTU124, OTU009, OTU027, OTU016, OTU113) with relative abundances below 10%. Three OTUs (OTU001, OTU002 and OTU003) dominated the samples from November and May 2011. They cluster within two subclades together with two cultured viruses, the prasinovirus *Micromonas pusilla virus*, and an unclassified virus shown to infect the haptophyte *Haptolina hirta* (HhV-Of01) [43] (Figure 7). Both clades also included environmental clones previously obtained at the same location [32]. The OTU016 and OTU113 also grouped together with an environmental clone from Raunefjorden [32]. OTU113 occurred only in May 2010 while OTU016 in addition were present on three other occasions (Figure 6, Table S5). OTU009, OTU124 and OTU027 did not match any viral sequences in GenBank. They were never abundant (relative abundances between 0.04–3.1%), but some were frequently observed (e.g., OTU009 and OTU027) (Figure 6, Table S5).

3. Discussion

3.1. Seasonal Patterns in Microbial Dynamics and Environmental Factors

The biological variables (Chl *a* concentration, phytoplankton, bacterial, and viral abundances) followed a seasonal and recurrent pattern that corroborated earlier descriptions of the microbial community in Raunefjorden [40,41,44–46]. The conditions in late winter and spring were nevertheless quite different in 2010 compared to 2011. The Chl *a* values were much higher in 2011 and nanoplankton, bacteria and virus abundances were stable or steadily increasing, although not fluctuating as in 2010. A deeper and more pronounced pycnocline in 2011 than in 2010 resulted in a more stable water column that sustained a longer-lasting bloom in 2011 than in 2010. In summer, fall and early winter, hydrographic conditions for the two years resembled each other, as did Chl *a* level and variation. High abundances of phototropic pico- and nanoeukaryotes and *Synechococcus* (Figure 2) matched peaks in Chl *a* concentration in August 2009, in June, September and October 2010, and in May 2011 (Figure 1). The concentration of Chl *a* in Raunefjorden is, however, largely determined by the abundance of larger phytoplankton forms like diatoms that were not counted in this study [40,44]. On several occasions (e.g., June 2009, June and September 2010, and March–May 2011) the decrease in Chl *a* concentrations may be related to a concurrent deep mixing and exchange of water masses (Figure 1).

3.2. Succession of Haptophytes and Co-Occurring DNA Viruses

454 pyrosequencing revealed a high diversity of haptophyte OTUs as well as of algal viruses. The diversity of haptophyte OTUs found in this study was larger than reported in earlier studies using microscopy (summarized in [47]). This demonstrates how high throughput sequencing of amplicon libraries is a powerful tool for detecting haptophyte species not yet morphologically and genetically characterized [48,49]. The level of haptophyte richness measured in Raunefjorden (93 different haptophyte OTUs) was at the same order of magnitude as the level previously found in Oslofjorden (156 haptophyte OTUs), a study for which the same sequencing technology and primers were used [47]. The loss of reads in the filtering process (Table S3) was high but can be explained by the fact that the multiplex identification *tag* was only present on the forward primer, leading to loss of nearly half of the reads.

Both haptophyte and viral communities varied substantially throughout the seasonal cycle. However, we could not distinguish a synchronization between their respective compositions. This may be explained by a complex virus-host relationship, or that our molecular approach was not sensitive enough to capture specific haptophyte viral-host pairs. Most empirical data on phytoplankton diversity resolve the host community diversity on a relatively coarse level, namely with host subgroups or species, typically defined by small subunit ribosomal RNA (SSU rRNA) gene sequences. Viral host-range, however, commonly dwells within strain or sub-species diversity levels [50]. Hence, 18S rRNA gene marker, as used in this and most other studies, might not be sensitive enough to capture the interaction dynamics between the viral sequences here observed and the true host to which they correspond.

Some viral OTUs resembled cultured haptophyte viruses but they were, in most cases, only distantly related. Others were more similar to viruses infecting other host groups. Due to the large diversity of haptophyte viruses and the paucity of isolated viruses infecting this important host group, there is, at present, no molecular approach available that allows us to target these viruses with specificity. Moreover, viral phylogeny does not necessarily reflect host phylogeny. Several algal viruses have been shown to cross host-species borders, and some infect hosts that are only distantly related [25]. Despite these challenges, our viral data did enable detection of successional patterns providing new insight into the interaction between viruses and their hosts. Some viral OTUs were highly abundant only at specific time-points, indicating a boom-bust relationship with their host, a pattern normally described for lytic viruses [11–13,15]. Surprisingly though, most of the viral OTUs were persistent indicating coexistence with their hosts, or alternatively an ability to exploit several host species.

Bloom communities normally comprise a few, and often recurrent, species [40,44]. Therefore, we were not surprised to find low haptophyte diversity, dominated by *E. huxleyi*, the common bloom-forming coccolithophores, and Chrysochromulinaceae (OTU001 and OTU004, respectively) in May both years. More to our surprise though, the diverse community of *Phycodneaviridae* and Megaviridae co-occurred with this recurrent, low-diversity haptophyte community, and the observed viral OTUs varied substantially between the two years. One possible explanation may be that genetically different viruses are exploiting the same hosts [25,51,52]. Even more surprising was the absence of the EhV in our flow cytometry (FCM) analysis. Blooms of *E. huxleyi* are frequently succeeded by large increases in this specific virus [53–55] which have a FCM characteristics that make them easy to distinguish and recognize [11,56] even without primers that capture their presence at the molecular level.

Our observation that the abundance of pico- and nanoeukaryotes and the diversity of haptophytes peaked in August is in accordance with the general narrative that relatively small forms typically dominate the diverse phytoplankton communities thriving in the temperate, stable, and nutrient-depleted summer water masses. The abundance of larger viruses (V2 and V3), i.e., viruses having a size typical of many dsDNA phytoplankton viruses [40,57], were also at their highest in the summer period. The *Phycodnaviridae* and Megaviridae communities were, however, dominated by a single OTU (OTU010), which was only distantly related to any known viral sequence. Thus, the diverse phytoplankton community in August seems to have sustained a high virus abundance with low diversity. One interpretation of this is that OTU010 represents a generalist virus type, able to infect several different species, a feature known for many viruses infecting members of Prymnesiales [25]. Other interpretations may be that phytoplankton viruses that are not targeted by our primers prevailed [25,32] or that the large viruses are related to other hosts groups.

Several of the haptophytes that we detected in our study are known to be susceptible to characterized algal viruses [24,25,58]. Viruses infecting *P. pouchetii*, *P. globosa*, *Haptolina ericina and Prymnesium kappa* have previously been isolated from Norwegian coastal waters and/or the North Sea [24,25,58,59], but none of the viral OTUs we found was similar to any of these characterized viruses within 95% aa similarity. Despite the low similarity, five of the OTUs clustered within the Megaviridae group together with several cultured representatives of viruses infecting the orders Prymnesiales and Phaeocystales. As very few algal viruses are cultured, our results may suggest that the diversity within this viral group is large. Alternatively, cultivated viruses may not represent the most abundant viral strains present in natural systems as current procedures for virus isolation [60] entail a strong selective pressure favoring lytic viruses with short replication cycles, a strategy perhaps not very common strategy in nature.

Some haptophyte and viral OTUs were remarkably persistent, considering that the samples were collected at different seasons, interspersed by mixing of the water-masses and changing environmental conditions. Haptophyte OTUs that were present in all samples may represent species that are able to tolerate a wide range of environmental conditions, either as actively dividing cells, or surviving periods of low activity. Most of these persistent OTUs were classified to Prymnesiales (eight OTUs), an order including several mixotrophic species known to survive even when light conditions are too low for photosynthesis [3]. A high degree of preservation and recurrence of virus-genotypes through the years have previously been observed for myovirus-like viruses [46], but this is the first observation for algal viruses. These year-long observations are contrasting and complementary to previous studies demonstrating clear boom and bust patterns for abundant algae and their viruses [15,53,61]. Viral particles are estimated to quickly degrade in seawater (inactivation rates of 5%–20% per h) [62–64], and should quickly disappear without co-occurring susceptible and active hosts. Persistent viral OTUs thus indicate either that they propagate and co-exist with a persisting host, or that they are able to infect various species. Virus-host coexistence [19,21,65] is regarded to be a paradox, since most cultured viruses quickly induce resistance in their hosts [21,66] and may be attributed to partial host resistance (strain specificity), low virus infectivity [21], or to chronic infections where only few cells in the host population produce the virus, while the rest grow normally [9,67]. Another possibility is that the persistent viruses have wide host ranges, which would allow them to proliferate on different host species [25]. The ability to infect several species may be especially beneficial in times when the phytoplankton community is very diverse or at low phytoplankton abundance and activity.

This inter-annual study of microbial communities in Raunefjorden is the first to apply high throughput sequencing to study seasonal variation in marine uncultured algal and viral communities. Five "snapshots" of the haptophyte and algal virus (*Phycodnaviridae* and Megaviridae) communities covering one year revealed a large diversity with many uncultured and unknown forms although we identified a stable "core" community of haptophyte and viral OTUs as well. Some abundant viral OTUs showed high relative abundance in several samples indicating virus-host coexistence or wider host-range than what we would expect from the existing isolates. The diversity varied a lot, and low haptophyte diversity in May was accompanied by high algal virus diversity whereas high haptophyte diversity in August co-occurred with low *Phycodnaviridae* and Megaviridae diversity. We suggest that several viruses may exploit the same hosts in the low-diversity spring communities, while a few viruses may be able to exploit several of the haptophytes in the high-diversity community in late summer. Notably, measured virus and host abundance illustrates the importance of viral caused mortality in the diverse late summer community.

4. Material and Methods

4.1. Sample Collection

Seawater samples were collected from 5 m depth in Raunefjorden (60°16.2′ N, 5°12.5′ E) Western Norway, between May 2009 and May 2011. The sampling interval was, with a few exceptions, 2–4 weeks. Temperature, salinity, density and Chl a fluorescence were determined using a CTD (Conductivity-Temperature-Depth) equipped with an in situ fluorometer (SD204 SAIV, SAIV A/S Environmental Sensors & Systems, Bergen, Norway). A 20 L aliquot of sampled water was filtered by peristaltic pumping through 3.00 μm and then through 0.45 μm pore-sized low-protein-binding filters (145 mm, Durapore, Millipore Corp., Billerica, MA, USA), within 3 h of collection. The filters were cut in two and immediately frozen in liquid N2 and thereafter stored at $-80\,°C$ until DNA-extraction for later use in PCR and 454 sequencing of the haptophyte community. Viruses in the 20 L 0.45 µm filtrate were concentrated 400 times (approximately 50 mL) using a tangential flow filtration system equipped with a 100,000 pore size (NMWC) hollow-fiber cartridge (QuixStand, GE Healthcare Bio-Sciences AB, Uppsala, Sweden). Aliquots of these viral concentrates were frozen at -80 °C for later use in PCR and 454 sequencing.

4.2. Microbial Abundance Measured by Flow Cytometry (FCM)

Phototrophic pico- and nano- plankton were counted in triplicate by FCM (Becton, Dickinson and Company, BD Biosciences, San Jose, CA, USA), using fresh, unpreserved samples, with the trigger set on red fluorescence and a flow rate giving 50–800 events per sec. Five different populations of phototrophs were defined in FCM-plots based on differences in side scatter, red and orange fluorescence: Synechococcus, picoeukaryotes, nanoeukaryotes, cryptophytes and E. huxleyi ([18,40,56,68], (Table S1)). Most haptophytes fall within the size class pico-and nanoeukaryotes [47,69,70].

Viruses and bacteria were counted in samples preserved with 1% glutaraldehyde (30 min at 4 °C) and snap frozen in liquid N_2. The samples were thawed, diluted, and stained with $1\times$SYBR Green I (Invitrogen, Carlsbad, CA, USA 10,000 × conc. in dimethyl sulfoxide (DMSO)) for 10 min at 80 °C [57], immediately before counting in triplicates. Bacteria and three different virus populations were defined based on side scatter properties and green fluorescence: low-, medium- and high-fluorescence viruses (V1, V2 and V3 respectively, [40]). Spearman rank order correlation analyses were calculated in Statistica 12 (StatSoft, Tulsa, OK, USA), to assess the relationship between abundance of different virus and algal groups, as well as their relationship with the measured abiotic factors. Missing values were pairwise deleted.

4.3. DNA Extraction, PCR and 454-Pyrosequencing

Five samples, collected on May 25, August 31, and November 30 (2010) and on February 22 and May 31 (2011), were used for targeted 454 pyrosequencing of haptophyte and Phycodnaviridae/Megaviridae communities. The samples were chosen based on pulsed field gel electrophoresis (PFGE) analysis to represent different seasonal community stages (Figure S1).

For haptophytes, DNA was extracted from $\frac{1}{2}$ of each 3 µm and 0.45 µm pore-sized filters (representing approximately 10 L of sea water) using DNeasy®Plant Mini kit (Qiagen, Hilden, Germany) according to the manufacturer's instructions. Initial re-suspension of cells was done by transferring the frozen filters into falcon tubes with 1 ml AP1 buffer (from the DNeasy®kit) and vortexing for 60 s. Extracted DNA from the two different size fractions was subjected to separate PCRs with tagged primers. The V4 region of the 18S rRNA gene (position 640–1060) was amplified using haptophyte-specific primers: 528Flong (5′GCGGTAATTCCAGCTCCAA3′) and PRYM01+7 (5′-GATCAGTGAAAACATCCCTGG-3′) [71]. Each PCR mixture contained 1 µL DNA template, 5 µL Phusion GC buffer (NEB Inc., Ipswich, MA, USA), 0.2 mM each deoxynucleoside triphosphate (dNTP), 400 nM each primer, 0.75 µL DMSO, 0.5 units Phusion®High-Fidelity DNA Polymerase (NEB Inc.), adjusted for a final volume of 25 µL. The cycling parameters were 98 °C for 30 s, and 30 cycles of 98 °C for 10 s, 55 °C for 30 s, and 72 °C for 30 s, with a final extension at 72 °C for 10 min [71].

For viruses, DNA was extracted from 0.5–1 mL of frozen viral concentrate (representing 200–400 mL of sea water). The concentrates were alternately heated to 90 °C and cooled on ice twice for 2 min. Ethylenediaminetetraacetic acid (EDTA, 20 mM) and proteinase K (100 µg/mL) were subsequently added before the samples were incubated for 10 min at 55 °C. Sodium dodecyl sulfate (SDS, final concentration 0.5%) was added and the samples incubated for 1 h at 55 °C. The extracted DNA was then purified with Zymo DNA Clean and Concentrator™ kit (Zymo Research, Irvine, CA, USA) following the manufacturer's protocols.

A segment of the major capsid protein (MCP) gene was amplified by the primers: MCPforwd (5′-GGY GGY CAR CGY ATT GA-3′) and MCPrev (5′-TGI ARY TGY TCR AYI AGG TA-3′) developed by Larsen et al. [32]. The PCR reactions (25 μL) contained: 0.625 U of HotStarTaq DNA polymerase (Qiagen), 1×PCR buffer, 0.2 mM of each dNTP, 0.5 μM of each of the primers and 1 μL template. The following PCR-program was used: Initial activation at 95 °C (15 min), a touchdown PCR of 20 cycles of denaturation at 94 °C (30 s), annealing at 60 °C and decreasing 0.5 °C per round (30 s) and extension at 72 °C (30 s), followed by 35 cycles with fixed annealing temperature at 45 °C, and a final elongation of 72 °C for 7 min. The PCR products were cleaned (Zymo DNA Clean and Concentrator™ kit) and amplified in new PCR reactions with tagged primers specific to each sample, 25 cycles with annealing at 45 °C, and otherwise as above.

For each haptophyte or viral sample, the products from eight replicate PCR reactions were cleaned, quantified and pooled before sequencing. The DNA amplicons were sent for Roche 454 (GS-FLX Titanium) library sequencing.

Two-directional amplicon sequencing using L chemistry was performed by LGC Genomics GmbH, Berlin, Germany. The following numbers of reads were obtained: haptophytes, 22588 (25 May 2010), 23261 (31 August 2010), 31959 (30 November 2010), 32540 (22 February 2011), 30380 (31 May 2011) (Table S2), viruses, 7909 (25 May 2010), 8195 (31 August 2010), 8558 (30 November 2010), 6791 (22 February 2011), 10353 (31 May 2011) (Table S4).

4.4. Sequence Analysis

All reads were filtered using AmpliconNoise [72], with default settings, and further analyzed using Mothur (www.mothur.org) [73] with the commands here provided in italics within brackets. Reads were trimmed (*trim.seqs*) and checked for chimeras by uchime (*chimera.uchime*) with Silva reference sequences for the haptophytes [74].

Prior to clustering, haptophyte reads were aligned to a reference alignment [47] (*align.seqs*) to ensure that the reads aligned in the targeted region, and to enable distance calculation by *dist.seqs*. Based on these distances, reads were clustered de novo into OTUs with 98% similarity (*cluster*), with the average neighbor-algorithm. An OTU definition of 98% nucleotide similarity was applied here in accordance with studies showing this to be a good threshold for delineating different species of most protists, while at the same time accounting for intra-species variation [75]. OTUs of different lengths, but which were otherwise identical, were clustered at 100% similarity by Uclust [76] and the longest sequence of each OTU was picked as representative. OTUs shorter than 250 bp were removed. OTUs that were represented by only a single read (singletons) were excluded from the analysis. Taxonomic classification was performed by MegaBLASTin Geneiuos v. 8.1.6 against the PIP Haptophyte 18S rDNA reference sequence database as described in [42] and available from figshare [77]. Diversity analyses were performed in Mothur (*collect.shared, summary.single*). To compare OTU richness between samples, all samples were subsampled to the number of reads in the smallest sample (1320), by the function *rrarefy* in the *vegan* package v. 2.4-1 [78] in R v. 3.3.2 (R Core Team 2016).

As many of the OTUs were found in both size fractions (>3 and 3–0.45 μm), the number of reads in the two size fractions were pooled, and the relative abundance of each OTU in the five samples was determined.

The filtered viral MCP reads were translated into the corresponding amino acid sequence in BioEdit [79]. Alignment of the amino acid reads was done with MAFFT v7) [80], with a gap opening penalty of 2.5, offset value of 0.1, and BLOSUM62 as the amino acid scoring matrix. Insertion/deletion errors were manually corrected. Reads that then did not align in the mid-conserved region (approx. position 100 in the alignment) or contained stop-codons, indicative of sequencing errors, were removed. The remaining reads were trimmed to equal length, i.e., position 117 in the alignment. A protein distance matrix was calculated by PROTDIST v3.5c (©1993 Joseph Felsenstein), and used to cluster the sequences in Mothur [73]. As the large number of PCR-cycles is prone to create artifacts, a 95% amino acid sequence identity threshold was applied [35]. To further decrease the number of spurious reads, only OTUs containing ten or more reads were used in the analysis. Mothur was used for downstream calculations of diversity indices. A representative sequence for each OTU containing more than 50 reads was used for phylogenetic analysis (together representing 84% of the reads). These OTUs were tentatively assigned a phylogenetic affiliation (BLAST-search closest hits) to the Megaviridae or *Phycodnaviridae* family. The tree was constructed, comprising the representative sequences together with reference sequences. Alignment and phylogenetic reconstructions were performed using the function "build" of ETE3 v3.0.0b32 [81] implemented on the GenomeNet, Tree [82] . The tree was constructed using FastTree v2.1.8 with default parameters [83]. Statistical support for the internal branches was calculated by an aLRTtest (SH-like), and through 100 bootstraps. Cluster diagrams were drawn for the haptophyte and virus samples separately. The cluster diagrams were based on Bray Curtis similarities of relative abundance of each virus or haptophyte OTU in the samples. A SIMPROF permutation test was applied to test if the samples could be differentiated at $p < 0.05$ (Primer 6, Primer-E Ltd., Ivybridge, UK).

Supplementary Materials: The follow supplementary materials can be found online at www.mdpi.com/1999-4915/9/4/84/s1, Table S1: Spearman Rank Order Correlations of the quantitative biological data, including the chl a measurements and the population abundances obtained by flow cytometry. Table S2: Heatmap and OTU table showing the relative abundance and percentage identity to nucleotide blast hits of the haptophyte OTUs in samples from Raunefjorden. Table S3: Results of the 454 sequencing and analysis of the V4-region of 18S rDNA 18S rDNA in haptophytes from Raunefjorden. Figure S1: Schematic representation of the relative abundance of distinct viral populations. Table S4: Result of the 454 sequencing of the viral MCP gene in samples from Raunefjorden. Table S5: Heatmap showing relative abundance of the different viral MCP OTUs in samples from Raunefjorden. Suplementary method and material describing viral diversity explored by pulsed-field gel electrophoresis (PFGE) and method precautions.

Acknowledgments: This work was funded by the Research Council of Norway through the project 190307/S54 "HAPTODIV" and project number 225956/E10 "MicroPolar", and by the European Research Council through the Advanced Grant project No. 250254 "MINOS". We are grateful to Knut Tomas Holden Sørlie for help with sampling, and to Hilde Marie Stabell, Jessica Ray and Jorunn Egge for help with seawater filtering

Author Contributions: Main contributor to analysis and writing has been Torill Vik Johannesen. Aud Larsen, Gunnar Bratbak, Bente Edvardsen and Ruth-Anne Sandaa have all contributed to scientific discussion of results, analysis, and writing. Elianne D. Egge has contributed to scientific discussion and analyses, while António Pagarete has performed some of the analysis included in the paper.

References

1. Field, C.B.; Behrenfeld, M.J.; Randerson, J.T.; Falkowski, P. Primary production of the biosphere: Integrating terrestrial and oceanic components. *Science* **1998**, *281*, 237–240. [CrossRef] [PubMed]
2. Eikrem, E.; Medlin, L.K.; Henderiks, J.; Rokitta, S.; Rost, B.; Probert, I.; Throndsen, J.; Edvardsen, B. Haptophyta. In *Handbook of the Protists*; Archibald, J.M., Simpson, A.G.B., Slamovits, C.H., Margulis, L., Melkonian, M., Chapman, D.J., Corliss, J.O., Eds.; Springer International Publishing: Cham, Switzerland, 2016; pp. 1–61.
3. Unrein, F.; Gasol, J.M.; Not, F.; Forn, I.; Massana, R. Mixotrophic haptophytes are key bacterial grazers in oligotrophic coastal waters. *ISME J.* **2014**, *8*, 164–176. [CrossRef] [PubMed]
4. Hallegraeff, G.M. Ocean climate change, phytoplankton community responses, and harmful algal blooms: A formidable predictive challenge. *J. Phycol.* **2010**, *46*, 220–235. [CrossRef]
5. Leadbeater, B.S.C. Identification, by means of electron microscopy, of flagellate nanoplankton from the coast of Norway. *Sarsia* **1972**, *49*, 107–124. [CrossRef]
6. Thomsen, H.A.; Buck, K.R.; Chavez, F.P. Haptophytes as components of marine phytoplankton. *Syst. Assoc. Spec. Vol. Ser.* **1994**, *51*, 187–208.
7. Egge, E.S.; Johannessen, T.V.; Andersen, T.; Eikrem, W.; Bittner, L.; Larsen, A.; Sandaa, R.-A.; Edvardsen, B. Seasonal diversity and dynamics of haptophytes in the Skagerrak, Norway, explored by high-throughput sequencing. *Mol. Ecol.* **2015**, *24*, 3026–3042. [CrossRef] [PubMed]

8. Fuhrman, J.; Noble, R. Viruses and protists cause similar bacterial mortality in coastal seawater. *Limnol. Oceanogr.* **1995**, *40*, 1236–1242. [CrossRef]

9. Short, S.M. The ecology of viruses that infect eukaryotic algae. *Environ. Microbiol.* **2012**, *14*, 2253–2271. [CrossRef] [PubMed]

10. Bouvier, T.; Del Giorgio, P.A. Key role of selective viral-induced mortality in determining marine bacterial community composition. *Envir. Microbiol.* **2007**, *9*, 287–297. [CrossRef] [PubMed]

11. Castberg, T.; Larsen, A.; Sandaa, R.A.; Brussaard, C.P.D.; Egge, J.K.; Heldal, M.; Thyrhaug, R.; van Hannen, E.J.; Bratbak, G. Microbial population dynamics and diversity during a bloom of the marine coccolithophorid *Emiliania huxleyi* (Haptophyta). *Mar. Ecol. Prog. Ser.* **2001**, *221*, 39–46. [CrossRef]

12. Brussaard, C.P.D.; Bratbak, G.; Baudoux, A.C.; Ruardij, P. *Phaeocystis* and its interaction with viruses. *Biogeochemistry* **2007**, *83*, 201–215. [CrossRef]

13. Martinez, J.M.; Schroeder, D.C.; Larsen, A.; Bratbak, G.; Wilson, W.H. Molecular Dynamics of *Emiliania huxleyi* and Cooccurring Viruses during Two Separate Mesocosm Studies. *Appl. Environ. Microbiol.* **2007**, *73*, 554–562. [CrossRef] [PubMed]

14. Santini, S.; Jeudy, S.; Bartoli, J.; Poirot, O.; Lescot, M.; Abergel, C.; Barbe, V.; Wommack, K.E.; Noordeloos, A.A.M.; Brussaard, C.P.D.; et al. Genome of *Phaeocystis globosa virus* PgV-16T highlights the common ancestry of the largest known DNA viruses infecting eukaryotes. *Proc. Natl. Acad. Sci. USA* **2103**, *110*, 10800–10805. [CrossRef] [PubMed]

15. Brussaard, C.P.D.; Kuipers, B.; Veldhuis, M.J.W. A mesocosm study of *Phaeocystis globosa* population dynamics: I. Regulatory role of viruses in bloom control. *Harmful Algae* **2005**, *4*, 859–874. [CrossRef]

16. Tomaru, Y.; Hata, N.; Masuda, T.; Tsuji, M.; Igata, K.; Masuda, Y.; Yamatogi, T.; Sakaguchi, M.; Nagasaki, K. Ecological dynamics of the bivalve-killing dinoflagellate *Heterocapsa circularisquama* and its infectious viruses in different locations of western Japan. *Environ. Microbiol.* **2007**, *9*, 1376–1383. [CrossRef] [PubMed]

17. Suttle, C.A.; Chan, A.M. Dynamics and distribution of cyanophages and their effect on marine *Synechococcus* spp. *Appl. Environ. Microbiol.* **1994**, *60*, 3167–3174. [PubMed]

18. Larsen, A.; Castberg, T.; Sandaa, R.A.; Brussaard, C.P.D.; Egge, J.; Heldal, M.; Paulino, A.; Thyrhaug, R.; van Hannen, E.J.; Bratbak, G. Population dynamics and diversity of phytoplankton, bacteria and viruses in a seawater enclosure. *Mar. Ecol. Prog. Ser.* **2001**, *221*, 47–57. [CrossRef]

19. Cottrell, M.T.; Suttle, C.A. Dynamics of a lytic virus infecting the photosynthetic marine picoflagellate *Micromonas pusilla*. *Limnol. Oceanogr.* **1995**, *40*, 730–739. [CrossRef]

20. Tarutani, K.; Nagasaki, K.; Itakura, S.; Yamaguchi, M. Isolation of a virus infecting the novel shellfish-killing dinoflagellate *Heterocapsa circularisquama*. *Aquat. Microb. Ecol.* **2001**, *23*, 103–111. [CrossRef]

21. Thyrhaug, R.; Larsen, A.; Thingstad, T.F.; Bratbak, G. Stable coexistence in marine algal host-virus systems. *Mar. Ecol. Prog. Ser.* **2003**, *254*, 27–35. [CrossRef]

22. Brussaard, C.P.D. Viral control of phytoplankton populations—A review. *J. Eukaryot. Microbiol.* **2004**, *51*, 125–138. [CrossRef] [PubMed]

23. Demory, D.; Arsenieff, L.; Simon, N.; Six, C.; Rigaut-Jalabert, F.; Marie, D.; Ge, P.; Bigeard, E.; Jacquet, S.; Sciandra, A.; et al. Temperature is a key factor in Micromonas-virus interactions. *ISME J.* **2017**. [CrossRef] [PubMed]

24. Sandaa, R.A.; Heldal, M.; Castberg, T.; Thyrhaug, R.; Bratbak, G. Isolation and characterization of two viruses with large genome size infecting *Chrysochromulina ericina* (Prymnesiophyceae) and *Pyramimonas orientalis* (Prasinophyceae). *Virology* **2001**, *290*, 272–280. [CrossRef] [PubMed]

25. Johannessen, T.V.; Bratbak, G.; Larsen, A.; Ogata, H.; Egge, E.S.; Edvardsen, B.; Eikrem, W.; Sandaa, R.-A. Characterisation of three novel giant viruses reveals huge diversity among viruses infecting Prymnesiales (Haptophyta). *Virology* **2015**, *476*, 180–188. [CrossRef] [PubMed]

26. Iyer, L.; Balaji, S.; Koonin, E.; Aravind, L. Evolutionary genomics of nucleo-cytoplasmic large DNA viruses. *Virus Res.* **2006**, *117*, 156–184. [CrossRef] [PubMed]

27. Wilson, W.H.; Etten, J.L.; Allen, M.J. The *Phycodnaviridae*: The story of how tiny giants rule the world. In *Lesser Known Large dsDNA Viruses*; Papers in Plant Pathology; Springer: Heidelberg, Germany, 2009; pp. 1–42.

28. La Scola, B.; Audic, S.; Robert, C.; Jungang, L.; de Lamballerie, X.; Drancourt, M.; Birtles, R.; Claverie, J.-M.; Raoult, D. A giant virus in amoebae. *Science* **2003**, *299*, 2033. [CrossRef] [PubMed]

29. Fischer, M.G.; Allen, M.J.; Wilson, W.H.; Suttle, C.A. Giant virus with a remarkable complement of genes infects marine zooplankton. *Proc. Natl. Acad. Sci. USA* **2010**, *107*, 19508–19513. [CrossRef] [PubMed]

30. Moniruzzaman, M.; LeCleir, G.R.; Brown, C.M.; Gobler, C.J.; Bidle, K.D.; Wilson, W.H.; Wilhelm, S.W. Genome of brown tide virus (AaV), the little giant of the Megaviridae, elucidates NCLDV genome expansion and host-virus coevolution. *Virology* **2014**, *466–467*, 59–69. [CrossRef] [PubMed]

31. Moniruzzaman, M.; Gan, E.R.; LeCleir, G.R.; Kang, Y.; Gobler, C.J.; Wilhelm, S.W. Diversity and dynamics of algal Megaviridae members during a harmful brown tide caused by the pelagophyte, *Aureococcus anophagefferens*. *FEMS Microbiol. Ecol.* **2016**. [CrossRef] [PubMed]

32. Larsen, J.B.; Larsen, A.; Bratbak, G.; Sandaa, R.A. Phylogenetic analysis of members of the *Phycodnaviridae* virus family, using amplified fragments of the major capsid protein gene. *Appl. Environ. Microbiol.* **2008**, *74*, 3048–3057. [CrossRef] [PubMed]

33. Monier, A.; Larsen, J.B.; Sandaa, R.-A.; Bratbak, G.; Claverie, J.M.; Ogata, H. Marine mimivirus relatives are probably large algal viruses. *Virol. J.* **2008**, *5*, 12. [CrossRef] [PubMed]

34. Kristensen, D.M.; Mushegian, A.R.; Dolja, V.V.; Koonin, E.V. New dimensions of the virus world discovered through metagenomics. *Trends Microbiol.* **2010**, *18*, 11–19. [CrossRef] [PubMed]

35. Park, Y.; Lee, K.; Lee, Y.S.; Kim, S.W.; Choi, T.J. Detection of diverse marine algal viruses in the South Sea regions of Korea by PCR amplification of the DNA polymerase and major capsid protein genes. *Virus Res.* **2011**, *159*, 43–50. [CrossRef] [PubMed]

36. Short, S.M.; Rusanova, O.; Staniewski, M.A. Novel phycodnavirus genes amplified from Canadian freshwater environments. *Aquat. Microb. Ecol.* **2011**, *63*, 61–67. [CrossRef]

37. Rozon, R.M.; Short, S.M. Complex seasonality observed amongst diverse phytoplankton viruses in the Bay of Quinte, an embayment of Lake Ontario. *Freshw. Biol.* **2013**, *58*, 2648–2663. [CrossRef]

38. Short, S.M.; Suttle, C.A. Sequence analysis of marine virus communities reveals that groups of related algal viruses are widely distributed in nature. *Appl. Environ. Microbiol.* **2002**, *68*, 1290–1296. [CrossRef] [PubMed]

39. Wang, M.-N.; Ge, X.-Y.; Wu, Y.-Q.; Yang, X.-L.; Tan, B.; Zhang, Y.-J.; Shi, Z.-L. Genetic diversity and temporal dynamics of phytoplankton viruses in East Lake, China. *Virol. Sin.* **2015**, *30*, 290–300. [CrossRef] [PubMed]

40. Larsen, A.; Flaten, G.A.F.; Sandaa, R.A.; Castberg, T.; Thyrhaug, R.; Erga, S.R.; Jacquet, S.; Bratbak, G. Spring phytoplankton bloom dynamics in Norwegian coastal waters: Microbial community succession and diversity. *Limnol. Oceanogr.* **2004**, *49*, 180–190. [CrossRef]

41. Sandaa, R.A.; Larsen, A. Seasonal variations in viral-host populations in Norwegian coastal waters: Focusing on the cyanophage community infecting marine Synechococcus species. *Appl. Environ. Microbiol.* **2006**, *72*, 4610–4618. [CrossRef] [PubMed]

42. Edvardsen, B.; Egge, E.S.; Vaulot, D. Diversity and distribution of haptophytes revealed by environmental sequencing and metabarcoding—A review. *Perspect. Phycol.* **2016**, *3*, 77–91. [CrossRef]

43. Johannesen, T.V. Marine Virus-phytoplankton Interactions. Ph.D. Thesis, University of Bergen, Bergen, Norway, 2015.

44. Erga, S.R.; Heimdal, B.R. Ecological studies on the phytoplankton of Korsfjorden, western Norway. The dynamics of a spring bloom seen in relation to hydrographical conditions and light regime. *J. Plankton. Res.* **1984**, *6*, 67–90. [CrossRef]

45. Bratbak, G.; Heldal, M.; Norland, S.; Thingstad, T.F. Viruses as partners in spring bloom microbial trophodynamics. *Appl. Environ. Microbiol.* **1990**, *56*, 1400–1405. [PubMed]

46. Pagarete, A.; Chow, C.-E.T.; Johannessen, T.; Fuhrman, J.A.; Thingstad, T.F.; Sandaa, R.A. Strong Seasonality and Interannual Recurrence in Marine Myovirus Communities. *Appl. Environ. Microbiol.* **2013**, *79*, 6253–6259. [CrossRef] [PubMed]

47. Egge, J.S.; Eikrem, W.; Edvardsen, B. Deep branching novel lineages and high diversity of haptophytes in Skagerak (Norway) uncovered by 454-pyrosequencing. *J. Eukaryot. Microbiol.* **2015**, *62*, 121–140. [CrossRef] [PubMed]

48. Moon-van der Staay, S.Y.; van der Staay, G.W.M.; Guillou, L.; Vaulot, D.; Claustre, H.; Medlin, L.K. Abundance and diversity of prymnesiophytes in the picoplankton community from the equatorial Pacific Ocean inferred from 18S rDNA sequences. *Limnol. Oceanogr.* **2000**, *45*, 98–109. [CrossRef]

49. Liu, H.; Probert, I.; Uitz, J.; Claustre, H.; Aris-Brosou, S.; Frada, M.; Not, F.; de Vargas, C. Extreme diversity in noncalcifying haptophytes explains a major pigment paradox in open oceans. *Proc. Natl. Acad. Sci. USA* **2009**, *106*, 12803–12808. [CrossRef] [PubMed]

50. Thingstad, T.F.; Våge, S.; Storesund, J.E.; Sandaa, R.-A.; Giske, J. A theoretical analysis of how strain-specific viruses can control microbial species diversity. *Proc. Natl. Acad. Sci. USA* **2014**, *111*, 7813–7818. [CrossRef] [PubMed]

51. Baudoux, A.; Brussaard, C. Characterization of different viruses infecting the marine harmful algal bloom species *Phaeocystis globosa*. *Virology* **2005**, *341*, 80–90. [CrossRef] [PubMed]

52. Nagasaki, K. Dinoflagellates, diatoms, and their viruses. *J. Microbiol.* **2008**, *46*, 235–243. [CrossRef] [PubMed]

53. Bratbak, G.; Egge, J.K.; Heldal, M. Viral mortality of the marine alga *Emiliania-huxleyi* (Haptophyceae) and termination of algal blooms. *Mar. Ecol. Prog. Ser.* **1993**, *93*, 39–48. [CrossRef]

54. Bratbak, G.; Levasseur, M.; Michaud, S.; Cantin, G.; Fernandez, E.; Heimdal, B.R.; Heldal, M. Viral activity in relation to *Emiliana huxleyi* blooms: A mechanism of DSMP release? *Mar. Ecol. Prog. Ser.* **1995**, *128*, 133–142. [CrossRef]

55. Wilson, W.H.; Tarran, G.A.; Schroeder, D.; Cox, M.; Oke, J.; Malin, G. Isolation of viruses responsible for the demise of an *Emiliania huxleyi* bloom in the English Channel. *J. Mar. Biol. Assoc. UK* **2002**, *82*, 369–377. [CrossRef]

56. Jacquet, S.; Heldal, M.; Iglesias-Rodriguez, D.; Larsen, A.; Wilson, W.; Bratbak, G. Flow cytometric analysis of an *Emiliana huxleyi* bloom terminated by viral infection. *Aquat. Microb. Ecol.* **2002**, *27*, 111–124. [CrossRef]

57. Brussaard, C.P.D. Optimization of procedures for counting viruses by flow cytometry. *Appl. Environ. Microbiol.* **2004**, *70*, 1506–1513. [CrossRef] [PubMed]

58. Jacobsen, A.; Bratbak, G.; Heldal, M. Isolation and characterization of a virus infecting *Phaeocystis pouchetii* (Prymnesiophyceae). *J. Phycol.* **1996**, *32*, 923–927. [CrossRef]

59. Brussaard, C.P.D.; Short, S.M.; Frederickson, C.M.; Suttle, C.A. Isolation and phylogenetic analysis of novel viruses infecting the phytoplankton *Phaeocystis globosa* (Prymnesiophyceae). *Appl. Environ. Microbiol.* **2004**, *70*, 3700–3705. [CrossRef] [PubMed]

60. Nagasaki, K.; Bratbak, G. Isolation of viruses infecting photosynthetic and nonphotosynthetic protists. In *Manual of Aquatic Viral Ecology*; Wilhelm, S.W., Weinbauer, M.G., Suttle, C.A., Eds.; ASLO: Waco, TX, USA, 2010; pp. 92–101.

61. Sandaa, R.-A. Burden or benefit? Virus-host interactions in the marine environment. *Res. Microbiol.* **2008**, *159*, 374–381. [CrossRef] [PubMed]

62. Suttle, C.A.; Chen, F. Mechanisms and Rates of Decay of Marine Viruses in Seawater. *Appl. Environ. Microbiol.* **1992**, *58*, 3721–3729. [PubMed]

63. Noble, R.T.; Fuhrman, J.A. Rapid virus production and removal as measured with fluorescently labeled viruses as tracers. *Appl. Environ. Microbiol.* **2000**, *66*, 3790–3797. [CrossRef] [PubMed]

64. Mojica, K.D.A.; Brussaard, C.P.D. Factors affecting virus dynamics and microbial host-virus interactions in marine environments. *FEMS Microbiol. Ecol.* **2014**, *89*, 495–515. [CrossRef] [PubMed]

65. Needham, D.M.; Chow, C.E.T.; Cram, J.A.; Sachdeva, R.; Parada, A.; Fuhrman, J.A. Short-term observations of marine bacterial and viral communities: Patterns, connections and resilience. *ISME J.* **2013**, *7*, 1274–1285. [CrossRef] [PubMed]

66. Ray, J.L.; Haramaty, L.; Thyrhaug, R.; Fredricks, H.F.; Van Mooy, B.A.S.; Larsen, A.; Bidle, K.D.; Sandaa, R.-A. Virus infection of *Haptolina ericina* and *Phaeocystis pouchetii* implicates evolutionary conservation of programmed cell death induction in marine haptophyte–virus interactions. *J. Plankton Res.* **2014**, *36*, 943–955. [CrossRef] [PubMed]

67. Rozenn, T.; Grimsley, N.; Escande, M.L.; Subirana, L.; Derelle, E.; Moreau, H. Acquisition and maintenance of resistance to viruses in eukaryotic phytoplankton populations. *Environ. Microbiol.* **2011**, *13*, 1412–1420.

68. Marie, D.; Brussaard, C.P.D.; Thyrhaug, R.; Bratbak, G.; Vaulot, D. Enumeration of marine viruses in culture and natural samples by flow cytometry. *Appl. Environ. Microbiol.* **1999**, *65*, 45–52. [PubMed]

69. Edvardsen, B.; Eikrem, W.; Throndsen, J.; Sáez, A.G.; Probert, I.; Medlin, L.K. Ribosomal DNA phylogenies and a morphological revision provide the basis for a revised taxonomy of the Prymnesiales (Haptophyta). *Eur. J. Phycol.* **2011**, *46*, 202–228. [CrossRef]

70. Not, F.; Siano, R.; Kooistra, W.H.C.F.; Simon, N.; Vaulot, D.; Probert, I. Diversity and Ecology of Eukaryotic Marine Phytoplankton. In *Genomic Insights into the Biology of Algae*; Piganeau, G., Ed.; Academic Press: London, UK, 2012; Volume 64, pp. 1–53.

71. Egge, E.; Bittner, L.; Andersen, T.; Audic, S.; de Vargas, C.; Edvardsen, B. 454 Pyrosequencing to Describe Microbial Eukaryotic Community Composition, Diversity and Relative Abundance: A Test for Marine Haptophytes. *PLoS ONE* **2013**, *8*, e74371. [CrossRef] [PubMed]

72. Quince, C.; Lanzen, A.; Davenport, R.J.; Turnbaugh, P.J. Removing Noise From Pyrosequenced Amplicons. *BMC Bioinform.* **2011**, *12*, 38. [CrossRef] [PubMed]

73. Schloss, P.D.; Westcott, S.L.; Ryabin, T.; Hall, J.R.; Hartmann, M.; Hollister, E.B.; Lesniewski, R.A.; Oakley, B.B.; Parks, D.H.; Robinson, C.J.; et al. Introducing mothur: Open-Source, Platform-Independent, Community-Supported Software for Describing and Comparing Microbial Communities. *Appl. Environ. Microbiol.* **2009**, *75*, 7537–7541. [CrossRef] [PubMed]

74. Quast, C.; Pruesse, E.; Yilmaz, P.; Gerken, J.; Schweer, T.; Yarza, P.; Peplies, J.; Glöckner, F.O. The SILVA ribosomal RNA gene database project: Improved data processing and web-based tools. *Nucleic Acids Res.* **2013**, *41*, D590–D596. [CrossRef] [PubMed]

75. Caron, D.A.; Countway, P.D. Hypotheses on the role of the protistan rare biosphere in a changing world. *Aquat. Microb. Ecol.* **2009**, *57*, 227–238. [CrossRef]

76. Edgar, R.C. Search and clustering orders of magnitude faster than BLAST. *Bioinformatics* **2010**, *26*, 2460–2461. [CrossRef] [PubMed]

77. Figshare. Available online: https://dx.doi.org/10.6084/m9.figshare.2759983.v1 (accessed on 22 March 2017).

78. Oksanen, J.; Blanchet, E.G.; Friendly, M.; Kindt, R.; Legendre, P.; McGlinn, D.; Minchin, P.R.; O'Hara, R.B.; Simpson, G.L.; Solymos, P.; et al. Vegan: Community Ecology Package, R package version 2.4-1. 2016. Available online: https://CRAN.R-project.org/package=vegan (accessed on 2 March 2017).

79. Hall, T. BioEdit: A user friendly biologicla sequence alignment editor and analysis program for windows 95/98/NT. *Nucl. Acids. Symp. Ser.* **1999**, *41*, 95–98.

80. Katoh, K.; Standley, D.M. MAFFT Multiple Sequence Alignment Software Version 7: Improvements in Performance and Usability. *Mol. Biol. Evol.* **2013**, *30*, 772–780. [CrossRef] [PubMed]

81. Huerta-Cepas, J.; Serra, F.; Bork, P. ETE 3: Reconstruction, Analysis, and Visualization of Phylogenomic Data. *Mol. Biol. Evol.* **2016**, *33*, 1635–1638. [CrossRef] [PubMed]

82. GenomeNet, Tree. Available online: http://www.genome.jp/tools/ete/ (accessed on 18 November 2016).

83. Price, M.N.; Dehal, P.S.; Arkin, A.P. FastTree: Computing Large Minimum Evolution Trees with Profiles instead of a Distance Matrix. *Mol. Biol. Evol.* **2009**, *26*, 1641–1650. [CrossRef] [PubMed]

5

Replication and Oncolytic Activity of an Avian Orthoreovirus in Human Hepatocellular Carcinoma Cells

Robert A. Kozak [1], Larissa Hattin [1], Mia J. Biondi [2], Juan C. Corredor [1], Scott Walsh [1], Max Xue-Zhong [3], Justin Manuel [3], Ian D. McGilvray [3], Jason Morgenstern [1], Evan Lusty [1], Vera Cherepanov [2], Betty-Anne McBey [1], David Leishman [1], Jordan J. Feld [2], Byram Bridle [1] and Éva Nagy [1,*]

[1] Department of Pathobiology, Ontario Veterinary College, University of Guelph, Guelph, ON N1G 2W1, Canada; rob.kozak@gmail.com (R.A.K.); larissa.hattin@medportal.ca (L.H.); corredor@uoguelph.ca (J.C.C.); scott.walsh22@gmail.com (S.W.); jason.d.morgenstern@gmail.com (J.M.); evanlusty@gmail.com (E.L.); bmcbey@uoguelph.ca (B.-A.M.); dleishma@uoguelph.ca (D.L.); bbridle@uoguelph.ca (B.B.)

[2] Sandra Rotman Centre for Global Health, University of Toronto, Toronto, ON M5G 1L7, Canada; mia.biondi@mail.mcgill.ca (M.J.B.); vera.cherepanov@uhn.ca (V.C.); Jordan.feld@uhn.ca (J.J.F.)

[3] Multi-Organ Transplant Program, Department of Surgery, University of Toronto, Toronto General Hospital, Toronto, ON M5G 2C4, Canada; maxuezhong@hotmail.com (M.X.-Z.); jmanuel@research.ca (J.M.); Ian.McGilvray@uhn.on.ca (I.D.M.)

* Correspondence: enagy@uoguelph.ca

Academic Editor: Andrew Mehle

Abstract: Oncolytic viruses are cancer therapeutics with promising outcomes in pre-clinical and clinical settings. Animal viruses have the possibility to avoid pre-existing immunity in humans, while being safe and immunostimulatory. We isolated an avian orthoreovirus (ARV-PB1), and tested it against a panel of hepatocellular carcinoma cells. We found that ARV-PB1 replicated well and induced strong cytopathic effects. It was determined that one mechanism of cell death was through syncytia formation, resulting in apoptosis and induction of interferon stimulated genes (ISGs). As hepatitis C virus (HCV) is a major cause of hepatocellular carcinoma worldwide, we investigated the effect of ARV-PB1 against cells already infected with this virus. Both HCV replicon-containing and infected cells supported ARV-PB1 replication and underwent cytolysis. Finally, we generated in silico models to compare the structures of human reovirus- and ARV-PB1-derived S1 proteins, which are the primary targets of neutralizing antibodies. Tertiary alignments confirmed that ARV-PB1 differs from its human homolog, suggesting that immunity to human reoviruses would not be a barrier to its use. Therefore, ARV-PB1 can potentially expand the repertoire of oncolytic viruses for treatment of human hepatocellular carcinoma and other malignancies.

Keywords: avian orthoreovirus; hepatocellular carcinoma; hepatitis C virus; oncolytic virus; syncytia

1. Introduction

Liver cancer is the third leading cause of cancer worldwide, and hepatocellular carcinoma (HCC) represents approximately 70%–80% of all cases [1]. The incidence of HCC is expected to increase markedly in North America over the next decades due to the high prevalence of hepatitis C virus (HCV) infection [2]. Additionally, as the prevalence of obesity and type 2 diabetes rises, metabolic diseases related to HCC will also continue to contribute to this burden [1]. Despite studies into the pathogenesis of HCC, and attempts to enhance therapy, improvement in patient outcomes has been marginal.

The use of oncolytic viruses (OVs) for therapy offers promise for cancers where survival outcomes are poor and treatment options are limited. The use of these viruses offers numerous advantages over conventional cancer therapies including reduced toxicity, treatments of relatively short duration, and the possibility of targeting micrometastases [3]. The list of candidates includes numerous DNA and RNA viruses [4–7] from a variety of viral families (extensively reviewed in [8,9]). Most recently, the oncolytic human herpes simplex virus 1 (HSV-1) carrying the granulocyte macrophage colony-stimulating factor (GM-CSF), known as T-VEC, was approved for melanoma treatment [10].

An ideal OV should eliminate cancer cells through a combination of three mechanisms: direct oncolysis, antiangiogenic or antivasculature effects, and activation of innate- and tumor-specific immune responses [8]. Currently, a number of human viruses are undergoing pre-clinical and clinical trials as OVs [11–14]. However, the presence of pre-existing neutralizing antibodies (such as those generated against human adenovirus serotype 5) reduces the potency of OVs. Even if a patient does not initially have pre-existing antibodies, they will be induced after treatment with an OV, thereby limiting the number of doses that can be administered [15]. Furthermore, while administration by alternative routes (e.g., intravenous) can be a potential means to circumvent this, it is highly likely that effective OV therapy will require multiple viruses, potentially in combination with chemotherapy or radiation. Animal-derived viruses that do not circulate extensively in the human population represent a potential source of OVs that can circumvent pre-existing immunity. Several animal viruses with demonstrated oncolytic properties, such as Newcastle disease virus, replicate and induce strong cytopathic effects (CPE) in human cancer cell lines [16].

Oncolytic animal viruses are often potent activators of the immune system, as their evolution in a non-human host has limited their ability to evade the immune response in humans [17,18]. Therefore, these viruses can potentiate immunocentric oncolytic virotherapy, where emphasis is placed on induction of immunological mechanisms that can target cancer cells [19]. Avian reoviruses (ARVs) are non-enveloped and some of them cause diseases in poultry. These viruses are part of the *Orthoreovirus* genus, and, although they share similarities with the mammalian reoviruses, they form a separate species, *Avian orthoreovirus*. In contrast to mammalian reoviruses, the receptor(s) for avian reoviruses has yet to be identified, although these viruses have a wide-range in terms of cellular tropism, as infections in poultry involve multiple systems and organs (including the liver), the most important disease in chickens is viral arthritis [20,21].

In general, reoviruses replicate in the cytoplasm of host cells, and their double-stranded (ds)RNA genome is composed of 10 segments [22]. Mammalian reoviruses are currently being investigated in the clinic and have demonstrated efficacy in conjunction with chemotherapy against head and neck cancer [23]. We postulate that ARVs offer several unique characteristics, and may provide a complementary platform for oncolytic virotherapy. First, ARVs are not known to be associated with human disease and pre-existing immunity would not hamper their clinical application. Second, unlike mammalian reoviruses, ARVs can induce syncytia through the fusion-associated transmembrane (FAST) protein [24–26] and thus it may facilitate virus spread and distribution within a tumor. Additionally, the ARV p17 protein induces autophagy through multiple pathways and activates protein kinase RNA-activated (PKR) signaling [27]. Both processes activate the innate immune system, which can induce immune responses against tumors.

In this work, we found that ARV-PB1 replicated and induced cytopathic effects in HCC cells. Mechanisms of cell death involved syncytia formation and apoptosis. Furthermore, ARV-PB1 induced the expression of interferon-stimulated genes (ISGs), which may be important in the induction of anti-tumoral immune responses.

2. Materials and Methods

2.1. Viruses

The original sample was obtained from the Animal Health Laboratory (AHL, University of Guelph, Guelph, ON, Canada) as a non-pathogenic serotype 11 fowl adenovirus (FAdV-11). Primary virus isolation at AHL was done from a chicken of a broiler flock presented with poorer performance and potential airsacculitis. Initially, the case was considered as FAdV associated inclusion body hepatitis (IBH), however the final diagnosis was non-IBH associated FAdV involvement. The presence and serotype of the isolated FAdV was demonstrated by polymerase chain reaction (PCR) and sequence analysis of the product, performed in the diagnostic lab. Isolation of more than one virus from chickens with performance issues is not uncommon. In this case, it is more the reovirus load in the tested tissues was very low and the subsequent passages in cell culture amplified the reovirus (ARV-PB1). The orthoreovirus ARV-PB1 (avian reovirus pathobiology 1) was initially detected in a field sample, and was subsequently isolated by multiple rounds of plaque purification in an avian hepatoma cell line (CH-SAH). Stocks of virus were generated in CH-SAH cells by successive freeze-thaw cycles and centrifugation. Virus propagation, titration and one-step growth curves were carried out in chicken hepatoma cells (CH-SAH cell line), as described [28]. To investigate the pre-existing immunity to ARV-PB1, plaque reduction assays were carried out using sera from HCC patients as described [29]. To assess the relative efficacy of ARV-PB1, it was compared to Reolysin^{TM} (Oncolytics Biotech Inc., Calgary, AB, Canada), a mammalian type 3 (Dearing strain) of reovirus, as a gold standard control, since it is undergoing extensive testing in human clinical trials (kindly provided by Dr. Matt Coffey, Oncolytics Biotech Inc.).

2.2. Cell Lines

Human and murine tumor cell lines were maintained in media supplemented with 10% fetal bovine serum (FBS), 2 mM L-glutamine, and penicillin (100 IU/mL)/streptomycin (100 µg/mL) (Sigma, Oakville, ON, Canada). The following cell lines were used: HepG2, Huh-7, Huh-7.5, Huh-7.5.1, BB7 as well as HeLa, L1210, PC-3, 22RV1, DU145, HCT 116, SK0V-3—obtained from the American Type Culture Collection (ATCC; Manassas, VA, USA)—and ID8 (kindly provided by Dr. James Petrik, University of Guelph, Guelph, ON, Canada). The CH-SAH cell line was maintained in DMEM/F12 (Dulbecco's Modified Eagle Medium: Nutrient Mixture F-12, Sigma) supplemented with 10% FBS, 2 mM L-glutamine, and penicillin (100IU/mL)/streptomycin (100 ug/mL).

2.3. Infection of Cells with Hepatitis C Virus (JFH-1 Strain)

Cells were seeded in 24-well plates and some of them were infected with JFH-1 virus (gift from Dr. Jake Liang, the National Institutes of Health (NIH), Liver Diseases Branch, Bethesda, MD, USA) at a multiplicity of infection (MOI) of 0.05 for 24 h. The cells were treated with ARV-PB1 at various MOIs, and after 48 h the cells were formalin-fixed and stained with crystal violet solution.

2.4. Preparation of Primary Hepatocytes

Human liver tissue was obtained from livers procured from multi-organ donors. The donor livers were perfused in situ with cold histidine-tryptophan-ketoglutarate (HTK) solution. The caudate lobe was resected and stored in HTK solution on ice before cell isolation. Briefly, the perfusion system consisted of a circulation pump which warmed the perfusion reagents to 41°C within a biological safety cabinet. Gas containing 95% O_2:5% CO_2 was used to oxygenate all reagents before perfuse into liver tissue. The average weight of the liver tissue was 10 to 20 g. Two or three irrigation cannulas with olive tips were inserted into the cut surface of the liver. Hank's balanced salt solution (HBSS) was used to flush out remaining blood followed by HBSS containing 0.1 mM EGTA (ethylene glycol-bis(β-aminoethyl ether)-N,N,N',N'-tetraacetic acid) for 10 min. Subsequently, 250 mL of HBSS containing 0.5 µM calcium chloride was added followed by fluid with collagenase containing neutral protease (VitaCyte, Indianapolis, IN, USA) according to the manufacturer's instructions. Perfusion was performed with recirculation of fluid for 15 to 20 min or until the liver appeared to break apart slightly under the Glisson's capsule. The digested liver was then placed in a crystallizing dish containing 100–200 mL of HBSS with 0.1% human albumin. The tissue was then cut to release cells contained. Cells were filtered through a 210 μm nylon mesh followed by a 70 μm nylon mesh. Filtered cells were centrifuged at $72 \times g$ for 5 min at 4 °C. The hepatocyte cell pellet was washed twice as above and HBSS with 0.1% human albumin was added to re-suspend cells. Approximately 8–12 million viable cells per gram of tissue were isolated as determined by Beckman ViCell trypan blue system.

Primary hepatocytes thawed and transferred into Williams E Medium supplemented (Life Technologies, Burlington, ON, Canada) with 5% FBS, 1 μM DMSO (dimethyl sulfoxide) and thawing plating cocktail A (Life Technologies) according to manufacturer's instructions. Subsequently, cells were re-suspended in Williams E Medium supplemented with 0.1 μM DMSO and Cell Maintenance Cocktail B (Life Technologies). Cells were added to collagen-coated plates (Life Technologies), and after 4 h the medium was replaced with fresh culture medium. Cells were incubated at 37 °C for 24 h prior to infection.

2.5. RNA Isolation and Sequencing

Viral RNA was extracted from infected CH-SAH cells with TRIzol (Life Technologies) according to manufacturer's protocol. In order to perform genomic sequencing, complementary DNA (cDNA) was generated following the method outlined by Jiang et al. [30]. Primers were designed to amplify specific viral genes, and the PCR products were sequenced at the University of Guelph Laboratory Services, Guelph, ON, Canada. Pairwise identity of the viral genes and comparison were performed with BLASTn [31].

2.6. Viral Growth and Cell Viability Assay

Survival of cancer cell lines after viral infection was determined by PrestoBlue™ Cell Viability Reagent (Life Technologies), a resazurin dye-based metabolic assay. Cells were plated at concentrations of 1×10^3 viable cells/well and allowed to adhere overnight. Cells were either uninfected or infected at various MOIs. At subsequent time points after viral infection, PrestoBlue™ Cell Viability Reagent was added according to the manufacturer's protocol. Cell viability was determined by comparing fluorescence readings of infected cells to uninfected controls. All samples were run in triplicate for each MOI, and each experiment was performed a minimum of three times.

To assess viral replication, cell monolayers were grown to 80%–90% confluency. Cells in six-well plates were infected with ARV-PB1 at an MOI of 5 for 1 h at room temperature. Subsequently, the inoculum was removed and the cells were washed with phosphate buffered saline (PBS, pH 7.4), and medium was added as described [28]. Cells were harvested at indicated time points and stored at −80 °C. Lysates were freeze-thawed three times to release viruses, and the samples were titrated in CH-SAH cells. Each viral growth curve was performed in duplicate.

2.7. Cell Staining

Cells were seeded in 35 mm cell culture dishes (5×10^5 cells/dish) containing sterile coverslips. After 24 h incubation at 37 °C, 5% CO_2, cells were infected with ARV-PB1 (MOI of 5) for 72 h. To study syncytia formation and cytopathic effects as well as to detect the viral genome in infected cells, medium was removed and cells were washed twice with PBS and fixed with 4% buffered-formalin (v/v) formalin for 5 min followed by another wash with PBS. Cells were permeabilized with 0.1% NP40-PBS for 10 min at room temperature followed by three washes with PBS. Cells were blocked with5% BSA (bovine serum albumin) in PBS for 1 h at room temperature followed by incubation with anti-dsRNA monoclonal antibody K1 (Scicons, Szirák, Hungary) for 2 h at room temperature. After three washes with PBS, cells were incubated with secondary goat anti-mouse immunoglobulin G (IgG) labeled with Nexa Fluor 594 (Invitrogen, Eugene, OR, USA). After three washes with PBS, cells were stained with 4′,6-diamidino-2-phenylindole (DAPI) according to manufacturer's protocol (Life Technologies). Analysis of cells was performed with a Zeiss Axio Cam MRm fluorescent microscope.

2.8. Flow Cytometry

Cells were seeded at a density of 1×10^6 cells per well in a six-well plate, and infected at an MOI of 5 and incubated for 72 h before staining. Medium was removed, and the cells were washed with PBS (pH 7.4) prior to trypsin treatment. Plates were incubated until cells appeared detached. 5×10^5 cells were centrifuged at $1000\times g$ for 5 min at room temperature, washed with PBS and stained with Annexin V-FITC (Calbiochem, Billerica, MA, USA) and 7-AAD (eBioscience, San Diego, CA, USA) according to the manufacturer's protocols. Samples were analyzed by flow cytometry using a FACS Aria IIu with FACSDiva™ Software V6 (BD Biosciences, Mississauga, ON, Canada), while data were analyzed with FlowJo software version X (Tree Star, Ashland, OR, USA).

2.9. In Silico Modeling

The nucleotide sequence of the viral S1 gene of ARV-PB1 was analyzed by I-TASSER (http://zhanglab.ccmb.med.umich.edu/I-TASSER/) to predict protein structure and function [32,33]. Based on the generated predictions, we identified the tertiary structure, which most closely resembled the S1 protein—identified as being avian reovirus strain S1133, Protein Data Bank (PDB) ID 2JJL. We next determined the structural similarity between ARV and a previously crystallized mammalian reovirus type 3 S1 protein. The analysis was performed using the mammalian reovirus type 3 (Dearing strain), 1KKE [34] as a reference. Modeling and tertiary alignments were carried out using USCF Chimera (https://www.cgl.ucsf.edu/chimera/).

2.10. Plaque Reduction Assays

Plaque reduction assays were performed as described [28] to investigate the presence of neutralizing antibodies in HCC patient sera. Whole blood was collected by venipuncture in a Vacutainer from patients with HCV-induced HCC at the Toronto Western Hospital Liver Clinic (Toronto, ON, Canada). Blood samples were centrifuged and sera stored at −80 °C. The experiments were approved and performed in accordance with the University Health Research Network Research Ethics Board. All patients provided written informed consent for the storage and use of their specimens for the purpose of research (University Health Network Research Ethics Board Biobank Protocol Number 13-6974). Patients tested negative for both human immunodeficiency virus (HIV) and hepatitis B virus (HBV), and were clinically diverse. HCV viral loads ranged from undetectable by the gold standard to 10^4 IU/mL; genotypes 2, 3, and 4 were represented; patients were either HCV treatment-naïve or experienced, and were either pre- or post-transplant.

2.11. Quantitative Polymerase Chain Reaction

Cells seeded in 24-well plates were infected with ARV-PB1 at an MOI of 5. At 6 hours post-infection (h.p.i.), total RNA was extracted using TRI Reagent® (Sigma) according to the manufacturer's instructions. Isolated RNA was quantified and subjected to DNaseI treatment to degrade genomic DNA. The reverse transcription reactions were performed with an IScript Reverse Transcription Kit (Bio-Rad, Mississauga, ON, Canada) according to the manufacturer's protocol. Real-time quantitative PCR (qPCR) reactions were carried out for all genes of interest in each sample using Light Cycler 480 SYBR Green1 (Roche Diagnostics, Mississauga, ON, Canada) Gene Expression Assays in a LightCycler 480 II (Roche Diagnostics). The sequences of the PCR primers are listed in Supplementary Table S3. The cycling conditions were: 1 cycle of denaturation at 95 °C for 5 min, followed by 45 three-segment cycles of amplification (95 °C for 10 s, 56–60 °C (gene depending), 72 °C/20 s) where the fluorescence was automatically measured during PCR and one three-segment cycle of product melting (95 °C for 5 s, 65 °C for 60 s, 98 °C continuous mode). The LightCycler480 Relative Quantification Software (Roche Applied Sciences, Penzberg, Germany), was used to determine the threshold cycle (Ct) in each reaction.

A melting curve was constructed for each primer pair to verify the presence of one gene-specific peak and the absence of primer dimer. Glyceraldehyde 3-phosphate dehydrogenase (GAPDH) was chosen as the reference housekeeping gene. Each experimental sample was repeated 12 times. For each cDNA sample, the Ct value of each target sequence was subtracted from the Ct value of the reference gene (GAPDH), to derive ΔCt. The level of expression of each target gene, normalized to GAPDH, was then calculated using the delta Ct method, where ΔCt value of each experimental sample was subtracted from ΔCt value of control samples, and $\log_2$ value was calculated.

2.12. Statistical Analyses

Figures were generated using GraphPad Prism 6.0 software (GraphPad Software, Inc., La Jolla, CA, USA). Statistical analyses were also performed using this software. Differences between means were evaluated using the Student's t-test and were deemed significant at $p \leq 0.05$.

3. Results

3.1. Isolation and Characterization of ARV-PB1

ARV-PB1 was isolated from a field sample. Virus propagation, titration and one-step growth curves were carried out in chicken hepatoma cells (CH-SAH cell line), as described [28]. ARV-PB1 infection induced cytopathic effect in CH-SAH cells through formation of syncytia, a hallmark of some orthoreoviruses (Figure S1). Reoviruses are known to have segmented genomes. We sequenced the sigma 1 (S1), sigma 2 (S2), sigma 4 (S4) and mu 2 (M2) genomic segments, and analysis by a nucleotide Basic Local Alignment Search Tool (BLASTn) [31] indicated that ARV-PB1 represented an avian orthoreovirus (Table S1).

3.2. ARV-PB1 Shows Cytolytic Activity in Cancer Cell Lines

Recent reports have highlighted the potential of reoviruses in oncolytic virotherapy [13]. Moreover, orthoreoviruses can infect a wide range of cells, and avian orthoreoviruses can cause hepatitis in chickens suggesting virus tropism to liver cells [21,28]. These led us to hypothesize that ARV-PB1 could have productive infection and induce cell death in HCC cells. To investigate this, ARV-PB1 was tested against four liver cancer cell lines: Huh-7, Huh-7.5, Huh-7.5.1 and HepG2 (Figure 1a). CPE and a decrease in cell viability were observed upon infection with ARV-PB1. To determine if viral replication was playing a role in oncolysis of HCC cells, one-step growth curves were performed. Each liver cell line supported virus replication with a one-log increase (or greater) over input viral titers, highlighted by a more than two-log increase in HepG2 cells by 72 h.p.i. (Figure 1b). Additionally, immunofluorescent staining of these cells with an antibody against dsRNA indicated the presence of dsRNA within the cells, further confirming active infection (Figure S2). Furthermore, crystal violet staining of infected cells revealed CPE (Figure 1c).

Additionally, we tested the efficacy of ARV-PB1 against a panel of cancer cell lines (Table S2) that represented a range of different tumor types and compared the cell viability between infected and uninfected cells. Slight to moderate decreases in viability were seen in the analyzed cell lines. More extensive decreases in cell viability upon infection were observed in HeLa, and 22Rv1 cells (Figure 1d). We used Reolysin^{TM}, an oncolytic human reovirus currently in clinical studies, as a reference virus to compare the cytolytic effects of ARV-PB1 in Huh-7.5 cells. As shown in Figure S3, the decrease of viability of cells infected with ARV-PB1 was comparable to those infected with Reolysin^{TM}.

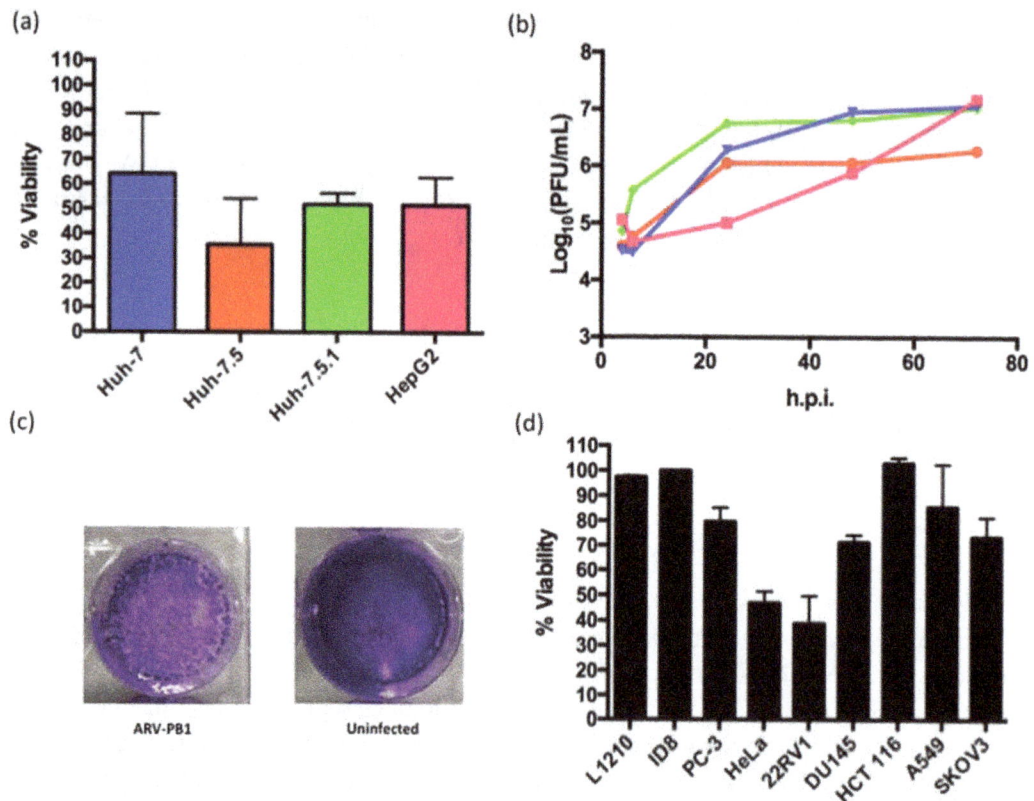

Figure 1. Evaluation of avian orthoreovirus (ARV-PB1) replication in cell lines. (**a**) Cell viability in human liver cell lines was measured at 96 hours post-infection (h.p.i.) with PrestoBlue™ Cell Viability Reagent (Life Technologies) and data was normalized to uninfected controls. Experiments for each cell line were performed a minimum of five times, and error bars represent the standard deviation. (**b**) ARV-PB1 titers at various time points, in cell lines following infection at a multiplicity of infection (MOI) of 5. The experiment was performed in duplicate and data are from one representative experiment. (**c**) Cytopathic effect following crystal violet staining in Huh-7.5 cells at 72 h.p.i. Similar results were observed in the other liver cell lines tested, which included Huh-7, Huh-7.5.1 and HepG2. Experiments were performed in triplicate, and one representative picture is shown. (**d**) Cell viability in other cancer cell lines infected at an MOI of 10 was measured at 96 h.p.i. with PrestoBlue™ Cell Viability Reagent (Life Technologies) and data was normalized to uninfected controls. The origin of the analyzed cell lines is described in Table S2. Experiments for the ID8 cell line was performed twice, all other cell lines were repeated a minimum of three times. Data represent the mean, and error bars represent the standard deviation.

3.3. ARV-PB1 Is Not Cytotoxic in Ex Vivo Hepatocytes

An important attribute of OVs is their selectivity for cancer cells. To test the specificity of ARV-PB1 to cancer cells, virus replication and CPE were examined ex vivo in primary hepatocyte cultures from biopsy samples. Upon virus infection, hepatocytes remained viable during the course of the experiment (Figure 2a) and virus titers decreased by 96 h.p.i. relative to the input infection dose (Figure 2b). Collectively, our results suggest that ARV-PB1 replicates and kills HCC cells while sparing normal hepatocytes.

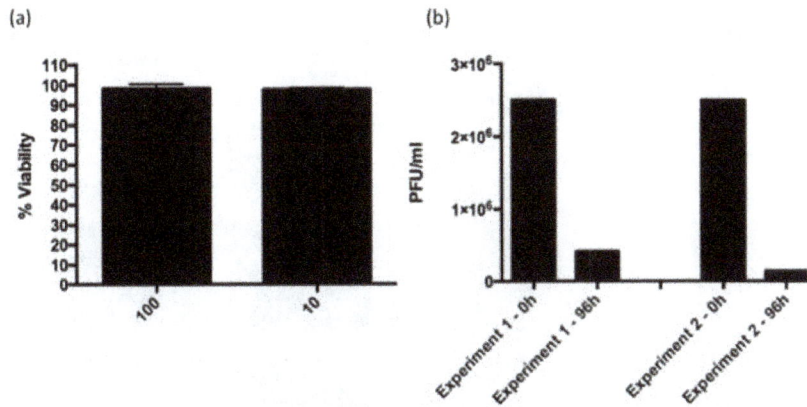

Figure 2. Cytotoxicity assessment of ARV-PB1. (**a**) Primary hepatocytes were infected at multiplicities of infection (MOI) of 10 and 100. After 1 h virus adsorption, the inoculum was removed and cells were washed with phosphate buffered saline (PBS) and fresh medium was added. Cell viability was determined at 96 h.p.i. by PrestoBlue™ Cell Viability Reagent (Life Technologies). Data were normalized to uninfected controls. Experiments were performed four times, and data represent the mean while error bars represent the standard deviation. (**b**) Primary hepatocytes were infected with ARV-PB1 at an MOI of 10 as described above and virus titers were examined at 0 and 96 h.p.i.

3.4. ARV-PB1 Induces Syncytia Formation and Apoptosis in Hepatocellular Carcinoma Cells

The presence of the FAST protein is unique amongst certain members of the family *Reoviridae*, resulting in characteristic formation of syncytia, and induction of apoptosis in infected cells [35,36]. To determine whether infection with ARV-PB1 induced syncytia, liver cells were stained with DAPI allowing visualization of multinucleated cells formed as a result of cell-to-cell fusion (Figure 3). Monoclonal antibody to dsRNA and DAPI staining of the analyzed HCC cell lines confirmed the presence of ARV-PB1 (red fluorescence signals) in multinucleated syncytial cells (Figure S2). Control cells, on the other hand, were negative for the presence of dsRNA and syncytia formation.

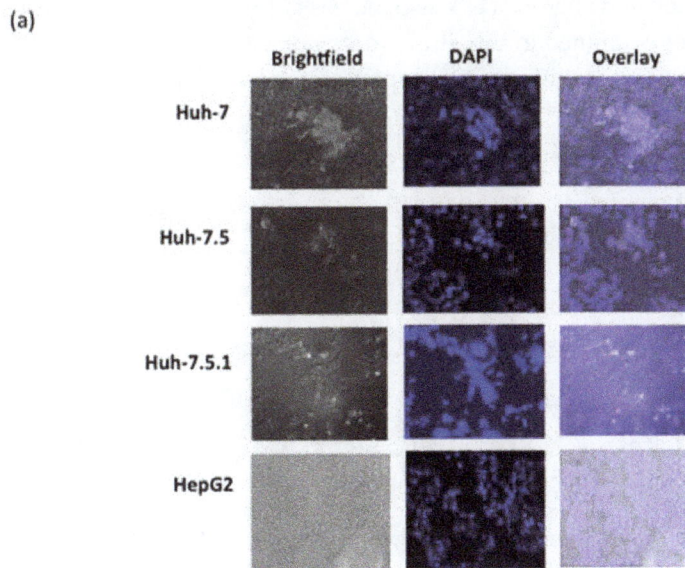

Figure 3. *Cont.*

(b)

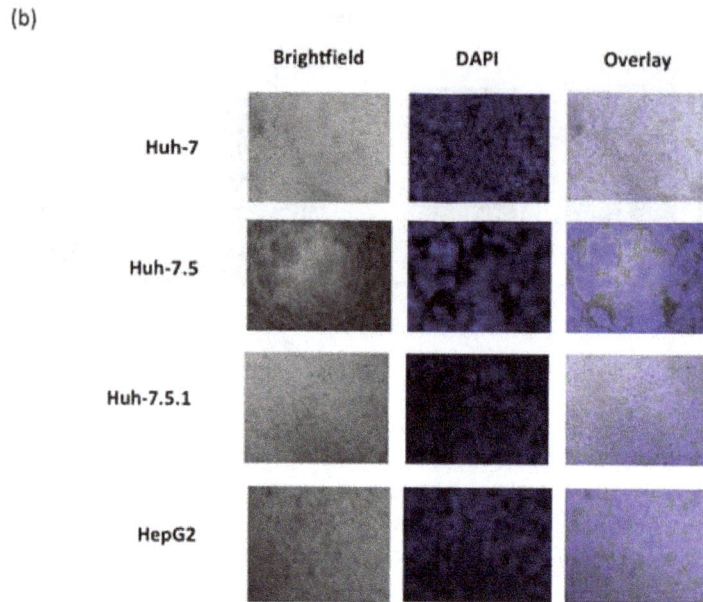

Figure 3. Syncytia formation in liver cell lines. (**a**) Cells were infected at an MOI of 5 for 72 h followed by fixation with formalin and 4′,6-diamidino-2-phenylindole (DAPI) staining. (**b**) No syncytial formation was observed in uninfected cells. Cells were analyzed by fluorescence and bright-field microscopy (100× magnification).

Syncytia formation mediated by FAST proteins has also been shown to activate cellular apoptotic pathways [35]. Therefore, to test whether ARV-PB1 induced apoptosis, Huh-7.5.1 cells were stained with annexin V and 7-aminoactinomycin D (7-AAD). During programmed cell death, phosphatidylserine (PS) is translocated to the extracellular membrane, and can be quantified by binding of fluorochrome-labelled annexin V. At 72 h.p.i, there were significantly fewer viable cells amongst the reovirus-infected groups compared to uninfected controls ($p = 0.005$). Furthermore, the percentages of apoptotic and necrotic cells also increased (Figure 4). Similar results were obtained in the other liver cell lines examined in this study (data not shown). These data suggest the correlation between syncytia formation and induction of apoptosis in liver cancer cells.

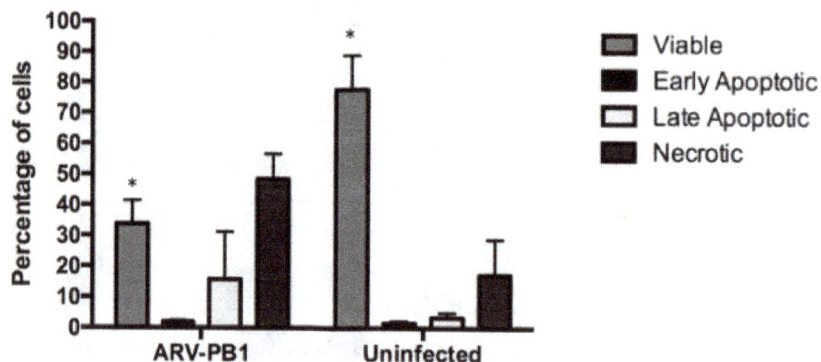

Figure 4. Analysis of apoptosis in hepatocellular carcinoma cells. Huh-7.5 cells were either mock infected or infected with ARV-PB1 at a MOI of 5 for 72 h. Subsequently, mechanisms of cell death were determined by annexin V and 7-AAD staining according to the manufacturer's instructions. The percentages of viable (annexin V− 7-AAD−), early apoptotic (annexin V+ 7-AAD−), late apoptotic (annexin V+ 7-AAD+) and necrotic (annexin V− 7-AAD+) cells were determined. Data represent the mean from experiments performed in triplicate. Error bars represent standard deviations. Samples were compared using Student's t-test with significant differences indicated by * $p \leq 0.05$.

3.5. ARV-PB1 Induces Expression of Interferon-Stimulated Genes

The induction of antitumor immune responses is important for achieving clearance of tumor cells. Additionally, recent work has shown that the oncolytic effect of OVs is largely determined by the anti-tumor immunity induced upon infection rather than virus-induced cell lysis [7,37,38]. We hypothesized that ARV-PB1 infection would induce the expression of ISGs that are linked to antitumor immunity. To test this hypothesis, Huh-7 cells were infected with ARV-PB1, and we examined the transcription of 11 ISGs involved in the production and induction of interferons at 6 h.p.i. This early time point was selected to ensure that any change in the expression was due to viral infection, rather than autocrine/paracrine effects of interferon production. We considered a >2-fold change in expression to be significant, and this was observed for the genes encoding IF144 (interferon (IFN)-induced protein 44), interleukin (IL)-8, IP10 (10 kDa interferon γ-induced protein), IFN-λ1 and SOCS3 (suppressor of cytokine signaling 3). Although, it was determined that while the expression of *IF144* was elevated, this failed to achieve statistical significance ($p = 0.06$). Notably, *IL8* and *IFN-λ1* showed very high levels of expression, with 5.3- and 16.7-fold increases, respectively (Figure 5).

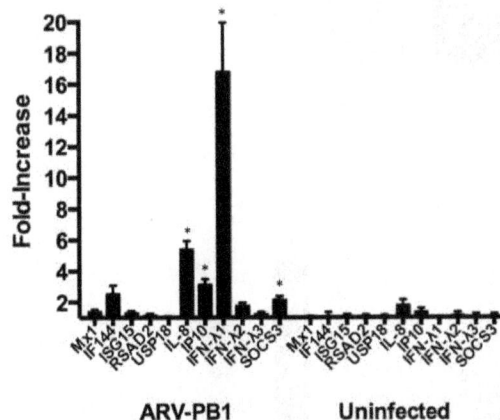

Figure 5. Expression of interferon-stimulated genes (ISGs). Huh-7 cells (5×10^5) were infected at a MOI of 5 and messenger RNA (mRNA) was collected for analysis at 6 hours post-infection. Transcription of the analyzed ISGs was analyzed by quantitative real-time polymerase chain reaction (PCR) with glyceraldehyde 3-phosphate dehydrogenase (GAPDH) as a housekeeping gene and expressed as a fold-change compared to uninfected cells (35). Data represent the mean from six replicate experiments, with error bars showing the standard error of the mean. Samples were compared using Student's *t*-test with significant differences indicated by * $p \leq 0.05$.

3.6. Activity in Hepatitis C Virus Replicon-Containing Cells

Infection with HCV is a major cause of HCC, and recurrent infection can occur even after liver transplantation [2]. Therefore, we investigated whether ARV-PB1 had cytolytic effect in an in vitro model of infection with HCV by using BB7 cells, a cell line derived from Huh-7.5.1 cells, which stably contains the HCV subgenomic genotype 1a H77 replicon. As shown in Figure 6, cell viability decreased as the MOI increased, and productive infection and titers were comparable to those in other liver cell lines. Likewise, virus infection induced syncytia and apoptosis in BB7 cells (Figure 6). Additionally, the decrease in cell viability upon infection with ARV-PB1 was comparable to that with Reolysin™ (Figure S3). Our results suggest that ARV-PB1 has cytolytic activity in HCC cell lines containing the HCV replicon, and that the presence of HCV does not prevent replication of ARV-PB1.

(a)

(b)

(c)

(d)

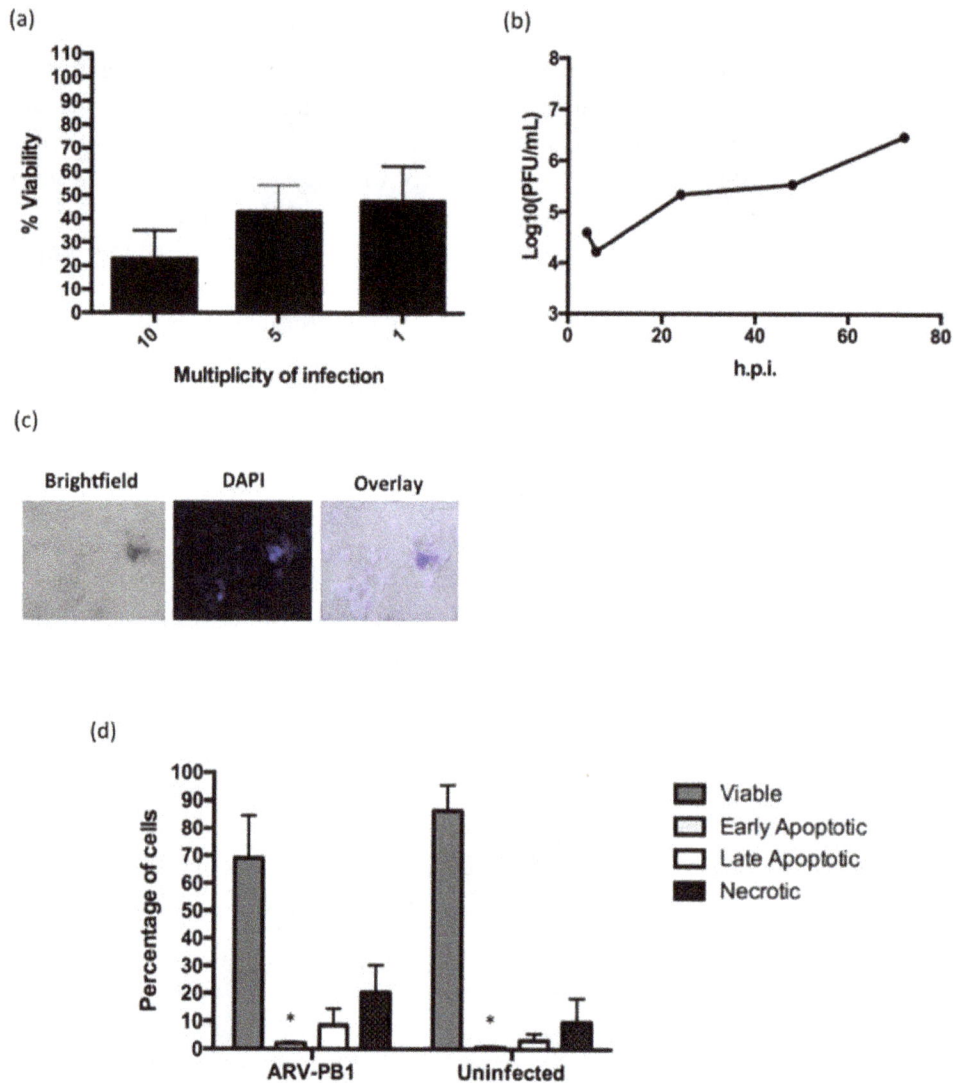

Figure 6. ARV-PB1 activity in HCV-replicon-containing cells. (**a**) Cell viability in BB7 cells infected at differing MOIs was measured at 96 hours post-infection (h.p.i.) using PrestoBlue^{TM} Cell Viability Reagent (Life Technologies). Data were normalized to uninfected controls. The experiment was performed at three times, and error bars represent the standard error of the mean. (**b**) ARV-PB1 viral titers in BB7 cells at various time points following infection at a MOI of 5. (**c**) Fluorescent and bright-field microscopy (100× magnification) to view DAPI staining and evaluate syncytia in BB7 cells infected with ARV-PB1 (MOI of 5) at 96 h.p.i. The experiment was performed in triplicate and data are from one representative experiment. (**d**) BB7 cells were either uninfected or infected with ARV-PB1 (MOI of 5) for three days. Subsequently, the cells were fixed and treated with annexin V and 7-AAD. The percentages of viable (annexin V− 7-AAD−), early apoptotic (annexin V+ 7-AAD−), late apoptotic (annexin V+ 7-AAD+) and necrotic (annexin V− 7-AAD+) cells were determined. Data represent the means from experiments performed in triplicate. Error bars show standard deviations. Samples were compared using Student's t-test with significant differences indicated by * $p \leq 0.05$.

3.7. ARV-PB1 Shows Cytolytic Activity in JFH-1 Infected Cells

The HCV genotype 1a H77 replicon contains only non-structural genes and thus virus replication and spread are ablated [39]. Therefore, we also tested ARV-PB1 against Huh-7 and Huh-7.5.1 cells infected with the lab-adapted infectious virus strain JFH-1. This virus, unlike the H77 replicon, contains the complete HCV genome, and is capable of producing infectious particles, making it a more physiologically relevant model of HCV infection [40]. The effect of ARV-PB1 on the viability of

JFH-1-infected cells was assessed at 48 h.p.i. CPE was observed upon infection with ARV-PB1 while absent in either uninfected cells or those infected with JFH-1 (Figure 7a,c). Moreover, DAPI staining showed large multinucleated cells and syncytia formation (Figure 7b), similar to those that were seen in cell lines containing the HCV replicon. This further demonstrates that infection with HCV does not inhibit replication and spread of ARV-PB1.

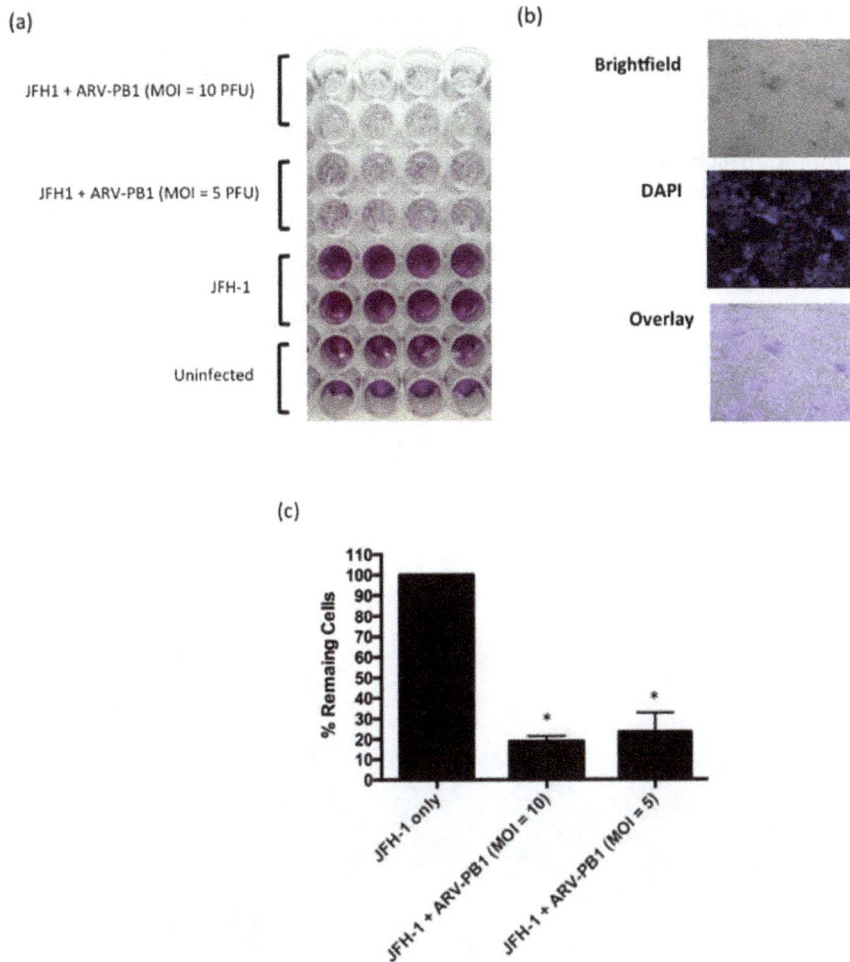

Figure 7. ARV-PB1 activity in JFH-1 -infected cells. (**a**) Huh-7.5.1 cells were infected with JFH-1 and 24 h later treated with ARV-PB1 at differing MOIs. At 48 hours post-treatment, cells were fixed and stained with crystal violet to evaluate cytopathic effect. The experiment was performed in triplicate. (**b**) Induction of syncytia in JFH-1-infected cells by ARV-PB1. Cells were infected with ARV-PB1 (MOI of 5) for 48 h. Subsequently, cells were fixed with formalin, stained with DAPI and analyzed by fluorescent and bright-field microscopy (100× magnification). (**c**) Measurement of remaining viable cells following treatment with various MOIs at 48h post-infection. The experiment was performed at four times, and error bars represent the standard deviation. Samples were compared using Student's t-test with significant differences indicated by * $p \leq 0.05$.

3.8. Serum Neutralization and In Silico Modeling

Neutralizing antibodies can limit the systemic distribution and therapeutic potential of OVs [15]. Mammalian reovirus type 1 σ1 adhesin fiber induces neutralizing antibodies [41]. In order to investigate whether antibodies to mammalian orthoreovirus would neutralize ARV-PB1, we performed plaque reduction assays using sera from patients with HCV-induced HCC and healthy volunteers. Sera from both HCC patients and healthy volunteers failed to reduce the number of virus induced plaques, which suggests the lack of pre-existing immunity to ARV-PB1

Using the online server I-TASSER we first analyzed the S1 gene to determine the most closely related crystallized structure. Although truncated, the server identified a structure of avian reovirus S1133 fiber confirming the identity of ARV-PB1 (Figure 8b). As demonstrated, the β-pleated sheets within this protein overlap with sequence homology of 87% in this region. Next, we used mammalian orthoreovirus type 3 (Dearing strain) as a reference to assess the tertiary structural similarity between human and avian orthoreovirus adhesion fibers. There is much less structural homology with the human protein in comparison to the crystallized avian protein, as evidenced by the minimal homology in beta-pleated sheets, and structural sequence alignments (Figure 8c,d). Furthermore, the structural alignment revealed significant amino acid difference in the epitope associated with antibody binding (highlighted in red) [41,42], suggesting that antibodies generated in patients who had received mammalian orthoreovirus type 3 (Dearing strain) (Reolysin™) would not neutralize ARV-PB1.

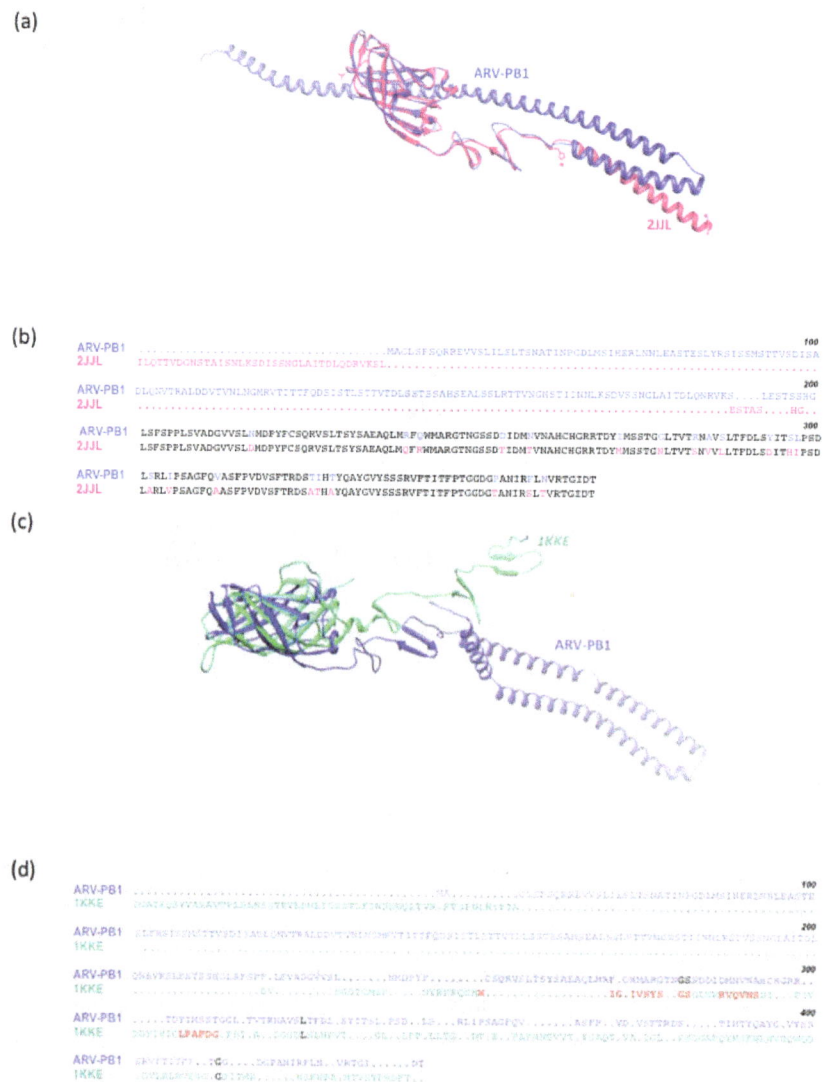

(a)

(b)

(c)

(d)

Figure 8. In silico structure generation of the S1 attachment protein of ARV-PB1. (**a**) I-TASSER-generated ARV-PB1 (blue), manually superimposed with avian reovirus σ C117-326, Protein Data Bank (PDB) ID 2JJL (pink), identified as the most closely related crystal structure. (**b**) Tertiary structural alignment of the S1 protein of ARV-PB1 and 2JJL. Conserved sequence regions are highlighted in black. (**c**) I-TASSER-generated ARV-PB1 (blue), manually superimposed with mammalian orthoreovirus 3 (Dearing strain), PDB ID 1KKE (green), using 1KKE as a generation template. (**d**) Tertiary structural alignment of the S1 protein of ARV-PB1 and 1KKE. Conserved sequence regions are highlighted in black, and the epitope targets of neutralizing antibodies are highlighted in red.

4. Discussion

The anticipated increase in worldwide HCC burden, combined with the poor outcomes, highlights the urgent need for new therapeutic options. We have isolated an avian orthoreovirus, ARV-PB1, and demonstrated its cytolytic activity in vitro against several liver cancer cell lines, including those infected with HCV. Therefore, our results suggest the oncolytic potential of ARV-PB1. Our results also suggest that the mechanisms of ARV-PB1-induced apoptosis are likely through syncytia formation by FAST proteins, as known for fusogenic reoviruses [36]. Interestingly, the presence of HCV did not impede viral replication, nor did it influence the oncolytic effects of ARV-PB1. Additionally, ARV-PB1 was an inducer of ISGs, which may have contributed to the apoptosis in Huh-7 cells. Therefore, we speculate that cytolytic properties of ARV-PB1 are likely determined by combination of direct cytopathic and immunostimulatory effects.

Recent clinical trials have highlighted the therapeutic potential of OVs, particularly reoviruses [9,13,23,42,43]. ARV-PB1 may provide a means of overcoming the challenges of pre-existing immunity that may otherwise limit the applicability of some of the identified OVs. The $\sigma1$ gene encodes the adhesin protein, which is the target of neutralizing antibodies [41,42]. It was noted that ARV-PB1 was not neutralized by a commercially available S1 adhesion protein-specific antibody, and that neutralizing antibodies against ARV-PB1 were not present in the serum of HCC patients infected with HCV. Further, significant structural and tertiary alignment sequence differences were found between the ARV-PB1 σ1 and its homolog in the mammalian reovirus serotype 3. Therefore, the presence of neutralizing antibodies in individuals treated with mammalian reoviruses (Reolysin™) would be low, and unlikely to inhibit ARV-PB1. Therefore, there is a potential for synergy with Reolysin™, or other OVs, or in combination with chemo drugs. Vaccinia virus and vesicular stomatitis virus are known to synergize in various tumor models [42]. Therefore, the use of ARV-PB1 in conjunction with another OV effective against HCC (such as JX-594 [38,44]) would be worth investigating.

This is the first report of the cytolytic effects of an avian reovirus against HCC, suggesting the potential oncolytic properties of ARV-PB1. We also demonstrated that ARV-PB1 neither replicates nor induces CPE in primary hepatocytes isolated from biopsy samples. Although treatments for HCV are rapidly improving and becoming more available, HCV continues to be a leading cause of liver cancer worldwide. To date, other OVs targeting HCC have shown efficacy against HBV and HCV-associated HCC [37,45]. Interestingly, recent work by Samson and colleagues has demonstrated the in vitro and in vivo oncolytic activity of a mammalian reovirus against HCC, including the ability to eradicate HCV from cells [45]. Moreover, induction of interferon was an important component of the antitumor response, as an effect was also seen with UV-inactivated virus. These findings support our data as we noted induction of IFN-λ1 as well oncolytic effect in HCV-infected cell lines. Future experiments will be performed to investigate if the oncolytic effect observed with ARV-PB1 is altered following UV-inactivation. However, it is tempting to speculate that the combination therapy with both mammalian and avian reoviruses could result in enhanced efficacy.

It has been shown that HCV inhibits protein kinase R (PKR), thereby preventing the induction of the antiviral state, which could promote replication of other viruses [46,47]. Additionally, some evidence suggests that inactive PKR may be unable to phosphorylate p53 and may also play a role in tumorigenesis through activation of signaling pathways involved in cell proliferation [47,48]. Therefore, activation of PKR may help in creating an anti-tumor environment in cells. Recent work demonstrated that ARVs may preferentially activate PKR and AMPK (AMP-activated protein kinase) as a means to aid viral replication [27,48]. Thus,we speculate that the CPE observed in cells co-infected with JFH-1 and ARV-PB1, may be due to HCV-mediated inhibition of PKR creating a favorable cellular environment for ARV-PB1 replication.

Our study demonstrated the ability of ARV-PB1 to form syncytia, and correlated this with the induction of apoptosis. Additionally, the sequence analysis revealed a putative homolog to the p10 gene found in ARV-138, known to encode a FAST protein (data not shown). It has been shown that genes encoding this family of proteins are found in some members of the family *Reoviridae*; however, they are not present in the mammalian reovirus currently being investigated as a virotherapeutic [35]. Moreover, it has been noted that vesicular stomatitis virus expressing the p14 FAST protein is able to induce cell fusion, and enhanced production of another oncolytic virus when administered together [43]. Thus, we hypothesize that the presence of a predicted FAST protein in ARV-PB1 will promote enhanced viral spread and oncolytic effect within the tumor environment, and we intend to test this in animal models in future studies.

An important goal of OV therapy is to achieve intratumoral spread to debulk the tumor while concurrently inducing an innate immune response. This assists in the removal of remaining tumor cells by direct mechanisms, or by activating adaptive immunity. As illustrated by our results, ARV-PB1 induced higher expression of IFN-λ1. This cytokine has been associated with induction of apoptosis and natural killer (NK) cell recruitment in the BNL cell line based model of HCC [49]. Three requirements have recently been highlighted for an OV to achieve a viroimmunotherapy response [50]: selectivity for cancer cells, induction of a potent immune response, and exposure of tumor-associated antigens to the immune system. This report demonstrates that ARV-PB1 selectively targets cancer cells, and induces genes associated with a potent innate immune response. We have proposed several mechanisms to explain its oncolytic activity in vitro. Taken together, these data provide rationale for in vivo tumor models to study the oncolytic effects of ARV-PB1.

Supplementary Materials: The following are available online at www.mdpi.com/1999-4915/9/4/90/s1, Figure S1: Cytopathic effect in response to viral infection, Figure S2: Presence of viral RNA in infected cells, Figure S3: Comparison of ARV-PB1 to Reolysin™ in Huh-7.5 cells, Figure S4: Comparison of ARV-PB1 to Reolysin™ in BB7 cells, Table S1: Homology of ARV-PB1 nucleotide sequences, Table S2: Characteristics of cancer cell lines, Table S3: Primers for qPCR.

Acknowledgments: This work was supported by the Natural Sciences and Engineering Research Council of Canada, and the Ontario Ministry of Agriculture, Food and Rural Affairs.

Author Contributions: Experiments were performed and data analyzed by R.A.K., L.H., M.J.B., S.W., J.C.C., M.X.-Z., J.M., E.L., J.M., I.D.M., V.C., B.-A.M., and D.L. Experimental design was done by R.A.K., M.J.B., J.J.F., B.B., I.D.M. and E.N. The manuscript was written by R.A.K., M.J.B., J.C.C, J.J.F., B.B. and E.N.

References

1. Bakiri, L.; Wagner, E.F. Mouse models for liver cancer. *Mol. Oncol.* **2013**, *7*, 206–223. [CrossRef] [PubMed]
2. Shah, H.A.; Heathcote, J.; Feld, J.J. A Canadian screening program for hepatitis C: Is now the time? *CMAJ* **2013**, *185*, 1325–1328. [CrossRef] [PubMed]
3. Russell, S.J.; Peng, K.W.; Bell, J.C. Oncolytic virotherapy. *Nat. Biotechnol.* **2012**, *30*, 658–670. [CrossRef] [PubMed]
4. Altomonte, J.; Marozin, S.; Schmid, R.M.; Ebert, O. Engineered newcastle disease virus as an improved oncolytic agent against hepatocellular carcinoma. *Mol. Ther.* **2010**, *18*, 275–284. [CrossRef] [PubMed]
5. Unno, Y.; Shino, Y.; Kondo, F.; Igarashi, N.; Wang, G.; Shimura, R.; Yamaguchi, T.; Asano, T.; Saisho, H.; Sekiya, S.; et al. Oncolytic viral therapy for cervical and ovarian cancer cells by Sindbis virus AR339 strain. *Clin. Cancer Res.* **2005**, *11*, 4553–4560. [CrossRef] [PubMed]
6. Evgin, L.; Vaha-Koskela, M.; Rintoul, J.; Falls, T.; Le Boeuf, F.; Barrett, J.W.; Bell, J.C.; Stanford, M.M. Potent oncolytic activity of raccoonpox virus in the absence of natural pathogenicity. *Mol. Ther.* **2010**, *18*, 896–902. [CrossRef] [PubMed]
7. Rintoul, J.L.; Lemay, C.G.; Tai, L.H.; Stanford, M.M.; Falls, T.J.; de Souza, C.T.; Bridle, B.W.; Daneshmand, M.; Ohashi, P.S.; Wan, Y.; et al. ORFV: A novel oncolytic and immune stimulating parapoxvirus therapeutic. *Mol. Ther.* **2012**, *20*, 1148–1157. [CrossRef] [PubMed]
8. Bartlett, D.L.; Liu, Z.; Sathaiah, M.; Ravindranathan, R.; Guo, Z.; He, Y.; Guo, Z.S. Oncolytic viruses as therapeutic cancer vaccines. *Mol. Cancer* **2013**, *12*, 103. [CrossRef] [PubMed]
9. Miest, T.S.; Cattaneo, R. New viruses for cancer therapy: Meeting clinical needs. *Nat. Rev. Microbiol.* **2014**, *12*, 23–34. [CrossRef] [PubMed]
10. Schmidt, C. Oncolytic Virus Approved To Treat Melanoma. *J. Natl. Cancer Inst.* **2016**, *108*. Print 2016 May. [CrossRef] [PubMed]

11. Comins, C.; Heinemann, L.; Harrington, K.; Melcher, A.; De Bono, J.; Pandha, H. Reovirus: Viral therapy for cancer "as nature intended". *Clin Oncol. (R. Coll. Radiol.)* **2008**, *20*, 548–554. [CrossRef] [PubMed]

12. Harrington, K.J.; Hingorani, M.; Tanay, M.A.; Hickey, J.; Bhide, S.A.; Clarke, P.M.; Renouf, L.C.; Thway, K.; Sibtain, A.; McNeish, I.A.; et al. Phase I/II study of oncolytic HSV GM-CSF in combination with radiotherapy and cisplatin in untreated stage III/IV squamous cell cancer of the head and neck. *Clin Cancer Res.* **2010**, *16*, 4005–4015. [CrossRef] [PubMed]

13. Morris, D.G.; Feng, X.; DiFrancesco, L.M.; Fonseca, K.; Forsyth, P.A.; Paterson, A.H.; Coffey, M.C.; Thompson, B. REO-001: A phase I trial of percutaneous intralesional administration of reovirus type 3 dearing (Reolysin(R)) in patients with advanced solid tumors. *Invest. New Drugs* **2013**, *31*, 696–706. [CrossRef] [PubMed]

14. Alemany, R. Viruses in cancer treatment. *Clin. Transl. Oncol.* **2013**, *15*, 182–188. [CrossRef] [PubMed]

15. Tatsis, N.; Tesema, L.; Robinson, E.R.; Giles-Davis, W.; McCoy, K.; Gao, G.P.; Wilson, J.M.; Ertl, H.C. Chimpanzee-origin adenovirus vectors as vaccine carriers. *Gene Ther.* **2006**, *13*, 421–429. [CrossRef] [PubMed]

16. Zamarin, D.; Palese, P. Oncolytic Newcastle disease virus for cancer therapy: Old challenges and new directions. *Futur. Microbiol.* **2012**, *7*, 347–367. [CrossRef] [PubMed]

17. Park, M.S.; Garcia-Sastre, A.; Cros, J.F.; Basler, C.F.; Palese, P. Newcastle disease virus V protein is a determinant of host range restriction. *J. Virol.* **2003**, *77*, 9522–9532. [CrossRef] [PubMed]

18. Lawrence, T.M.; Wanjalla, C.N.; Gomme, E.A.; Wirblich, C.; Gatt, A.; Carnero, E.; Garcia-Sastre, A.; Lyles, D.S.; McGettigan, J.P.; Schnell, M.J. Comparison of Heterologous Prime-Boost Strategies against Human Immunodeficiency Virus Type 1 Gag Using Negative Stranded RNA Viruses. *PLoS ONE* **2013**, *8*, e67123. [CrossRef] [PubMed]

19. Alemany, R.; Cascallo, M. Oncolytic viruses from the perspective of the immune system. *Futur. Microbiol.* **2009**, *4*, 527–536. [CrossRef] [PubMed]

20. Kibenge, F.; Gwaze, G.; Jones, R.; Chapman, A.; Savage, C. Experimental reovirus infection in chickens: Observations on early viraemia and virus distribution in bone marrow, liver and enteric tissues. *Avian Pathol.* **1985**, *14*, 87–98. [CrossRef] [PubMed]

21. Jones, R.C. Reovirus infections. In *Diseases of poultry*, 13th ed.; Swayne, D.E., Glisson, J.R., McDougald, L.R., Nolan, L.K., Suarez, D.L., Nair, V.L, Eds.; Wiley & Sons: Ames, IA, USA, 2013; pp. 351–373.

22. Benavente, J.; Martinez-Costas, J. Avian reovirus: Structure and biology. *Virus Res.* **2007**, *123*, 105–119. [CrossRef] [PubMed]

23. Roulstone, V.; Twigger, K.; Zaidi, S.; Pencavel, T.; Kyula, J.N.; White, C.; McLaughlin, M.; Seth, R.; Karapanagiotou, E.M.; Mansfield, D.; et al. Synergistic cytotoxicity of oncolytic reovirus in combination with cisplatin-paclitaxel doublet chemotherapy. *Gene Ther.* **2013**, *20*, 521–528. [CrossRef] [PubMed]

24. Schnitzer, T.J.; Ramos, T.; Gouvea, V. Avian reovirus polypeptides: Analysis of intracellular virus-specified products, virions, top component, and cores. *J. Virol.* **1982**, *43*, 1006–1014. [PubMed]

25. Duncan, R.; Chen, Z.; Walsh, S.; Wu, S. Avian reovirus-induced syncytium formation is independent of infectious progeny virus production and enhances the rate, but is not essential, for virus-induced cytopathology and virus egress. *Virology* **1996**, *224*, 453–464. [CrossRef] [PubMed]

26. Nibert, M.L.; Duncan, R. Bioinformatics of recent aqua- and orthoreovirus isolates from fish: Evolutionary gain or loss of FAST and fiber proteins and taxonomic implications. *PLoS ONE* **2013**, *8*, e68607. [CrossRef] [PubMed]

27. Chi, P.I.; Huang, W.R.; Lai, I.H.; Cheng, C.Y.; Liu, H.J. The p17 nonstructural protein of avian reovirus triggers autophagy enhancing virus replication via activation of phosphatase and tensin deleted on chromosome 10 (PTEN) and AMP-activated protein kinase (AMPK), as well as dsRNA-dependent protein kinase (PKR)/eIF2α signaling pathways. *J. Biol. Chem.* **2013**, *288*, 3571–3584. [CrossRef] [PubMed]

28. Tran, A.; Berard, A.; Coombs, K.M. Avian reoviruses: Propagation, quantification, and storage. *Curr. Protoc. Microbiol.* **2009**, *Chapter 15*, Unit15C 2. [CrossRef]

29. Russell, P.K.; Nisalak, A.; Sukhavachana, P.; Vivona, S. A plaque reduction test for dengue virus neutralizing antibodies. *J. Immunol.* **1967**, *99*, 285–290. [PubMed]

30. Jiang, J.; Hermann, L.; Coombs, K.M. Genetic characterization of a new mammalian reovirus, type 2 Winnipeg (T2W). *Virus Genes* **2006**, *33*, 193–204. [CrossRef] [PubMed]

31. Blast. Bethesda (MD): National Library of Medicine (US), National Center for Biotechnology Information. 2004. Available online: http://blast.ncbi.nlm.nih.gov/Blast.cgi (accessed on 20 April 2017).

32. Roy, A.; Xu, D.; Poisson, J.; Zhang, Y. A protocol for computer-based protein structure and function prediction. *J. Vis. Exp.* **2011**, e3259. [CrossRef] [PubMed]

33. Roy, A.; Kucukural, A.; Zhang, Y. I-TASSER: A unified platform for automated protein structure and function prediction. *Nat. Protoc.* **2010**, *5*, 725–738. [CrossRef] [PubMed]

34. Chappell, J.D.; Prota, A.E.; Dermody, T.S.; Stehle, T. Crystal structure of reovirus attachment protein sigma1 reveals evolutionary relationship to adenovirus fiber. *EMBO J.* **2002**, *21*, 1–11. [CrossRef] [PubMed]

35. Shmulevitz, M.; Duncan, R. A new class of fusion-associated small transmembrane (FAST) proteins encoded by the non-enveloped fusogenic reoviruses. *EMBO J.* **2000**, *19*, 902–912. [CrossRef] [PubMed]

36. Salsman, J.; Top, D.; Boutilier, J.; Duncan, R. Extensive syncytium formation mediated by the reovirus FAST proteins triggers apoptosis-induced membrane instability. *J. Virol.* **2005**, *79*, 8090–8100. [CrossRef] [PubMed]

37. Liu, T.C.; Hwang, T.; Park, B.H.; Bell, J.; Kirn, D.H. The targeted oncolytic poxvirus JX-594 demonstrates antitumoral, antivascular, and anti-HBV activities in patients with hepatocellular carcinoma. *Mol. Ther.* **2008**, *16*, 1637–1642. [CrossRef] [PubMed]

38. Melcher, A.; Parato, K.; Rooney, C.M.; Bell, J.C. Thunder and lightning: immunotherapy and oncolytic viruses collide. *Mol. Ther.* **2011**, *19*, 1008–1016. [CrossRef] [PubMed]

39. Lohmann, V.; Korner, F.; Koch, J.; Herian, U.; Theilmann, L.; Bartenschlager, R. Replication of subgenomic hepatitis C virus RNAs in a hepatoma cell line. *Science* **1999**, *285*, 110–113. [CrossRef] [PubMed]

40. Wakita, T.; Pietschmann, T.; Kato, T.; Date, T.; Miyamoto, M.; Zhao, Z.; Murthy, K.; Habermann, A.; Krausslich, H.G.; Mizokami, M.; et al. Production of infectious hepatitis C virus in tissue culture from a cloned viral genome. *Nat. Med.* **2005**, *11*, 791–796. [CrossRef] [PubMed]

41. Helander, A.; Miller, C.L.; Myers, K.S.; Neutra, M.R.; Nibert, M.L. Protective immunoglobulin A and G antibodies bind to overlapping intersubunit epitopes in the head domain of type 1 reovirus adhesin sigma1. *J. Virol.* **2004**, *78*, 10695–10705. [CrossRef] [PubMed]

42. Dietrich, M.; Ogden, K.; Katen, S.; Reiss, K.; Sutherland, D.; Carnhan, R.; Goff, M.; Cooper, T.; Dermody, T.; Stehle, T. Structural insights into Reovirus σ1 interactions with two neutralizing antibodies. *J. Virol.* **2017**, *91*, e01621–16. [CrossRef] [PubMed]

43. Kyula, J.N.; Roulstone, V.; Karapanagiotou, E.M.; Melcher, A.A.; Harrington, K.J. Oncolytic reovirus type 3 (Dearing) as a novel therapy in head and neck cancer. *Expert Opin. Biol. Ther.* **2012**, *12*, 1669–1678. [CrossRef] [PubMed]

44. Heo, J.; Breitbach, C.J.; Moon, A.; Kim, C.W.; Patt, R.; Kim, M.K.; Lee, Y.K.; Oh, S.Y.; Woo, H.Y.; Parato, K.; et al. Sequential therapy with JX-594, a targeted oncolytic poxvirus, followed by sorafenib in hepatocellular carcinoma: preclinical and clinical demonstration of combination efficacy. *Mol. Ther.* **2011**, *19*, 1170–1179. [CrossRef] [PubMed]

45. Le Boeuf, F.; Diallo, J.S.; McCart, J.A.; Thorne, S.; Falls, T.; Stanford, M.; Kanji, F.; Auer, R.; Brown, C.W.; Lichty, B.D.; et al. Synergistic interaction between oncolytic viruses augments tumor killing. *Mol. Ther.* **2010**, *18*, 888–895. [CrossRef] [PubMed]

46. Samson, A.; Bentham, M.; Scott, K.; Nuovo, G.; Bloy, A.; Appleton, E.; Adair, R.; Dave, R.; Peckham-Cooper, A.; Toogood, G.; et al. Oncolytic reovirus as a combined antiviral and anti-tumour agent for the treatment of liver cancer. *Gut* **2016**, *15*, 2016–312009. [CrossRef] [PubMed]

47. Taylor, D.R.; Shi, S.T.; Romano, P.R.; Barber, G.N.; Lai, M.M. Inhibition of the interferon-inducible protein kinase PKR by HCV E2 protein. *Science* **1999**, *285*, 107–110. [CrossRef] [PubMed]

48. Dabo, S.; Meurs, E.F. dsRNA-dependent protein kinase PKR and its role in stress, signaling and HCV infection. *Viruses* **2012**, *4*, 2598–2635. [CrossRef] [PubMed]

49. Abushahba, W.; Balan, M.; Castaneda, I.; Yuan, Y.; Reuhl, K.; Raveche, E.; de la Torre, A.; Lasfar, A.; Kotenko, S. V Antitumor activity of type I and type III interferons in BNL hepatoma model. *Cancer Immunol. Immunother.* **2010**, *59*, 1059–1071. [CrossRef] [PubMed]

50. Lichty, B.D.; Breitbach, C.J.; Stojdl, D.F.; Bell, J.C. Going viral with cancer immunotherapy. *Nat. Rev. Cancer* **2014**, *14*, 559–567. [CrossRef] [PubMed]

6

Nasal Infection of Enterovirus D68 Leading to Lower Respiratory Tract Pathogenesis in Ferrets (*Mustela putorius furo*)

Hui-Wen Zheng, Ming Sun, Lei Guo, Jing-Jing Wang, Jie Song, Jia-Qi Li, Hong-Zhe Li, Ruo-Tong Ning, Ze-Ning Yang, Hai-Tao Fan, Zhan-Long He * and Long-Ding Liu *

Institute of Medical Biology, Chinese Academy of Medical Sciences & Peking Union Medical College, Kunming 650118, China; zhenghuiwen12@126.com (H.-W.Z.); sunming@imbcams.com.cn (M.S.); gl2011@imbcams.com.cn (L.G.); wangjingjing@imbcams.com.cn (J.-J.W.); songjie@imbcams.com.cn (J.S.); lijiaqipumc@yahoo.com (J.-Q.L.); lhz@imbcams.com.cn (H.-Z.L.); ruotongning@126.com(R.-T.N.); zenningyang@163.com (Z.-N.Y.); a779091958@163.com (H.-T.F.)

* Correspondence: hzl@imbcams.com.cn (Z.-L.H.); longdingl@gmail.com (L.-D.L.)

Academic Editor: Andrew Mehle

Abstract: Data from EV-D68-infected patients demonstrate that pathological changes in the lower respiratory tract are principally characterized by severe respiratory illness in children and acute flaccid myelitis. However, lack of a suitable animal model for EV-D68 infection has limited the study on the pathogenesis of this critical pathogen, and the development of a vaccine. Ferrets have been widely used to evaluate respiratory virus infections. In the current study, we used EV-D68-infected ferrets as a potential animal to identify impersonal indices, involving clinical features and histopathological changes in the upper and lower respiratory tract (URT and LRT). The research results demonstrate that the EV-D68 virus leads to minimal clinical symptoms in ferrets. According to the viral load detection in the feces, nasal, and respiratory tracts, the infection and shedding of EV-D68 in the ferret model was confirmed, and these results were supported by the EV-D68 VP1 immunofluorescence confocal imaging with α2,6-linked sialic acid (SA) in lung tissues. Furthermore, we detected the inflammatory cytokine/chemokine expression level, which implied high expression levels of interleukin (IL)-1a, IL-8, IL-5, IL-12, IL-13, and IL-17a in the lungs. These data indicate that systemic observation of responses following infection with EV-D68 in ferrets could be used as a model for EV-D68 infection and pathogenesis.

Keywords: enterovirus D68; animal models; ferret; lower respiratory tract pathogenesis

1. Introduction

Recently, there have been many epidemiological studies reporting on the continuous circulation and infection of enterovirus D68 (EV-D68) [1,2]. This virus is associated with severe diseases, including acute respiratory distress syndrome (ARDS) and central nervous system (CNS) clinical signs [3–5], and identifying the virulence factors contributing to these manifestations has been the focus of many recent studies [6,7]. EV-D68 was first identified in California in 1962. However, EV-D remains a poorly characterized species of the family *Picornaviridae* [8]. Notably, the mechanisms leading to increased pathogenesis of EV-D68, particularly the tropism infection in the human upper and lower respiratory tract, remain to be discovered. Clinical data has shown that pathology in the patients' lower respiratory tract (LRT) are mainly featured by serious respiratory diseases in children and acute flaccid myelitis [9,10]. Some studies have shown that the most important pathway of EV-D68 infection is the respiratory tract by binding sialic acid on the membrane [11,12], but the shedding of the virus from the upper respiratory tracts and the early immune response of patient cytokine secretions and excretions have not been well documented. In translational medicine research, the cotton rat models for studying the infection and transfer of EV-D68 can be helpful for characterizing the behavior of the virus as well as physiological responses to it [13]. However, the use of cotton rat-adapted EV-D68 strains for direct intranasal or intraperitoneal inoculation cannot mimic the natural route of infection in humans.

It is well documented that the sialic acid on the surface of respiratory tract can mediate influenza viral receptor binding protein attachment, which is believed to be an important determinant in tissue tropism of this virus [14].For instance, hemagglutinin (HA) of human influenza viruses have a binding preference for α2,6-linked sialic acids (SAs) dominated in upper respiratory tract (URT), supposedly indicating the pathology characteristics in the upper respiratory tract [15]. As an animal model, the domestic ferret (*Mustela putorius furo*) is a conventional model for studying the pathogenesis of viral respiratory infection, including influenza and Severe Acute Respiratory Syndrome (SARS) coronavirus, because there is no need for virus adaptation [16]. Advantageously, ferrets exhibit a similar distribution of sialic acid in respiratory tract as in humans [8,14,17,18]. Due to the respiratory-infecting characteristics of EV-D68 in humans, the infection efficiency of the virus in ferrets could be a key animal model for conducting studies of the EV-D68 infection mechanisms and immune responses via the respiratory tract. In this study, we focused on determining whether or not ferrets have the potential to serve as a small animal model for EV-D68. A nose aerosol spray was used to induce EV-D68 infection in a ferret model, and serial nasal, throat, and feces samples were collected from the ferrets during and after EV-D68 infection. The pathological changes and expressive levels of inflammatory cytokines/chemokines in URT and LRT of ferrets correlated with viral shedding. These results could be useful supporting data for the evaluation of EV-D68-induced manifestations. Collectively, these data provide the rationale to assess the utility of the ferret as a model for EV-D68 infection.

2. Materials and Methods

2.1. Ethics Statement and Animal Experiments

The animal experiments in this study comply with the Replacement, Refinement, & Reduction (3R) principles. The study was also in accordance with the Animal Research: Reporting of In Vivo Experiments (ARRIVE) guidelines. The animal procedures were approved by the Institutional Animal Care and Use Committee (IACUC) of the Institute of Medical Biology, Chinese Academy of Medical Sciences. Ferret experiments were conducted, and a total of 18 male ferrets (weight: 800–1000 g) were used in this study. Serum samples from the ferrets were tested by neutralization assay to ensure seronegativity for EV-D68 before the experiments. Fifteen ferrets were infected with $10^{4.5}$ 50% cell culture infectious doses ($CCID_{50}$) of EV-D68 via the nostrils dropwise, three ferrets of which were included for immune evaluation. Three ferrets used as mock controls were inoculated with 0.5 mL phosphate-buffered saline (PBS). All ferrets were housed in a high-efficiency particulate air-filtered individual isolation unit in an Animal Biosafety Level 2-enhanced (ABSL-2+) facility, which complied with the requirements for ferret housing, environment, and comfort as described in the Guide for Laboratory Animals Care issued by the Institute of Medical Biology.

The animals were visually inspected daily. Body weight and temperature were measured daily for two weeks post-infection. The onset and duration of all visible changes, such as abnormal respiration and excretions, were recorded, as well as any observed sneezing and nasal discharge. Blood samples were collected before and after viral infection. Animal feces, nasal washes, and throat swabs were collected on days 1, 3, 5, 7, 9, 11, 13, and 15 after the challenge and frozen (−80 °C) until analysis. After infection, on days 3, 5, 7, 9, the ferrets were euthanized with an intramuscular injection of a ketamine (20 mg/kg), then the organs or tissues were harvested for histopathology and viral distribution analysis. After the last sacrifice, three ferrets were followed for 28 days to determine the neutralization antibody titer.

2.2. Cells and Virus

The EV-D68 Fermon strain used in this study was preserved in the Institute of Medical Biology. EV-D68 seed stocks with a titer of $10^{6.5}CCID_{50}$ units/mL were propagated in Vero cells (ATCC). The seed stocks were diluted to the appointed titer and used for EV-D68 $CCID_{50}$ as well as neutralization antibody assays.

2.3. Real-Time PCR Test for Viral Load Quantity

Total RNA was extracted from fresh tissue, feces, bronchoalveolar lavage fluid (BALF), and blood from the experimental animals using the TRNzol-A+ Reagent mini kit (TianGen Biotech, Co., Ltd., Beijing, China) according to the manufacturer's instructions. The total RNA was eluted in a final volume of 20 μL. For quantification, a single-tube, real-time Taqman RT-PCR assay was performed using the Taqman one-step RT-PCR Master Mix in the CFX96 Touch™ Real-Time PCR Detection system (Bio-Rad, Laboratories, Hercules, CA, USA). The experiments were carried out by adding the primer (200 nm), FAM/TAMRA probe (100 nm) (TAKARA Biotechnology Co., Ltd., Dalian, China), and 2 μL of RNA into the Taqman PCR mater mix, of which total reaction volume is 20μL. The following sequences including EV-D68-specific primer and probe: forward primer, 5′-CACCATACTCACAACTGTGGCAG-3′; reverse primer 5′-CTAGCATTACTGCCTGATTGCCAATG-3′ and the probe 5′-TGACTTGACACTCCAAGCAATG TTTG-3′; The following reaction conditions were applied for all PCR experiments: 5 min at 42 °C and 10 s at 95 °C, followed by 40 cycles at 95 °C for 5 s, and 60 °C for 30 s. A standard reference curve was established by measuring the serially-diluted concentrations of the EV-D68 RNA standards generated from the in vitro transcription of a DNA gene fragment containing the EV-D68 p1 gene region.

2.4. Histopathological and Immunohistochemical (IHC) Staining

Tissue samples were obtained from the infected ferrets and were fixed with 10% formalin in PBS and embedded in paraffin. Paraffin-embedded sections were stained with hematoxylin and eosin (H&E). For immunohistochemical detection of the VP1 antigen of EV-D68, slides of paraffin-embedded sections were detected by anti-EV-D68 monoclonal antibodies (GeneTex, Inc., Irvine, CA, USA) and horseradish peroxidase (HRP)-conjugated anti-rabbit IgG antibodies (Cell Signaling Techonology, Shanghai, CST-US subsidiary in China). Peroxidase activity was detected with an Enhanced HRP-DAB Chromogenic Substrate Kit (TianGen Biotech, Co., Ltd., Beijing, China) [19,20].

2.5. Laser Confocal Microscopy Analysis of the Infected Ferret Lung Sections

The lung sections were blocked for two hours in 10% normal goat serum to reduce nonspecific antibody binding. The lung was then incubated with anti-EV-D68VP1 antibody 5 μg/mL (GeneTex, Inc., Irvine, CA, USA) and with biotinylated MAA (α2,3-linkage) and SNA (α2,6-linkage) 5 μg/mL (Vector Laboratories, Inc., CA, USA) at 4 °C overnight. The sections were rinsed extensively in Tris-buffered saline. The cell nuclei were strained by DAPI (Beyotime Biotech, Co., Ltd., Shanghai, China). The primary antibody was detected using 5 μg/mL of Fluorescein Avidin DCS (Vector Laboratories, Inc., Burlingame, CA, USA) and 2 μg/mL Texas-Red-conjugated goat anti-rabbit IgG antibodies (Molecular Probes, Carlsbad, CA, USA) [20,21]. The stained slides were analyzed under a Leica TCS SP8Laser Confocal microscope (Leica Microsystems, Wetzlar, Germany).

2.6. Cytokine Quantification by Magnetic Beads-Based Bio-Plex Assay

In this study, the Non-Human Primate Cytokine Magnetic Bead Panel Kit (Millipore Corporation, Billerica, MA, USA) is attempted for simultaneous quantification of the infected ferret serum of the following 23 kinds of cytokines: G-CSF, GM-CSF, IFN-γ, IL-1B, IL-1a, IL-2, IL-4, IL-5, IL-6, IL-8, IL-10, IL-12/23, IL-13, IL-15, IL-17A, IL-18, MCP-1, MIP-1B, MIP-1a, sCD40L, TGF-a, TNF-a, and VEGF. In brief, 25μL aliquots of standard, control and sample were diluted 1:4 with diluent, following incubation with antibody-coupled beads, and thoroughly washed, the detection antibodies were added for co-incubation and testing. Finally, the plate was run on a Bio-PLEX (Bio-Rad, Laboratories, Inc.). A five-parameter logistic method was used to calculate analytic concentrations in samples.

2.7. Quantification of Cytokine mRNA

RNA isolations were performed on ferret lung tissue samples with the TRNzol-A+ Reagent kit (TianGen Biotech, Co., Ltd., Beijing, China) according to the manufacturer's protocols. Then, cytokine expression levels were normalized to Beta-actin (β-actin) and are reported as the fold change compared with mock-infected animals. Primer sequences for IL-1a, IL-5, IL-8 [22], IL-12 [23], IL-13, IL-17A, and β-actin [23] were published elsewhere. Primers sequences for the remaining genes are as follows: IL-1a: forward, 5′-GAGATGCCTGAGACACCCAAA-3′; reverse, 5′-TGTGCACCAGTTTTCGTTCC-3′; IL-5: forward, 5′-GGAGGCTGTGGATAAACTATTCC-3′; reverse, 5′-CCGGTGTCCACTCAGTGTTTAT-3′; IL-13: forward, 5′-AGAATCAGGCATCCCTCTGC-3′; reverse, 5′-CTTACTGGAGATCCCTGCCG-3′; IL-17A: forward, 5′-GTGCTGACGGGACGGTAAA-3′; reverse, 5′-ACCAGCATCTTTTCCAACCG-3′; Quantitative real-time PCR (qRT-PCR) was performed by using aCFX96 Touch^{TM} Real-Time PCR Detection system (Bio-Rad, Laboratories), and a One Step SYBR PrimeScript^{TM} RT-PCR Kit (TAKARA Biotechnology Co., Ltd.).Each reaction consisted of 1 cycle of 42 °C for 5 min, 95 °C for 10 s, followed by 40 cycles of 95 °C for 5 s and 60 °C for 30 s.

2.8. Neutralization Antibody Titer Test

In brief, ferret serum was heat-inactivated for 30 min at 56 °C, then diluted 1:2 in minimum essential medium containing 2% fetal bovine serum (FBS)(Gibco, LifeTechbologies, Shanghai, China). After cooling, the medium to room temperature, diluted serum was transferred in triplicate to the first row of one 96-well plate and then diluted two-fold from 1:4 to 1:512. One hundred $CCID_{50}$ were combined with the diluted sera in a 96-well, white, opaque-bottom plate and incubated at 35 °C for 3 h before adding 10,000 Vero cells/well. After incubating the samples for six days at 35 °C and 5% CO_2, the reciprocal measurement of the highest serum dilution that inhibited 50% of the viral cytopathic effect was defined as the neutralization antibody (NA) titer against the relative EV-D68. Neutralization titers were estimated with the Spearman–Karber method and expressed in log2 form (e.g., 4 is a titer of 1:16) [24,25].

2.9. Statistics

GraphPad Prism 5 (Version 5.0, La Jolla, CA, USA) was used to graph data and perform statistical analyses. To compare blood cell count and cytokine expressed level between groups, the Mann–Whitney U test was used. Mean ± SEM (standard error of the mean) were graphed and * ($p < 0.05$) was considered to be statistically significant.

3. Results

3.1. EV-D68 Infection in Ferrets Caused Normal, Cold-Like Clinical Signs

Generally, patients infected with this virus can appear to have various disease severities ranging from mild respiratory illnesses, such as cold-like clinical signs, to severe lower respiratory tract infections (LRTI), including pneumonia, wheezing, and bronchiolitis [26]. In our study, the clinical signs of respiratory illness, including cough, nasal discharge (Figure S1A) and dry nose (Figure S1B) were present in 4 of 15 infected animals. No significant increase in body temperature (Figure 1A) was observed in the ferrets with virus infection, while this phenomenon is in accordance with the clinical report that some patients with EV-D68 infection were characterized by low-grade or absent fever [27,28]. Although, there is an increase in all ferret body weight (Figure 1B) during the period of experimental observation, the uninfected ferrets gained more weight than the infected ferrets (a mean of 13.7% vs. 2.7%) at 14 days post-infection, indicating that the overall health of the mock ferrets was better. In addition, during early infection (5–7 days post-infection), some ferrets with EV-D68 infection had a slightly increase of neutrophils (from mean 42.3% to 44.6%) and monocytes (from mean 4.6% to 6.5%) when compared with the three days post-infection ferrets, but there is no change in the number of the lymphocytes and eosinophils (Figure 1C).

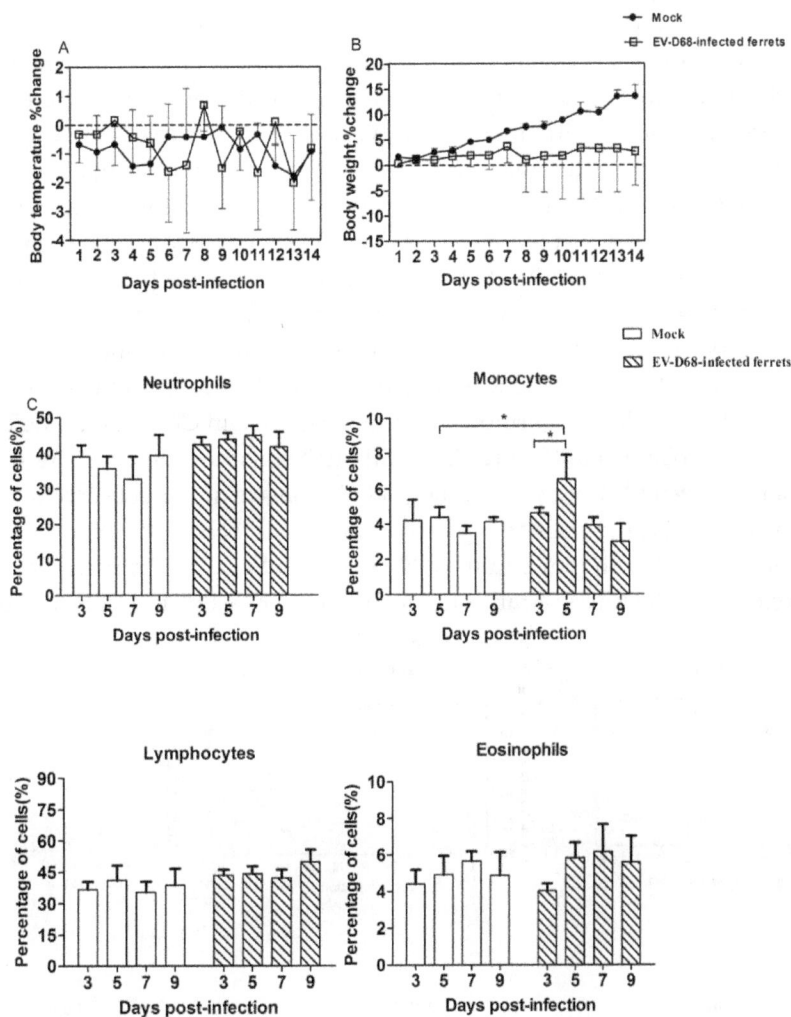

Figure 1. Clinical features of ferrets infected with EV-D68 virus. 12 EV-D68-infected ferrets ($10^{4.5}$CCID$_{50}$ per animal) and 3 mock-infected ferrets (equivalent volumes of virus-free DMEM) were monitored daily for clinical features. The data were recorded as percentage changes compared with value of zero days post-infection. (**A**) The body temperature changes of infected and uninfected animals. Data is mean of rectal temperatures of each ferret groups of 3 to 12 ferrets per time point. The error bars show the SEM of temperature changes at different time; (**B**) The body weight changes of infected and uninfected animals. Data is mean of each groups of 3 to 12 ferrets per time point. The error bars show the SEM of temperature changes at different time; (**C**) The complete blood count (CBC) analysis of mock and EV-D68-infected ferrets. The percentages of lymphocytes, monocytes, neutrophils, and eosinophils were plotted respectively. Blood samples were collected at three, five, seven, and nine days post-infection. All samples were run in triplicate, with mean value and SEM. Monocytes: *, $p < 0.05$, EV-D68-infectedferrets compared with mock on five days post-infection; five days post-infected ferrets compared with three days post-infected ferrets.

3.2. Virus Shedding Potential and Distribution in Different Tissues of EV-D68-Infected Ferrets

Similar to other enteroviruses, EV-D68 has the ability to infect lymphocytes [29]. In addition, several epidemiological studies have demonstrated that EV-D68 is associated with severe lower respiratory tract infection and central nervous system (CNS) pathogenicity, including acute focal limb weakness, paralysis, and acute cranial nerve dysfunction [4,27,30]. In our research, the partial VP1 gene (120 nt) was detected using RT-PCR analysis with sequence-specific primers and probes in the feces, nasal washes, throat swabs, blood, lung, BALF, trachea, and CNS samples that were collected at different times post-infection from the ferrets. The virus was detected in the feces and nasal washes from the third day post-infection, and a peak level of over 70,000 copies per 100 mg sample (Figure 2A), was reached on the fifth day, and 90,000 copies per 1 mL of nasal wash (Figure 2B) was detected on the ninth day, but no virus was detected in the throat swabs (Figure 2C). For viremia of EV-D68 infection, the viral load in blood samples is less than 50 copies/mL (Figure 2D). Several studies have indicated that the enterovirus invasion lead to a transient minor viremia that delivers the virus to lymphoid tissues [31]. In this study, the virus replication in the axillary lymph nodes was detected at five and seven days post-infection. With levels of 18,000–50,000 copies per 100 mg of tissue (Figure 2E). To measure the profile of virus replication in lower respiratory systems, the EV-D68 virus was detected in lung, trachea, and BALF. Infectious viruses were detectable from three to seven days post-infection. With a peak level of 89,000 copies per 100 mg of lung tissue on the fifth day post-infection (Figure 2F). However, viral load in BALF and in the trachea was less than 40 copies/mL (Figure S2A,B). Detection of EV-D68 Fermon strain was negative in CNS (including brain, midbrain, cerebellum, and medulla oblongata) (Figure S2C). Although some clinical results report that there is a possible association between EV-D68 and neurological disease, we did not observe the central nervous system symptoms in ferrets [4], which was in accordance with Schieble's experimental result from sucking mice by inoculation with four strains of EV-D68, in which CNS impairments were observed only in mice infected with the Rhyne strain—not in Fermon, Franklin, and Robison strains [32].

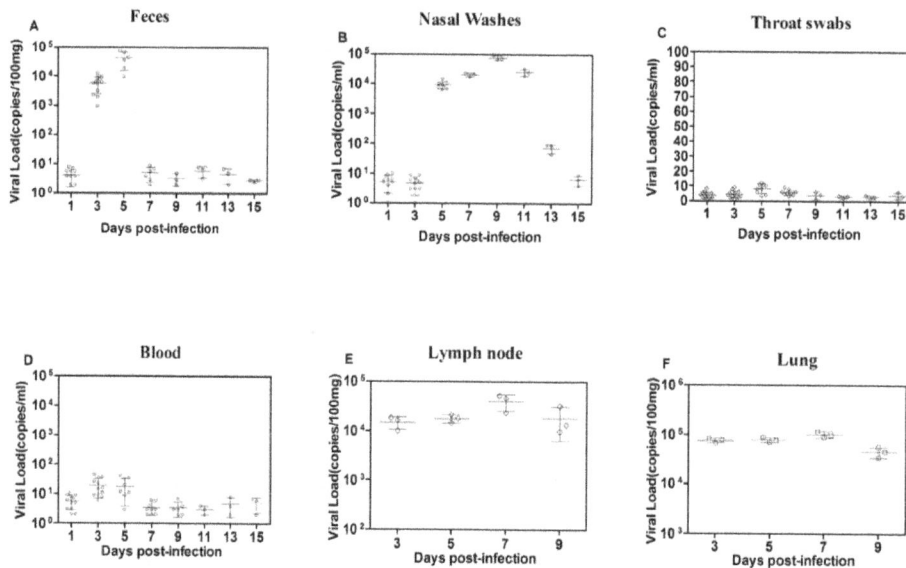

Figure 2. Dynamic distribution of EV-D68 virus in infected ferrets through respiratory route. Twelve ferrets were infected with EV-D68 ($10^{4.5}$CCID$_{50}$/ferret) via nostrils dropwise and three ferrets set as mock control. Viral load in feces (**A**) and nasal washes (**B**) and throat swabs (**C**) of infected ferrets were detected in the whole course of EV-D68 infection (1, 3, 5, 7, 9, 11, 13, 15 days post-infection); (**D**) Viral RNA was picked up from blood samples and analyzed by Taqman real-time quantitative PCR assay, according to the protocols provided by the qPCR kit. Each blood samples were gathered during the whole EV-D68 infection course (1, 3, 5, 7, 9, 11, 13, 15 days post-infection); (**E**)Viral load in lung of ferret at different day post-infection (3, 5, 7, 9 days post-infection); (**F**) Viral load in lymph nodes at different day post-infection(3, 5, 7, 9 days post-infection). The vitro-synthesized RNA was applied to quantify the viral copies of each RNA sample, and viral RNA's relative copies for each sample was calculated by the mathematical formula as follow: [(µg of RNA/µL)/(molecular weight)] × Avogadro's number = viral copy number/µL. A viral load which is less than 10 copies is regarded as negative.

3.3. Histopathological Examinationand Immunohistochemical Analysis

Two clinical cases have reported that diffusion and patchy alveolar infiltration were present in lung tissues of the EV-D68-infected patients [10,33] according to computed tomography angiograms of the chest. In this study, the tissue samples from the trachea and lungs were histopathologically examined in order to observe the pathological change of the respiratory apparatus from the ferret infected with the EV-D68. Inflammation and diffuse alveolar hemorrhage were found in lungs where the viral load level was very high from EV-D68-inoculated ferrets on three and seven days post-infection (Figure 3A), but there were no pathological changes in the trachea (Figure 3B). These findings suggest that ferret infection might have a remarkable pathogenesis in the lower respiratory tract at three to seven days post-infection (Figure 3A). At the same time, immunolabeling of the VP1 antigen was observed in lung cells around the pulmonary alveolus cells (Figure3C), while VP1 antigen is undetectable in the trachea (Figure 3D). Together with the virus detection and histopathological analyses of the lung tissues, all EV-D68-infectedanimals likely demonstrated pulmonary manifestations at three to seven days post-infection.

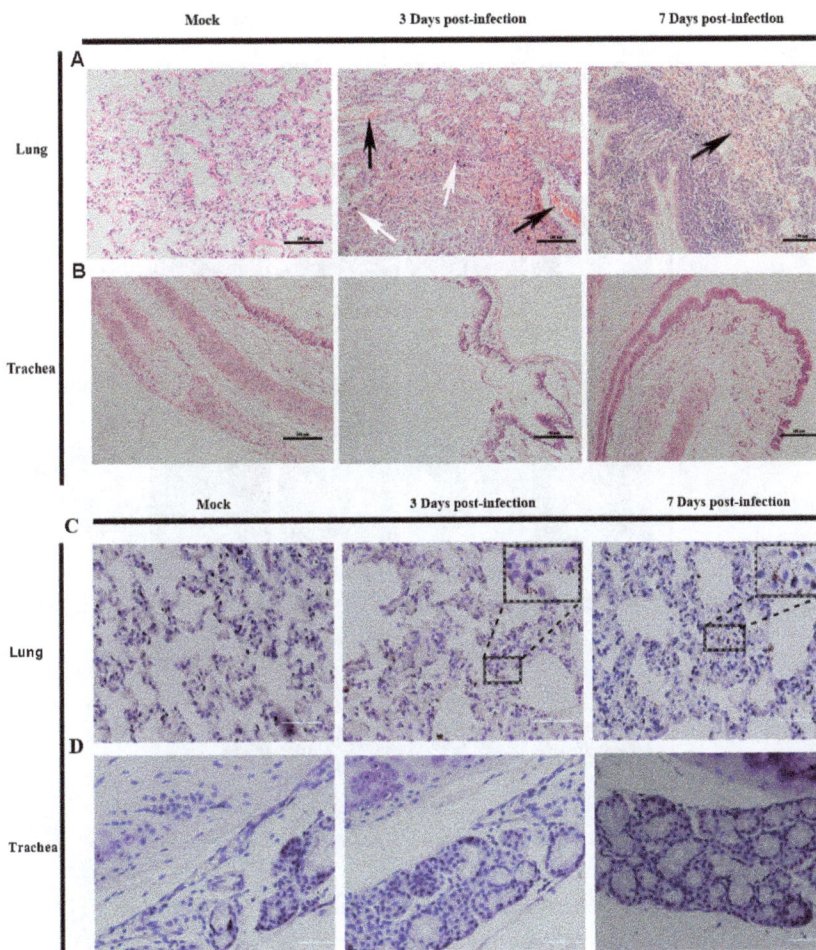

Figure 3. Pathological manifestations in respiratory apparatus of EV-D68-infected ferrets. (**A**) The clinical pathological features of EV-D68 infection in ferret lung tissues; (**B**) There is no pathological change in trachea. In these images, the white arrow represents the inflammatory cell infiltration, the black arrow points the diffuse alveolar hemorrhage. The microscope magnified the images 200 times, Bar, 100 μm; (**C**) The viral antigen expression in lung tissue from ferrets infected by $10^{4.5}$CCID$_{50}$ EV-D68. The black dashed boxes indicate the EV-D68 VP1 antigen expression and the insert picture enlarge the antigen (on the top right corner); (**D**) There is no VP1 antigen detectable in trachea. Images are magnified by 400 times; Bar, 50 μm.

3.4. Confocal Imaging of EV-D68 VP1 with the α2,6-Linked SAs in the Lung Tissue of EV-D68-Infected Ferrets

Sialic acid (SA) has been reported as the receptor for EV-D68, and it has been shown that EV-D68 has a stronger affinity for α2,6-linked SAs than for α2,3-linked SAs [12,34]. To compare the distribution of the α2,3-linked SAs and α2,6-linked SAs in the cells of lungs of viral-infected animals, the lung samples were stained by anti-EV-D68 virus VP1 protein monoclonal antibodies, then were detected by a Texas-Red-conjugated secondary antibody (Figure 4). The biotinylated lectins SNA and MAA were used to detect the α2,6-linked SAs and α2,3-linked SAs, respectively. The bound lectins were distinguished by avidin DCS labeled by FITC (Figure 4 and Figure S3). In the ferrets, we found that α2,3-linked SAs are more dominantly expressed in the lung tissue (Figure S3) than α2,6-linked SAs (Figure 4), However, both three days post-infection and seven days post-infection, the overlaid image of the two colors (green and red) exhibit intense co-location of the α2,6-linked SAs on EV-D68-positive lung cells (Figure 4). Remarkably, α2,3-linked SAs were not observed on the EV-D68 positive lung cells according to the co-localized image (Figure S3).These results indicated EV-D68 infection may prefer α2,6-linked SAs over α2,3-linked SAs in the ferrets' lower respiratory tracts (Figure 4).

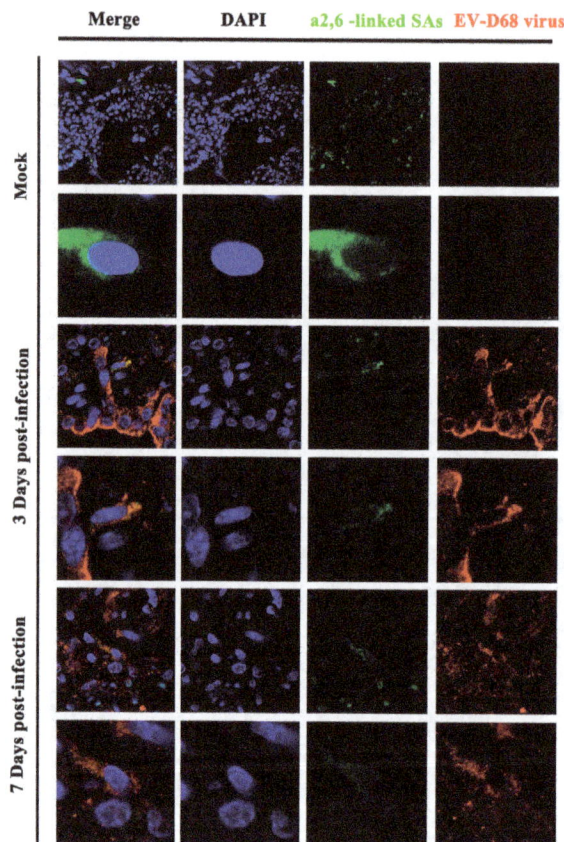

Figure 4. Confocal imaging of α2,6-linked SAs and viral antigen in the lung tissue of EV-D68-infected ferrets. The samples were obtained at three and seven days post-infection. EV-D68 VP1 antigen was labeled with Texas-Red-conjugated anti-IgG antibody (red), while α2,6-linked SAs were labeled with FITC (green). Images are shown at 630× magnification.

3.5. Inflammatory Cytokines Increased in Pulmonary Pathogenesis

To determine the possibility of the pathological progress in the LRT of EV-D68-infected ferrets due to acute lung injury of inflammatory immune responses, we intended to evaluate the inflammatory cytokine levels in the lung during the early and late stages of infection. Despite the prominent use of ferrets in medical research, the immune system of ferrets remains poorly characterized. There are insufficient ELISA reagents for the detection of cytokines in ferrets, so we attempted to use the Bio-Plex Suspension Array System to evaluate serum cytokine levels. Numerous studies have reported that many animal species' antibodies specific for cytokines can be used to detect ferret cytokines [35,36]. Therefore, in this study, the non-human primate cytokine Th1/Th2 assay was used to cross-test 23 cytokines and chemokines in ferrets following the protocol in the user manual (Figure S4). For verification of this test, the levels of gene expression were also measured by qRT-PCR. The results indicated differential expression of the inflammatory cytokines/chemokines in the LRT, showing high expression levels of interleukin-1α (IL-1α), IL-5, IL-8 (Figure 5A,B), IL-12, IL-13, and IL-17a in serum (Figure 5C,D), which correlated with the peak levels of viral shedding and pathological changes in the lungs. These data suggest that infection increased the release of inflammatory cytokines in the infected ferret lungs during early infection and lung pathology was continuously exacerbated during the progression of infection and caused lung edema as well as lung injury at the middle and late stages of infection.

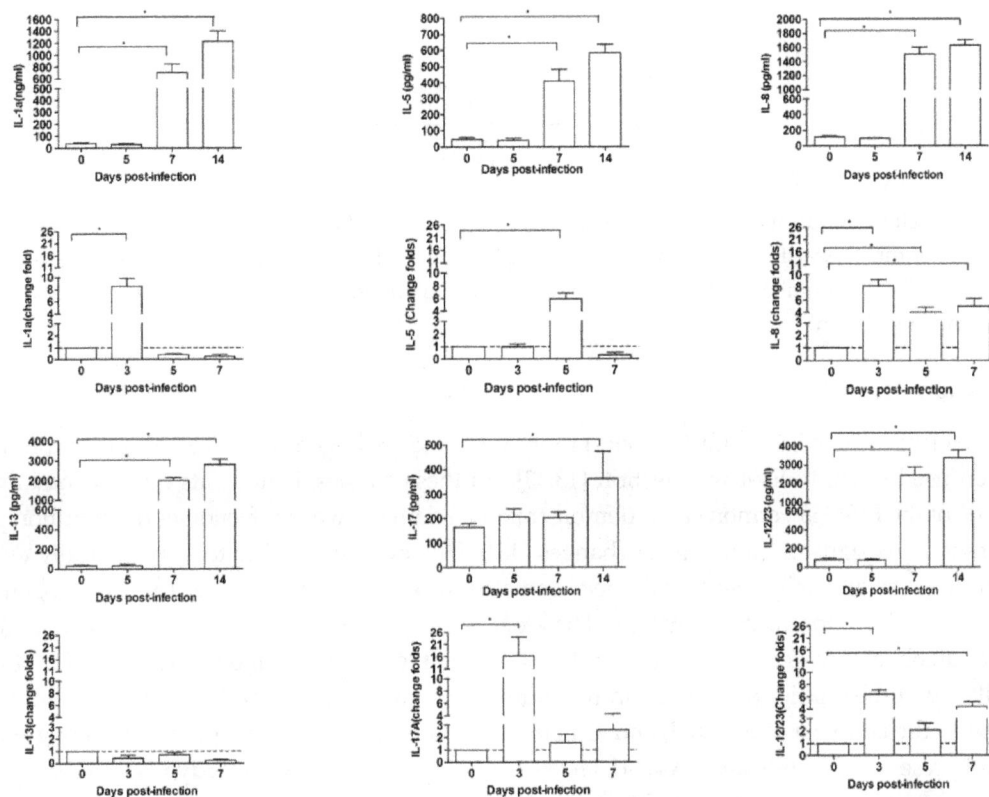

Figure 5. Inflammatory cytokines detection of the ferrets after EV-D68 infection. (**A**) Cytokines (IL-1a, IL-5, IL-8) of serum were determined using multiplex bead-based Bio-Plex assay and are detected at different day (0, 5, 7, 14) post-infection; (**B**) The average transcription levels of interleukin-1α (IL-1α), IL-5, IL-8 of lungs were determined by qPCR and are plotted graphically for various time points (0, 5, 7, 9) following infection. Increases of mRNA levels were relative to β-actin and then normalized to the PBS control groups; (**C**) Cytokines (IL-12, IL-13, IL-17a) of serum were determined using multiplex bead-based Bio-Plex assay and are detected at different days (0, 5, 7, 14) post-infection; (**D**) The average transcription levels of IL-12, IL-13, IL-17a of lungs were determined by qPCR and are plotted graphically for various time points (0, 5, 7, 9) following infection. Increases of mRNA levels were relative to β-actin and then normalized to the PBS control groups. Average data were obtained from three independent experiments, and the error bars indicate SEM. The dot line represents the mRNA base level. Horizontal bars show the statistical analysis performed between the selected two groups. *, $p < 0.05$. The expressed level of different cytokine on 7 & 14 days post-infection compared with 0 days post-infection.

3.6. Immunological Response of Ferrets after EV-D68 Infection

To detect the ability of viral infection to elicit an antibody response, three ferrets were followed for up to 28 days after inoculation. The immunological analysis of the EV-D68-infected animals showed a typical antibody response of viral-induced characteristics. The neutralization antibody GMT exhibited a slight and slow increase of 1:16 levels at week 4 after inoculation (Figure 6). With respect to the report by Patel et al., in the EV-D68 Cotton rat model, Fermon infection resulted in no detectable neutralization antibody (NA) response [13]. Findings in this study demonstrate that the EV-D68 Fermon strain infection may induce faint immunological response in the form of the neutralization antibody titer in ferrets.

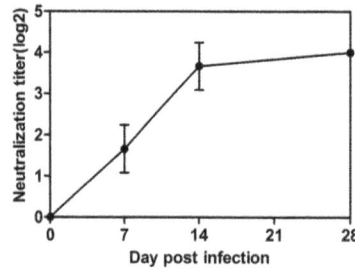

Figure 6. The neutralization antibody titer of ferrets against the EV-D68 virus. Each plot represents the average of three logarithmically transformed measurements (log2). Vertical bars represent the mean value ± SEM. The number under each figure represents the timing of blood collection at different days after the inoculation.

4. Discussion

To determine the in vivo effects of viral replication and pathogenesis, mice and rat cotton models have been used to study EV-D68 infection [13,32]. In these studies, cotton rats infected nasally with $10^6 CCID_{50}$ of the EV-D68 Fermon strain demonstrated a relatively weak infection and replication profile with non-obvious signs of pathological changes [13]. However, the ability to systematically evaluate infection progress in small rodents is limited, and such models do not yield sufficient pathogenic and pathological evidence for infection analysis. Therefore, the implementation of effective ferret models now represents an obligatory step for the preclinical evaluation of respiratory pathogen infections [37,38]. With the similarity of the distributing profile in the human respiratory tract α2,6-linked SAs of ferrets were dominant in the epithelia of the URT, and a relative increase of α2,3-linked SAs in LRT, along with the continued presence of α2,6-linked SAs [14], ferrets have largely been used to study respiratory infections, such as influenza, respiratory syncytial virus, and SARS virus. The infection pathways of EV-D68 were reported to be in the respiratory tract [39,40], indicating that it would be easier to induce respiratory infections in ferret models, which would display symptoms similar to those in humans.

In this study, we used a ferret model to assess clinical symptoms, viral replication, and shedding, pathogenicity, and expression of inflammatory cytokines in the respiratory tracts of ferrets infected with the EV-D68 virus. This model utilized a commonly-employed intranasal challenge route for ferrets to assess the natural mucosal routes of infection. In the results, we found that there was a clear profile in viral replication rates in ferrets compared to results with cotton rats described by Patel [13]. Notably, despite having direct access to the nose upon infection, we observed less viral shedding from nasal discharges compared to fecal matter. Instead, the most impressive viral burdens were noted in the lung, which were similar to the experimental results for cotton rats [13]. While the EV-D68 infection was not measured directly by viral titration in our study, the results of qRT-PCR for viral genome and immunohistochemical analysis for viral antigens can support the EV-D68 virus replication in infected ferrets. Moreover, since EV-D68 usually targets the respiratory tract for infection and induces severe pneumonia among children who have asthma or a history of wheezing [41,42], we also focused on analyzing the pathogenesis in the URT and LRT tissues. Our results clearly showed inflammation and diffuse alveolar hemorrhage occurred in the lungs of EV-D68-inoculated ferrets at three and seven days post-infection, suggesting that the ferret infections might have remarkable pathogenesis in the lower respiratory tract.

Despite of initializing the common cold-like symptoms (e.g., a runny nose, sore throat, and cough) in the upper respiratory tract, more serious pneumonia-like symptoms related with the lower respiratory tract may occur in some cases [26,43]. Thus, EV-D68 attaching and invading in the LRT could contribute to above outcomes. Furthermore, we investigated the relationship between lectin receptors on the surface of lower respiratory tract cells and EV-D68 invasion. In our findings, α2,6-linked SAs staining was more apparent with viral infection, but less of a signal appeared in the α2,3-linked SAs, which indicated that EV-D68 specifically recognized the α2,6-linked SAs of cell tropism in the ferret lower respiratory tract, inducing LRT EV-D68 infection and pathogenic changes. In this case, ferrets are proposed to be a good small-animal model of EV-D68-related lower respiratory manifestations.

Inflammatory immune responses have been reported to be related to acute lung injury after viral infection. We further assessed the inflammatory cytokine and chemokine levels during pathological changes in the LRT of EV-D68-infectedferrets. In the serum and lung during the early stages of infection, there was increased expression of inflammatory cytokines/chemokines, including interleukin-1α (IL-1α), IL-5, IL-8, IL-12, IL-13, and IL-17a, which correlated with the pathological changes in the lungs and may play a key role in causing lung edema as well as lung injury in the middle and late stages of infection.

Taken together, these data imply that the ferret models have the potential to be used for characterizing key events in the pathogenesis of EV-D68, even for the Fermon strain, which has lower antigenicity. This model may be useful for evaluating the newly-isolated EV-D68 strains and potential candidate medical interventions, including vaccines. Further studies are needed to more fully characterize the transmission of infection in ferrets to expand the predictive efficacy of this model for intensive study.

Acknowledgments: This work was supported by CAMS Innovation Fund for Medical Sciences (2016-I2M-1-014), National Natural Sciences Foundations of China (81373142).The funders had no role in the study, design, data collection, and analysis, decision to publish, or preparation of the manuscript.

Author Contributions: L.-D.L. and H.-W.Z. conceived and designed the experiments; H.-W.Z., L.G., and Z.-L.H. performed the experiments; M.S., J.-J.W., and J.S. analyzed the data; J.-Q.L., H.-Z.L., R.-T.N., H.-T.F., and Y.-Z.N. contributed reagents/materials/analysis tools; L.-D.L. and H.-W.Z. wrote the paper.

References

1. Tan, Y.; Hassan, F.; Schuster, J.E.; Simenauer, A.; Selvarangan, R.; Halpin, R.A.; Lin, X.; Fedorova, N.; Stockwell, T.B.; Lam, T.T.; et al. Molecular evolution and intraclade recombination of enterovirus D68 during the 2014 outbreak in the United States. *J. Virol.* **2015**, *90*, 1997–2007. [CrossRef] [PubMed]

2. Bottcher, S.; Prifert, C.; Weissbrich, B.; Adams, O.; Aldabbagh, S.; Eis-Hubinger, A.M.; Diedrich, S. Detection of enterovirus D68 in patients hospitalised in three tertiary university hospitals in Germany, 2013 to 2014. *Eurosurveillance* **2016**, *21*. [CrossRef] [PubMed]

3. Farrell, J.J.; Ikladios, O.; Wylie, K.M.; O'Rourke, L.M.; Lowery, K.S.; Cromwell, J.S.; Wylie, T.N.; Melendez, E.L.; Makhoul, Y.; Sampath, R.; et al. Enterovirus D68-associated acute respiratory distress syndrome in adult, United States, 2014. *Emerg. Infect. Dis.* **2015**, *21*, 914–916. [CrossRef] [PubMed]

4. Messacar, K.; Schreiner, T.L.; Maloney, J.A.; Wallace, A.; Ludke, J.; Oberste, M.S.; Nix, W.A.; Robinson, C.C.; Glodé, M.P.; Abzug, M.J.; et al. A cluster of acute flaccid paralysis and cranial nerve dysfunction temporally associated with an outbreak of enterovirus D68 in children in Colorado, USA. *Lancet* **2015**, *385*, 1662–1671. [CrossRef]

5. Greninger, A.L.; Naccache, S.N.; Messacar, K.; Clayton, A.; Yu, G.; Somasekar, S.; Federman, S.; Stryke, D.; Anderson, C.; Yagi, S.; et al. A novel outbreak enterovirus D68 strain associated with acute flaccid myelitis cases in the USA (2012–2014): A retrospective cohort study. *Lancet Infect. Dis.* **2015**, *15*, 671–682. [CrossRef]

6. Xiang, Z.; Liu, L.; Lei, X.; Zhou, Z.; He, B.; Wang, J. 3C protease of enterovirus D68 inhibits cellular defense mediated by interferon regulatory factor 7. *J. Virol.* **2015**, *90*, 1613–1621. [CrossRef] [PubMed]

7. Xiang, Z.; Li, L.; Lei, X.; Zhou, H.; Zhou, Z.; He, B.; Wang, J. Enterovirus 68 3C protease cleaves TRIF to attenuate antiviral responses mediated by Toll-like receptor 3. *J. Virol.* **2014**, *88*, 6650–6659. [CrossRef] [PubMed]

8. Holm-Hansen, C.C.; Midgley, S.E.; Fischer, T.K. Global emergence of enterovirus D68: A systematic review. *Lancet Infect. Dis.* **2016**, *16*, e64–e75. [CrossRef]

9.   Aliabadi, N.; Messacar, K.; Pastula, D.M.; Robinson, C.C.; Leshem, E.; Sejvar, J.J.; Nix, W.A.; Oberste, M.S.;
      Feikin, D.R.; Dominguez, S.R. Enterovirus D68 infection in children with acute flaccid myelitis, Colorado,
      USA, 2014. *Emerg. Infect. Dis.* **2016**, *22*, 1387–1394. [CrossRef] [PubMed]

10.   Matsumoto, M.; Awano, H.; Ogi, M.; Tomioka, K.; Unzaki, A.; Nishiyama, M.; Toyoshima, D.;
      Taniguchi-Ikeda, M.; Ishida, A.; Nagase, H.; et al. A pediatric patient with interstitial pneumonia due
      to enterovirus D68. *J. Infect. Chemother.* **2016**, *22*, 712–715. [CrossRef] [PubMed]

11.   Liu, Y.; Sheng, J.; Baggen, J.; Meng, G.; Xiao, C.; Thibaut, H.J.; van Kuppeveld, F.J.; Rossmann, M.G. Sialic
      acid-dependent cell entry of human enterovirus D68. *Nat. Commun.* **2015**, *6*, 8865. [CrossRef] [PubMed]

12.   Imamura, T.; Okamoto, M.; Nakakita, S.; Suzuki, A.; Saito, M.; Tamaki, R.; Lupisan, S.; Roy, C.N.;
      Hiramatsu, H.; Sugawara, K.E.; et al. Antigenic and receptor binding properties of enterovirus 68. *J. Virol.*
      2014, *88*, 2374–2384. [CrossRef] [PubMed]

13.   Patel, M.C.; Wang, W.; Pletneva, L.M.; Rajagopala, S.V.; Tan, Y.; Hartert, T.V.; Boukhvalova, M.S.; Vogel, S.N.;
      Das, S.R.; Blanco, J.C. Enterovirus D-68 infection, prophylaxis, and vaccination in a novel permissive animal
      model, the cotton rat (*Sigmodon hispidus*). *PLoS ONE* **2016**, *11*, e0166336. [CrossRef] [PubMed]

14.   Belser, J.A.; Blixt, O.; Chen, L.M.; Pappas, C.; Maines, T.R.; van Hoeven, N.; Donis, R.; Busch, J.; McBride, R.;
      Paulson, J.C.; et al. Contemporary North American influenza H7 viruses possess human receptor specificity:
      Implications for virus transmissibility. *Proc. Natl. Acad. Sci. USA* **2008**, *105*, 7558–7563. [CrossRef] [PubMed]

15.   Matrosovich, M.N.; Gambaryan, A.S.; Teneberg, S.; Piskarev, V.E.; Yamnikova, S.S.; Lvov, D.K.; Robertson, J.S.;
      Karlsson, K.A. Avian influenza A viruses differ from human viruses by recognition of sialyloligosaccharides
      and gangliosides and by a higher coservation of the HA receptor-binding site. *Virology* **1997**, *233*, 224–234.
      [CrossRef] [PubMed]

16.   Sutton, T.C.; Subbarao, K. Development of animal models against emerging coronaviruses: From SARS to
      MERS coronavirus. *Virology* **2015**, *479–480*, 247–258. [CrossRef] [PubMed]

17.   Shinya, K.; Ebina, M.; Yamada, S.; Ono, M.; Kasai, N.; Kawaoka, Y. Avian flu: Influenza virus receptors in the
      human airway. *Nature* **2006**, *440*, 435–436. [CrossRef] [PubMed]

18.   Leigh, M.W.; Connor, R.J.; Kelm, S.; Baum, L.G.; Paulson, J.C. Receptor specificity of Influenza virus
      influences severity of illness in ferrets. *Vaccine* **1995**, *13*, 1468–1473. [CrossRef]

19.   Liu, L.; Zhao, H.; Zhang, Y.; Wang, J.; Che, Y.; Dong, C.; Zhang, X.; Na, R.; Shi, H.; Jiang, L.; et al. Neonatal
      rhesus monkey is a potential animal model for studying pathogenesis of EV71 infection. *Virology* **2011**, *412*,
      91–100. [CrossRef] [PubMed]

20.   Wang, J.; Zhang, Y.; Zhang, X.; Hu, Y.; Dong, C.; Liu, L.; Yang, E.; Che, Y.; Pu, J.; Wang, X.; et al. Pathologic
      and immunologic characteristics of coxsackievirus A16 infection in rhesus macaques. *Virology* **2017**, *500*,
      198–208. [CrossRef] [PubMed]

21.   Eash, S.; Tavares, R.; Stopa, E.G.; Robbins, S.H.; Brossay, L.; Atwood, W.J. Differential distribution of the JC
      virus receptor-type sialic acid in normal human tissues. *Am. J. Pathol.* **2004**, *164*, 419–428. [CrossRef]

22.   Carolan, L.A.; Butler, J.; Rockman, S.; Guarnaccia, T.; Hurt, A.C.; Reading, P.; Kelso, A.; Barr, I.; Laurie, K.L.
      TaqMan real time RT-PCR assays for detecting ferret innate and adaptive immune responses. *J. Virol. Methods*
      2014, *205*, 38–52. [CrossRef] [PubMed]

23.   Fang, Y.; Rowe, T.; Leon, A.J.; Banner, D.; Danesh, A.; Xu, L.; Ran, L.; Bosinger, S.E.; Guan, Y.; Chen, H.; et al.
      Molecular characterization of in vivo adjuvant activity in ferrets vaccinated against influenza virus. *J. Virol.*
      2010, *84*, 8369–8388. [CrossRef] [PubMed]

24.   Sun, M.; Ma, Y.; Xu, Y.; Yang, H.; Shi, L.; Che, Y.; Liao, G.; Jiang, S.; Zhang, S.; Li, Q. Dynamic profiles
      of neutralizing antibody responses elicited in rhesus monkeys immunized with a combined tetravalent
      DTaP-Sabin IPV candidate vaccine. *Vaccine* **2014**, *32*, 1100–1106. [CrossRef] [PubMed]

25.   Zhang, Y.; Moore, D.D.; Nix, W.A.; Oberste, M.S.; Weldon, W.C. Neutralization of Enterovirus D68 isolated
      from the 2014 US outbreak by commercial intravenous immune globulin products. *J. Clin. Virol.* **2015**, *69*,
      172–175. [CrossRef] [PubMed]

26.   Linsuwanon, P.; Puenpa, J.; Suwannakarn, K.; Auksornkitti, V.; Vichiwattana, P.; Korkong, S.;
      Theamboonlers, A.; Poovorawan, Y. Molecular epidemiology and evolution of human enterovirus serotype
      68 in Thailand 2006–2011. *PLoS ONE* **2012**, *7*, e35190. [CrossRef] [PubMed]

27.   Vazquez-Perez, J.A.; Ramirez-Gonzalez, J.E.; Moreno-Valencia, Y.; Hernandez-Hernandez, V.A.;
      Romero-Espinoza, J.A.; Castillejos-Lopez, M.; Hernandez, A.; Perez-Padilla, R.; Oropeza-Lopez, L.E.;

Escobar-Escamilla, N.; et al. EV-D68 infection in children with asthma exacerbation and pneumonia in Mexico City during 2014 autumn. *Influenza Other Respir. Viruses* **2016**, *10*, 154–160. [CrossRef] [PubMed]

28. Esposito, S.; Chidini, G.; Cinnante, C.; Napolitano, L.; Giannini, A.; Terranova, L.; Niesters, H.; Principi, N.; Calderini, E. Acute flaccid myelitis associated with enterovirus-D68 infection in an otherwise healthy child. *Virol. J.* **2017**, *14*, 4. [CrossRef] [PubMed]

29. Smura, T.; Ylipaasto, P.; Klemola, P.; Kaijalainen, S.; Kyllonen, L.; Sordi, V.; Piemonti, L.; Roivainen, M. Cellular tropism of human enterovirus D species serotypes EV-94, EV-70, and EV-68 in vitro: Implications for pathogenesis. *J. Med. Virol.* **2010**, *82*, 1940–1949. [CrossRef] [PubMed]

30. Thongpan, I.; Wanlapakorn, N.; Vongpunsawad, S.; Linsuwanon, P.; Theamboonlers, A.; Payungporn, S.; Poovorawan, Y. Prevalence and phylogenetic characterization of enterovirus D68 in pediatric patients with acute respiratory tract infection in Thailand. *Jpn. J. Infect. Dis.* **2016**, *69*, 426–430. [CrossRef] [PubMed]

31. Enterovirus D68: Background, Pathophysiology, Etiology. Available online: http://emedicine.medscape.com/article/2236902-overview (accessed on 11 January 2016).

32. Schieble, J.H.; Fox, V.L.; Lennette, E.H. A probable new human picornavirus associated with respiratory diseases. *Am. J. Epidemiol.* **1967**, *85*, 297–310. [CrossRef] [PubMed]

33. Ward, N.S.; Hughes, B.L.; Mermel, L.A. Enterovirus D68 infection in an adult. *Am. J. Crit. Care* **2016**, *25*, 178–180. [CrossRef] [PubMed]

34. Uncapher, C.R.; Dewitt, C.M.; Colonno, R.J. The major and minor group receptor families contain all but one human rhinovirus serotype. *Virology* **1991**, *180*, 814–817. [CrossRef]

35. Martel, C.J.; Aasted, B. Characterization of antibodies against ferret immunoglobulins, cytokines and CD markers. *Vet. Immunol. Immunopathol.* **2009**, *132*, 109–115. [CrossRef] [PubMed]

36. Pedersen, L.G.; Castelruiz, Y.; Jacobsen, S.; Aasted, B. Identification of monoclonal antibodies that cross-react with cytokins from different animal species. *Vet. Immunol. Immunopathol.* **2002**, *88*, 111–112. [CrossRef]

37. Camp, J.V.; Bagci, U.; Chu, Y.K.; Squier, B.; Fraig, M.; Uriarte, S.M.; Guo, H.; Mollura, D.J.; Jonsson, C.B. Lower respiratory tract infection of the ferret by 2009 H1N1 pandemic influenza A virus triggers biphasic, systemic, and local recruitment of neutrophils. *J. Virol.* **2015**, *89*, 8733–8748. [CrossRef] [PubMed]

38. Gustin, K.M.; Belser, J.A.; Veguilla, V.; Zeng, H.; Katz, J.M.; Tumpey, T.M.; Maines, T.R. Environmental conditions affect exhalation of H3N2 seasonal and variant influenza viruses and respiratory droplet transmission in ferrets. *PLoS ONE* **2015**, *10*, e0125874. [CrossRef] [PubMed]

39. Huang, S.S.; Banner, D.; Paquette, S.G.; Leon, A.J.; Kelvin, A.A.; Kelvin, D.J. Pathogenic influenza B virus in the ferret model establishes lower respiratory tract infection. *J. Gen. Virol.* **2014**, *95*, 2127–2139. [CrossRef] [PubMed]

40. Imamura, T.; Oshitani, H. Global reemergence of enterovirus D68 as an important pathogen for acute respiratory infections. *Rev. Med. Virol.* **2015**, *25*, 102–114. [CrossRef] [PubMed]

41. Xiao, Q.; Ren, L.; Zheng, S.; Wang, L.; Xie, X.; Deng, Y.; Zhao, Y.; Zhao, X.; Luo, Z.; Fu, Z.; et al. Prevalence and molecular characterizations of enterovirus D68 among children with acute respiratory infection in China between 2012 and 2014. *Sci. Rep.* **2015**, *5*, 16639. [CrossRef] [PubMed]

42. Orvedahl, A.; Padhye, A.; Barton, K.; O'Bryan, K.; Baty, J.; Gruchala, N.; Niesen, A.; Margoni, A.; Srinivasan, M. Clinical characterization of children presenting to the hospital with enterovirus D68 infection during the 2014 outbreak in St. Louis. *Pediatr. Infect. Dis. J.* **2016**, *35*, 481–487. [CrossRef] [PubMed]

43. Imamura, T.; Suzuki, A.; Lupisan, S.; Okamoto, M.; Aniceto, R.; Egos, R.J.; Daya, E.E.; Tamaki, R.; Saito, M.; Fuji, N.; et al. Molecular evolution of enterovirus 68 detected in the Philippines. *PLoS ONE* **2013**, *8*, e74221. [CrossRef] [PubMed]

7

Microbial Natural Product Alternariol 5-*O*-Methyl Ether Inhibits HIV-1 Integration by Blocking Nuclear Import of the Pre-Integration Complex

Jiwei Ding [1,†], Jianyuan Zhao [1,†], Zhijun Yang [2,†], Ling Ma [1], Zeyun Mi [1], Yanbing Wu [1], Jiamei Guo [3], Jinmin Zhou [1], Xiaoyu Li [1], Ying Guo [3], Zonggen Peng [1], Tao Wei [4], Haisheng Yu [5], Liguo Zhang [5], Mei Ge [2] and Shan Cen [1,*]

[1] Institute of Medicinal Biotechnology, Chinese Academy of Medical Sciences and Peking Union Medical School, Beijing 100050, China; jiweiding1223@aliyun.com (J.D.); zjyuan815@163.com (J.Z.); maling26@163.com (L.M.); mizeyun@126.com (Z.M.); wyblily@sina.com (Y.W.); zhou_jim@hotmail.com (J.Z.); xiaoyulik@hotmail.com (X.L.); pumcpzg@126.com (Z.P.)
[2] School of Pharmacy, Shanghai Jiaotong University, Shanghai 200040, China; yangzjun80@sina.com (Z.Y.); hccbred@gmail.com (M.G.)
[3] Institute of Materia Medica, Chinese Academy of Medical Sciences and Peking Union Medical School, Beijing 100050, China; guojiamei@imm.ac.cn (J.G.); yingguo6@imm.ac.cn (Y.G.)
[4] Department of Food Science, Beijing Union University, Beijing 100101, China; weitao@buu.edu.cn
[5] Institute of Biophysics, Chinese Academy of Sciences, Beijing 100101, China; yuhaisheng@moon.ibp.ac.cn (H.Y.); zhanglgf@hotmail.com (L.Z.)
*   Correspondence: shancen@imb.pumc.edu.cn
†   These authors contributed equally to this work.

Academic Editor: Andrew Mehle

Abstract: While Highly Active Antiretroviral Therapy (HAART) has significantly decreased the mortality of human immunodeficiency virus (HIV)-infected patients, emerging drug resistance to approved HIV-1 integrase inhibitors highlights the need to develop new antivirals with novel mechanisms of action. In this study, we screened a library of microbial natural compounds from endophytic fungus *Colletotrichum* sp. and identified alternariol 5-*O*-methyl ether (AME) as a compound that inhibits HIV-1 pre-integration steps. Time-of addition analysis, quantitative real-time PCR, confocal microscopy, and WT viral replication assay were used to elucidate the mechanism. As opposed to the approved integrase inhibitor Raltegravir, AME reduced both the integrated viral DNA and the 2-long terminal repeat (2-LTR) circular DNA, which suggests that AME impairs the nuclear import of viral DNA. Further confocal microscopy studies showed that AME specifically blocks the nuclear import of HIV-1 integrase and pre-integration complex without any adverse effects on the importin α/β and importin β-mediated nuclear import pathway in general. Importantly, AME inhibited Raltegravir-resistant HIV-1 strains and exhibited a broad anti-HIV-1 activity in diverse cell lines. These data collectively demonstrate the potential of AME for further development into a new HIV inhibitor, and suggest the utility of viral DNA nuclear import as a target for anti-HIV drug discovery.

Keywords: HIV integration; pre-integration complex; alternariol 5-*O*-methyl ether; nuclear import; natural product

---

1. Introduction

An essential step in the replication cycle of human immunodeficiency virus (HIV)-1 is integration of viral DNA into host chromosome DNA, which is catalyzed by viral integrase (IN). Since there are no cellular homologs in human cells and the reactions which INs catalyze are unique, they avoid off-target effects and adverse drug actions. Additionally, IN is necessary for HIV-1 replication and several key residue mutations render it inactive, thus, it represents an ideal target for the discovery of HIV-1 inhibitors [1]. HIV integration includes three essential steps: 3'-processing, nuclear import of pre-integration complex (PIC), and strand transfer. Early IN inhibitor (INI) design and discovery have focused on directly inhibiting IN enzymatic activities. The β-Diketo acid derivatives (DKAs) are the first compounds that were shown to specifically impede strand transfer [2]. DKAs inspired the development of the first-generation INIs including Raltegravir (RAL) and Elvitegravir (EVG), which are also referred to as integrase strand transfer inhibitors (INSTIs) [3]. These INSTIs show a low genetic barrier to resistance development and IN resistance mutations show cross-resistance to RAL and EVG [4]. The second-generation INSTIs including MK2048 and dolutegravir (DTG) inhibit most RAL-resistant HIV-1 mutants. However, development of cross-resistant viral strains appears to be inevitable because both the first and second generations of INSTIs share an overlapping binding site in the IN catalytic domain. Indeed, DTG resistance mutations including G118R and R263K were recently identified, and some of them conferred cross-resistance to RAL and EVG [5]. Development of allosteric IN inhibitors (ALLINIs) provides an alternative approach to discover compounds that are effective against INSTI-resistant HIV-1 [6]. Yet, their effectiveness in clinics awaits further studies.

In addition to targeting the catalytic activity of HIV-1 IN, nuclear entry of HIV-1 pre-integration complex (PIC) represents a promising target. PIC is generated in the cytoplasm following the reverse transcription of viral RNA into viral DNA [7]. The viral components include IN, nucleocapsid (CA), matrix (MA), viral protein R (Vpr), and reverse transcriptase (RT) [8]. Several host proteins including lens epithelium-derived growth factor (LEDGF/p75), barrier-to-autointegration factor (BAF), and integrase interactor 1 have also been found in PIC [9]. Transport of PIC into the nucleus is the prerequisite for the integration of HIV-1 DNA into cellular chromosomes [7]. Yet, it is still uncertain which nuclear import pathway is adopted by PIC. Several lines of evidence support the involvement of importin α/β [10–13]. Other studies suggest a key role of a non-classical nuclear import localization signals (NLS) of IN in the nuclear accumulation of PIC [14,15]. Mutations in MA and CA also block PIC nuclear transport, suggesting that MA and CA also promote the nuclear import of PIC [16–18]. In addition to these viral proteins, cellular import factors are also believed to assist PIC nuclear import. For example, nuclear pore complex component nucleoporin 153 (Nup153) and transportin-3 (TNPO3) operate synergistically in nuclear import of PIC [19,20]. Importin α3 (Imp α3), importin 7 (Imp7), and transportin-SR2 (TRN-SR2) have been shown to interact with HIV-1 IN and affect viral nuclear import [13,21–23].

Although the detailed mechanisms and the nature of host/viral factors behind nuclear import of HIV-1 PIC are still being investigated, some efforts have been made to find small compounds that can target this process. For example, styrylquinoline derivatives show post-entry, pre-integration antiviral activities [24]. An importin α/β pathway inhibitor called ivermectin inhibits IN nuclear transport and HIV-1 replication [25]. Mifepristone was reported to specifically block nuclear import of HIV-1 IN [26]. These studies support the idea that inhibiting IN nuclear import is a feasible approach for discovering new HIV-1 inhibitors. In the present study, we have identified alternariol 5-O-methyl ether (AME), a natural product from endophytic fungus *Colletotrichum* sp, as a novel anti-HIV compound. Further studies revealed that AME specifically blocks nuclear import of HIV-1 PIC and does not affect the importin α/β and importin β-mediated nuclear import pathway in general. Importantly, AME inhibits both wild-type HIV-1 and RAL-resistant viruses, suggesting the potential of AME as a prototype of IN nuclear import inhibitors with the potential to be further developed into a new generation of anti-HIV compounds.

2. Materials and Methods

2.1. Cell Culture and Transfection

HEK293T cells and HeLa cells were maintained in Dulbecco's modified Eagle's Medium supplemented with 10% Fetal Bovine Serum (FBS). MT-2, MT-4, SupT1, and Jurkat cells were maintained in RPMI1640 medium supplemented with 10% FBS. Peripheral blood mononuclear cells (PBMC) from healthy donors were isolated by Ficoll-Hypaque centrifugation and incubated in RPMI1640 medium containing 5 µg/mL phytohemagglutinin (PHA) and 50 U/mL human recombinant IL-2 for 72 h prior to anti-viral assays. A total of 1×10^6 HEK293T cells were transfected with 0.6 µg pNL4-3luc.R-E- in the presence of 60 µM AME or DMSO using Lipofectamine2000 (Invitrogen, Carlsbad, CA, USA). Two days post-transfection, cells were harvested with RIPA buffer (50 mM Tris-HCl (pH 7.4), 150 mM NaCl, 1% NP-40, 0.1% SDS) and the levels of Gag in transfected cells were assessed by Western blotting. For immunostaining assay, 0.1 µg IN-EGFP or Rev-HA was transfected into 2×10^4 HEK293T cells using Lipofectamine2000 in NuncLab-TekII (Thermo Fisher Scientific, Pittsburgh, PA, USA).

2.2. Plasmids and Reagents

The HIV pNL4-3.Luc.R-E- vector contains a full-length HIV-1 proviral DNA in which *env* is defective and *nef* was replaced by luciferase. The vesicular stomatitis virus glycoprotein (VSV-G) expressing vector pHIT/G was provided by Johnny He [15]. To construct IN-EGFP, the coding region of HIV-1 IN was amplified by PCR and inserted into the pEGFP-C1 expression vector (Clontech Laboratories, Palo Alto, CA, USA) at EcoR I and Bgl II. The primers used to amplify the HIV-1 IN are as follows: 5′TAG GAA TTC ATG TTT TTA GAT GGA ATA GAT AAG 3′ (sense) and 5′TAG GGA TCC ATC CTC AT C CTG TCT ACT TGC CAC 3′ (antisense). RAL-resistant mutant V151L was kindly provided by Yong Xiao (McGill University, Montreal, QC, Canada). Another three Q148 pathway mutants G140SQ148H, Q148H, and Q148S, were kindly provided Matthew D. Marsden (University of California, Los Angeles, CA, USA) [27]. To construct subtype C founder/transmitted and chronic infection Luciferase Reporter pseudovirus, the fragment from 5′LTR to the initiation site of env CDS of pNL4-3.Luc.R-E- was replaced by the counterpart of pZM247Fv1 (NIH AIDS REAGENT PROGRAM Catalog #11941) or pIndie-C1 [28] using an IN-Fusion HD cloning kit (Clontech), these chimera constructs were named pZM247Fv1Luc [29] (founder/transmitted reporter virus) and pIndie-C1-Luc (chronic infection virus), respectively.

2.3. Screening for Anti-HIV Compounds from a Library of Microbial Natural Compounds

Microbial natural products were obtained from fungus *Colletotrichum* sp. Briefly, the fermented substrate was extracted with AcOEt and evaporated to get crude extract. Then, the crude extract was fractionated by silica gel and purified by semi-preparation RP-HPLC to obtain the compounds, including AME. The screening was performed as previously described [30]. Briefly, 2×10^5 HEK293T cells were co-transfected with 0.6 µg of pNL4-3Luc.R-E- and 0.4 µg of pHIT/G. After 48 h, the VSV-G pseudotyped HIV-1 viruses were harvested by filtration through a 0.45 mm filter and the concentration of viral capsid protein was determined by p24 antigen capture ELISA. A total of 1×10^5 SupT1 cells were subject to VSV-G pseudotyped HIV-1 infection (MOI = 1) in the absence or presence of test compounds (Efavirenz used as positive control). The inhibition rate was determined by a firefly Luciferase Assay System (Promega, Madison, WI, USA) at 48 h post-infection.

2.4. Assay for Measuring the Inhibitory Activity of Compounds on Different HIV-1 Strains

The inhibitory activity of AME on infection by a typical HIV-1 strain, NL4-3luc.R-E-, and three RAL-resistant strains, G140SQ148H, Q148H, and Q148S, were tested in SupT1 cells. Briefly, 1×10^5 cells were infected by VSV-G-pseudotyped HIV-1 viruses, followed by addition of compounds at serial dilutions. After further incubation at 37 °C for 48 h, cells were harvested and luciferase activities were measured by 960 luminometer. The concentration of the compound for inhibiting 50% viral replication (IC50) was determined by Origin 8.0 software.

2.5. Cytotoxicity Assay

AME was added to HEK293T cells at 1×10^5 per well, followed by incubation at 37 °C for 48 h. Ten microliters of CCK-8 reagent were added to the cells. After incubation at 37 °C for 4 h to allow color development of the XTT formazan product, the absorbance of each well was read at 450 nm. The 50% cytotoxicity concentration (CC50) was generated by Origin 8.0 software.

2.6. Time of Addition Experiment

A total of 1×10^5 SupT1 cells/well were infected with VSV-G-pseudotyped NL4-3Luc.R-E- viruses and incubated at 4 °C for 1 h in order to synchronize infection. The unbound viruses were washed off with phosphate-buffered saline (PBS) and then the cells were incubated at 37 °C. Four nM efavirenz (EFV), 25 nM RAL, 76 nM 3TC, and 60 μM AME were added at 0, 0.5, 1, 2, 4, 6, 7, 12, 15, 18, and 21 h. Two days post-infection, cells were lysed with cell culture lysis buffer (Promega), luciferase activities were measured, and the time-response curves were generated using Origin 8.0 software.

2.7. Semi-Quantitative Real-Time PCR

A total of 5×10^6 SupT1 cells were infected with VSV-G pseudotyped NL4-3Luc.R-E- viruses in the treatment of 4 nM EFV, 25 nM RAL, or 60 μM AME. Cells were harvested at 3, 8, or 24 h post-infection. After washing with PBS, total DNA was extracted using DNeasy Blood&Tissue Kit (Qiagen, Hilden, Germany). An aliquot of each sample was analyzed by PCR. The PCR program was a relative quantitative procedure in an Mx3000P real time PCR system (Agilent Technologies Inc., Palo Alto, CA, USA). The primers used in the PCR were: U5-Gag sense, 5′TGT GTG CCC GTC TGT TGT GTG A3′; U5-Gag antisense, 5′TCA GCA AGC CGA GTC CTG CGT3′; Alu-LTR sense, 5′TCC CAG CTA CTC GGG AGG CTG AGG3′; Alu-LTR antisense, 5′AGG CAA GCT TTA TTG AGG CTT AGC3′; 2-long terminal repeat (2-LTR) sense, 5′AAC TAG GGA ACC CAC TGC TTA AG3′; and 2-LTR antisense, 5′ TCC ACA GAT CAA GGA TAT CTT GTC 3′. Total viral DNA, integrated DNA, and 2-LTR were expressed as copy numbers per cell, with DNA template normalized by the GAPDH gene that was amplified using GAPDH primers.

2.8. Immunostaining

HeLa cells were grown on coverslips in the treatment of DMSO, 25 nM RAL, 60 μM AME, or 25 μM ivermectin for 48 h, and then fixed in 4% paraformaldehyde and permeabilized in 0.4% TritonX-100 followed by incubation in primary and secondary antibodies. The primary antibody was anti-IN at 1:500 (Abcam, Cambridge, MA, USA) [31]. The secondary antibody was TRITC-conjugated goat anti-mouse immunoglobulin (Ig)G or FITC-conjugated goat anti-mouse IgG (Sigma-Aldrich, St Louis, MO, USA). Fluorescence images were acquired on an Olympus FV1000 confocal fluorescence microscope.

2.9. Nuclear and Cytoplasmic Fractionation

HEK293T cell cytosol and nucleus were prepared according to the manual of the Nuclear and Cytoplasmic Protein Extraction Kit (Sangon Biotech, Shanghai, China). In brief, 1×10^6 HEK293T cells were collected by centrifugation at 3000 rpm at 4 °C. Cells were washed twice with PBS and once with Solution A supplemented with protease inhibitors, DTT, phosphatase inhibitors and PMSF. After sonication, cell lysate were centrifuged at 12,000 rpm for 30 min at 4 °C. The supernatant contained cell cytosol. The pellet was incubated with Solution B supplemented with protease inhibitors, DTT, phosphatase inhibitors and PMSF, and then centrifuged at 12,000 rpm for 30 min at 4 °C; the supernatant contained cell nuclei. Each sample was added with 5× SDS lysis buffer and boiled, then detected by SDS-PAGE and Western blotting. The PVDF membranes were immunoblotted with mouse anti-IN, anti-β-actin, and anti-MCM2 antibodies (Abcam) and then with complementary horseradish peroxidase (HRP)-conjugated secondary antibodies.

2.10. Viral Preparation and Infection Assay

HIV-Luc Reporter viruses were produced from HEK293T cells by transfecting with VSV-G and pNL4-3luc.R-E-. Founder/transmitted Reporter viruses were produced by transfecting with VSV-G and pZM247Fv1Luc or pIndie-C1-Luc as a chronic infection control. The supernatants were filtered and stored at $-80\,^{\circ}$C for future use. For single round infection assays, 5×10^5/mL SupT1 cells were infected with pseudotyped viruses. Two days post-infection, cells were lysed with cell culture lysis buffer (Promega) for 20 min at $37\,^{\circ}$C, 6 μL of the lysate was added with 40 μL substrate, luciferase activities were measured by 960 luminometer.

For detection of viral entry, 1×10^4 SupT1 cells per well were seeded in 96-well plates and divided into two populations. In one population, the cells were treated with T-20, AME, or DMSO as vehicle control at the concentrations of 2 fold IC_{50}, immediately followed by infection with NL4-3Luc pseudoviruses. In another population, cells were incubated with pseudoviruses at $4\,^{\circ}$C for 1 h and then the cells were washed with PBS followed by treatment of T-20, AME or DMSO. Forty-eight h later, cells were harvested and Luc activity was measured. Luc activity of each sample was normalized by the vehicle control.

For WT HIV infection, HIV-1 NL4-3 viruses were produced from HEK293T cells by transfecting with pNL4-3. One million PBMC were seeded in each well in a 96-well plate and incubated with DMSO, 2 nM EFV, 35 μM or 180 μM AME for 5h. Then, the medium was refreshed and cultured for a further 72 h. Supernatants were harvested at day 3 and day 5 and assayed for HIV-1 p24 content by an enzyme-linked immunosorbent assay (ELISA, ZeptoMetrix Corp., Buffalo, NY, USA).

2.11. In Vitro Integrase Assay

The wells of a microtiter plate were coated with a 30 base pair (bp) stretch of the sequence encoding U5-LTR. HIV-1 integrase and tested compounds were added and incubated in reaction buffer (25 mM MOPS, pH 7.2, 20 mM Tris-HCl, pH 7.8, 20 mM DTT, 15 mM $MnCl_2$ and 0.2%TritonX-100) at $37\,^{\circ}$C for 60 min. Then, a 20 bp oligonucleotide of 3′-biotin-conjugated target substrate was added into the microplate and incubated for another 60 min. After washing with PBS, the microplate was added with horseradish peroxidase (HRP) dilution buffer (10 mM Tris-HCl, pH7.5, 0.15M NaCl, 1 mM EDTA, 1%BSA) and incubated at $37\,^{\circ}$C for 30 min. Then, the plate was washed with PBS. One hundred μL of TMB buffer (3,3′,5,5′-tetramethylbenzidine) was dispensed into each well, incubated for 15 min in the dark, and an equal volume of stopping solution (2M H_2SO_4) was added. The optical density was read at 450 nm.

3. Results

3.1. AME Inhibits HIV Infection

In an effort to discover new anti-HIV compounds, we examined a library of microbial natural products as previously described [30]. The results revealed a number of hits including alternariols from endophytic fungus *Colletotrichum* sp., among which AME showed antiviral activity (Figure 1A). In contrast, its derivative alternariol (AOH) (Figure 1B) had little effect on HIV-1 infection with an inhibition rate ~6% even at a concentration of 300 μM. AME inhibited the HIV-1 infection with an EC_{50} of 30.9 ± 0.5 μM ($n = 3$) in a one-cycle HIV-1 infection assay (Figure 1C), which is much lower than a CC_{50} of 392.3 ± 7.2 μM ($n = 3$). The result excludes the possibility that the anti-HIV activity was due to the cytotoxicity of AME.

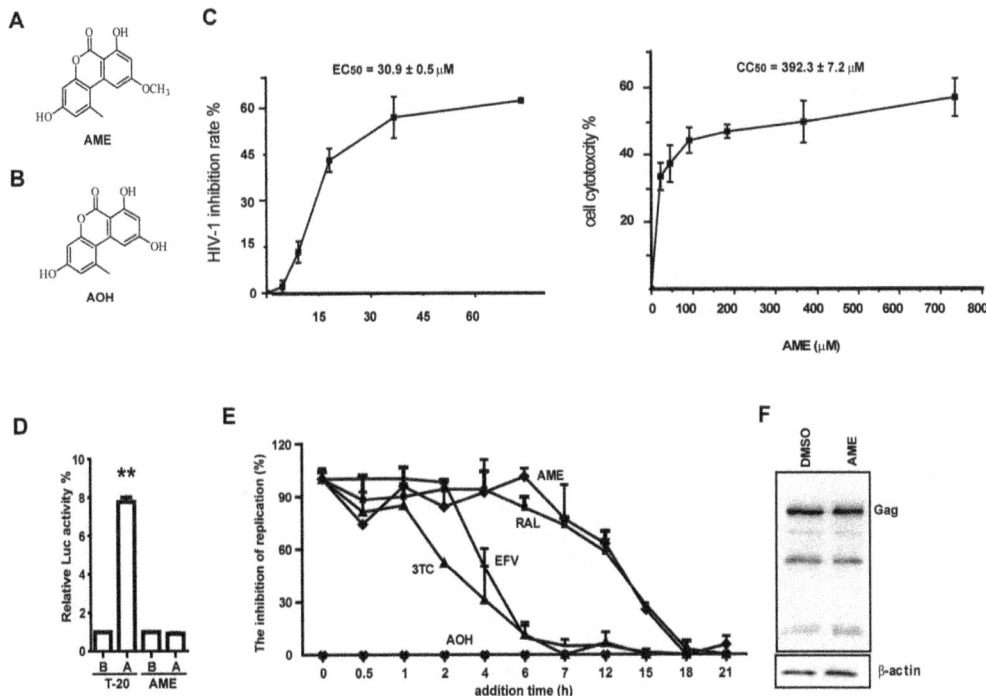

Figure 1. Alternariol 5-O-methyl ether (AME) potentially inhibits early stage of human immunodeficiency virus (HIV)-1 replication. (**A**) The structure of AME was solved through spectroscopic analysis; (**B**) The structure of alternariol (AOH); (**C**) SupT1 cells were infected with HIV-1 NL4-3luc.R-E- pseudoviruses in the presence of different concentrations of AME. 48 h post-infection, cells were lysed and luciferase activity was measured. A dose-response curve was generated using the Origin 8.0 software; (**D**) SupT1 cells were infected with HIV-1 NL4-3luc.R-E- pseudoviruses before or after treatment with either AME or T-20 at 2 fold IC50. Luciferase activity was measured before (B) or after (A) viral infection. Relative luciferase activity was a ratio of the luciferase activity in cells infected before AME (or T20) treatment over that in cells infected after drug treatment (set as 1); (**E**) A time of addition (TOA) experiment was carried out to determine the step of HIV-1 infection that AME inhibits. Briefly, after infection of SupT1 cells with HIV-1 NL4-3luc.R-E- pseudoviruses, inhibitors of the indicated concentrations were added at different time points ranging from 1 to 24 h post-infection. The relative inhibition rate was calculated by dividing the inhibition rate at 0 h for each compound. White diamond: AME; black diamond: efavirenz (EFV); triangle: 3TC; rectangular: Raltegravir (RAL); cross: AOH. Viral infection was determined by measuring luciferase activity at 48 h post-infection. As controls, reverse transcriptase (RT) inhibitors EFV and 3TC, integrase strand transfer inhibitors (INSTIs) RAL, as well as inactive congener AOH were also tested; (**F**) HEK293T cells were transfected with pNL4-3luc.R-E-. Forty-eight h post-transfection, cells were lysed, and Gag expression was determined by Western blotting. Data represent the mean ± SD of three independent experiments.

To explore the anti-HIV mechanism of AME, we first examined whether AME impairs viral entry. T-20, a synthetic peptide which blocks viral entry, was used as positive control. As shown in Figure 1D, the antiviral effect of HIV-1 entry inhibitor T-20 was significantly reduced when the compound was added at post-infection, while no difference in the inhibitory effect of AME was observed regardless of the addition of the compound before or after viral infection. This indicates that AME acts, at least in part, downstream of the viral entry step. Next, we further conducted the time-of-addition experiment (TOA), which has been used to pinpoint which step of HIV-1 replication cycle is blocked by antiretroviral compounds. Briefly, AME was added at the different time points after infection of SupT1 cells with pseudotyped HIV-1 NL4-3luc.R-E-, and 48 h after infection, luciferase activities were measured to determine the anti-HIV effect of AME. As controls, nucleoside reverse transcription inhibitor Lamivudine (3TC), non-nucleoside reverse transcription inhibitor, efavirenz (EFV), integrase inhibitor RAL, and inactive congener AOH were used to treat infected SupT1 cells. 3TC, EFV, and RAL exhibited their antiviral activities within 1 h, 2 h, and 7 h post-infection, respectively, which reflect the time required for completion of reverse transcription and integration. AME still inhibited HIV-1 when added 7 h after infection, and lost its inhibitory activity when added 15 h after infection, which corresponds to the time-activity curve of RAL (Figure 1E). Thus, this suggests that AME likely inhibits HIV at a step in the replication cycle similar to that of RAL. As expected, AOH showed no anti-HIV activity at all the time points we tested. To further rule out the possibility that the reduced viral infectivity resulted from an inhibitory effect of AME on transcription or translation of the viral gene, HEK293T cells were transfected with pNL4-3luc.R-E- in the presence of 60 μM AME or DMSO. Levels of Gag in transfected cells were assessed by Western blotting. The results showed that AME had no effect on HIV-1 Gag expression (Figure 1F). Taken together, these data suggest that AME may inhibit HIV-1 replication by blocking viral integration.

3.2. AME Causes Defective HIV-1 Integration

We next performed relative quantitative PCR (RQ-PCR) to directly assess the effect of AME on viral reverse transcription and integration. Primers were designed to detect the U5 and Gag (U5-Gag) sequences indicative of the following reverse transcription products: integrated viral DNA (Alu-LTR) and 2-LTR-containing DNA circles. A total of 1.5×10^6 SupT1 cells per well were infected with pseudotyped HIV-1 NL4-3luc.R-E- in the treatment of DMSO, EFV, RAL, AOH, or AM. Cells were harvested at 3, 8, and 24 h after infection. DNA was extracted to quantify late reverse transcripts, 2-LTR circles, and integrated proviruses via qPCR. As shown in Figure 2A, among three compounds (EFV, RAL, and AME), only EFV caused a significant reduction in U5-Gag products, suggesting that AME had no effect on reverse transcription. Importantly, both RAL and AME severely reduced Alu-LTR DNA that represents viral integration product at 8 h post-infection (Figure 2B). Together with the results of the TOA analysis, we conclude that AME blocks the generation of integrated HIV-1 DNA.

2-LTR circles form at a low level in the nucleus through the action of cellular nonhomologous DNA end joining. Post nuclear entry blockage to HIV-1 infection is expected to result in an increase in 2-LTR circles. Therefore, the abundance of 2-LTR circles is an indirect measure for the nuclear import of the PIC. The decline of 2-LTR and subsequent block of proviral integration pinpoint a block of HIV replication between the late reverse transcription and the nuclear import steps. Indeed, the amount of 2-LTR circles was 3.9-fold higher in RAL-treated cells compared with DMSO-treated control cells at 8 h post-infection (Figure 2C). In contrast, AME reduced the level of 2-LTR circles. This suggests that, as opposed to RAL that impairs integration, AME appears to interfere with the nuclear import of HIV-1 DNA. In support of this result, AME did not affect 3′ processing and strand transfer reactions in vitro by ELISA at the concentration of 100 μM. (Figure 2D).

3.3. AME Blocks Nuclear Transport of PIC

Next, we measured the effect of AME upon nuclear import of HIV-1 PIC by performing an indirect immunofluorescence assay. HeLa cells were infected with HIV-1 in the presence of AME or DMSO. Subcellular localization of integrase was monitored by an indirect immunofluorescence assay using anti-integrase monoclonal antibody. IN was detected in the nucleus in DMSO-treated cells as opposed to the dominant cytoplasmic presence of IN in AME-treated cells (Figure 3A). When the ratio of nuclear fluorescence and total cellular fluorescence were measured, which represents the relative efficiency of IN nuclear import, 30% of IN was found in the nucleus of AME-treated cells as compared to about 60% of IN in the nucleus of DMSO-treated cells (Figure 3B). We, therefore, conclude that AME inhibits nuclear transport of HIV-1 IN that is an integral component of viral PIC.

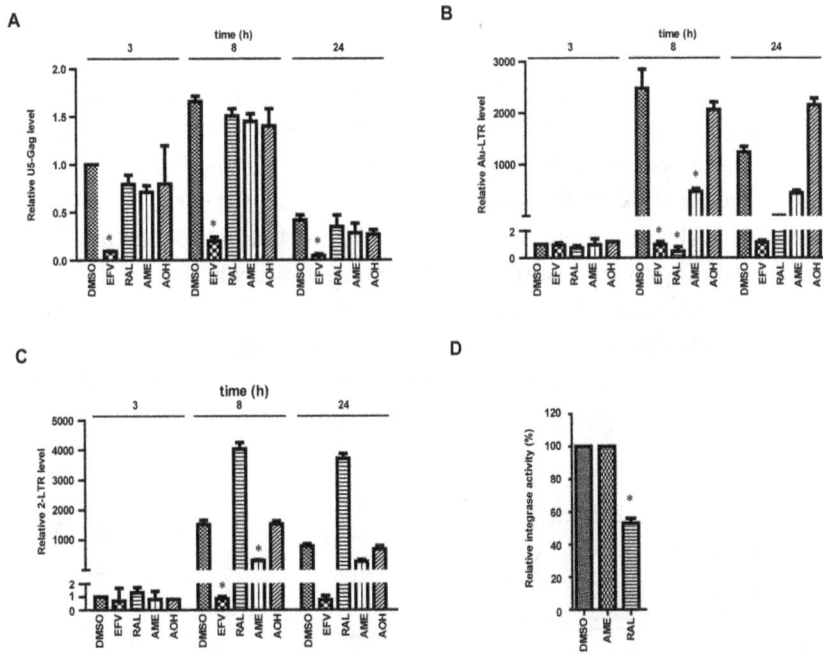

Figure 2. AME inhibits HIV integration. (**A–C**) SupT1 cells were infected with vesicular stomatitis virus glycoprotein (VSV-G) pseudotyped NL4-3Luc R-E- viruses with the treatment of DMSO, 4 nM EFV, 25 nM RAL, 60 μM AOH, or AME of indicated concentrations as described in the Materials and Methods section. DNA was extracted from infected cells at 3 h, 8 h, and 24 h post-infection. HIV early reverse transcription products (**A**), integrated DNA (**B**), and 2-LTR circles (**C**) were analyzed by RQ-PCR using the corresponding primers mentioned in the Materials and Methods section; (**D**) The activity of integrase was measured by an ELISA-based in vitro assay in the treatment of DMSO, 100 μM AME, and 100 nM RAL. * $p < 0.05$.

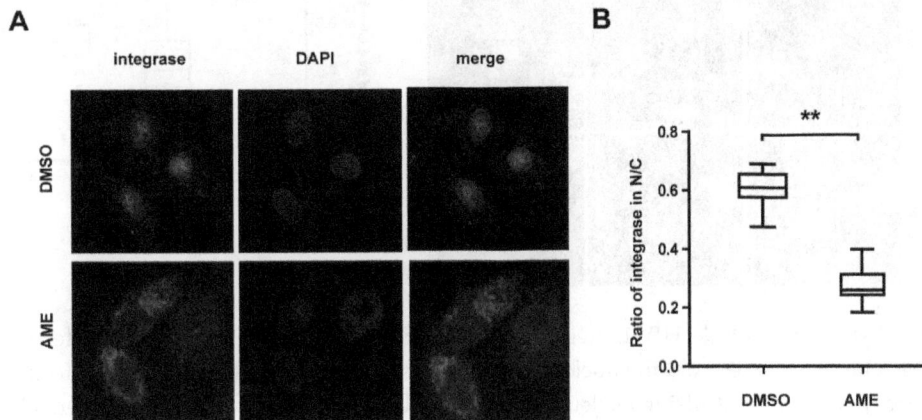

Figure 3. AME blocks pre-integration complex (PIC) nuclear import. (**A**) HeLa cells were infected with VSV-G pseudotyped NL4-3Luc R-E- viruses in the presence of 60 μM AME or DMSO. Twelve h post-infection, cells were fixed and labelled with mouse anti-integrase (anti-IN) antibody followed by FITC-conjugated goat anti-mouse secondary antibody, and the nuclei were stained with 4′6′-diamidino-2-phenylindole (DAPI). Cells were analyzed by confocal microscopy (with a 60× objective lens); (**B**) Statistical analysis for the ratio of integrase in nuclei and cell cytosol. Ratios of fluorescence in nucleus and whole cells at 488 nm were measured from 25 DMSO-treated cells and 25 AME-treated cells randomly selected from three independent experiments. The fluorescence density at 488 nm was quantified by ImageJ and statistical analysis was performed in Graphpad Prism 5.0 software. The data represent the percentage of PIC in the nucleus. Error bars represent standard deviation (SD) of all cells in each group. ** $p < 0.01$ (determined by Student's t test).

We further tested whether AME directly targets IN rather than other viral components in PIC [16]. To this end, HEK293T cells were transfected with a plasmid expressing IN-EGFP, and then treated with DMSO, RAL, and AME. Levels of IN-EGFP in the nucleus and the cytoplasm were assessed by cytoplasma-nucleus separation followed by Western blotting. A successful isolation of the nucleus and cytoplasm was corroborated by Western blotting of β-actin and MCM-2 in its corresponding lysates. The result showed that IN-EGFP was predominantly localized in the cytoplasm in AME-treated cells as compared with RAL-treated and control cells (Figure 4A). AME inhibits the nuclear entry of IN-EGFP by 14-fold according to gray value of each band quantified by ImageJ (Figure 4B). These data were corroborated by the results of fluorescence imaging (Figure 4C). Like an importin α/β inhibitor ivermectin, AME caused the diffusion of IN into the cytoplasm compared with the diffusion caused by DMSO-treated cells (Figure 4C). Together, these data demonstrated that AME blocked PIC transport into the nucleus by directly by targeting the import of HIV integrase.

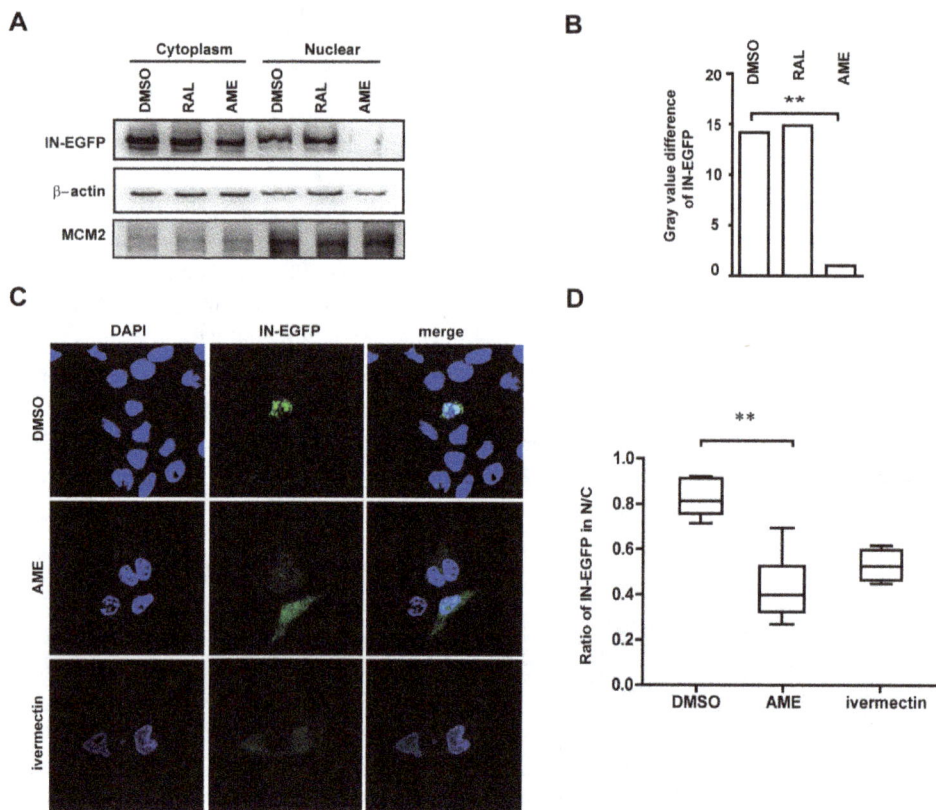

Figure 4. AME directly blocks HIV-1 integrase nuclear entry. (**A**) HEK293T cells were transfected with IN-EGFP. Cells were lysed, and the nucleus and cytoplasm were isolated 40 h post-transfection. IN expression, β-actin, and MCM-2 in nucleus (N) and cytoplasm (C) were measured by Western blotting (left panel); (**B**) The gray value of each band in (A) was quantified and analyzed by ImageJ (right panel); (**C**) HEK293T cells were transfected with IN-EGFP in the presence of DMSO, ivermectin, or AME. Cells were fixed, and the nuclei were stained with 4′6′-diamidino-2-phenylindole (DAPI) after 24 h. Cells were analyzed by confocal microscopy (with a 60× objective lens); (**D**) Ratios of fluorescence in nucleus and whole cells at 488 nm were measured from more than 25 cells per condition randomly selected from three independent experiments. The fluorescence density at 488 nm was quantified by ImageJ and statistical analysis was performed in Graphpad Prism 5.0 software. The data represent the percentage of IN-EGFP in the nucleus. Error bars represent standard deviation (SD) of all cells in each group. ** $p < 0.01$ (determined by Student's t test).

We next asked whether AME specifically inhibited the nuclear import of PIC, or exerted a general effect on nuclear transport. Nuclear import of cellular proteins mostly depends on nuclear import localization signals (NLS). Classic NLS comprise either a short stretch of basic amino acids, such as SV40 large T antigen (PKKKRKV) or a bipartite NLS including two interdependent stretches of basic amino acids with a spacer cluster in between, such as nucleoplasmin (KRPAATKKAGQAKKKK). Karyphilic proteins containing the classic NLS require both importin α and β for their nuclear entry. Alternatively, other proteins such as Rev, cyclin B1, and hTAP only require importin β. These proteins have arginine rather than lysine in the NLS. In this context, we assessed the effect of AME upon the nuclear transportation of SV40 large T antigen and Rev that are imported into the nucleus by the importin α/β and importin β-mediated pathway, respectively. HeLa cells were transfected with plasmids expressing either HIV-1 Rev or SV40 large T antigen, and treated with DMSO, ivermectin, or AME. Immunofluorescence analysis revealed that AME had no effect upon the nuclear localization of Rev (Figure 5A) or SV40 large T antigen (Figure 5B). In contrast, ivermectin caused cytoplasmic accumulation of SV40 large T antigen, confirming the experimental system is sound (Figure 5B), whereas, HIV-1 Rev, which is dependent on import β alone during nuclear import, was not affected (Figure 5A). These data suggest that AME specifically inhibits the nuclear import of HIV-1 IN without interrupting cellular importin α/β and importin β-mediated nuclear transport pathways.

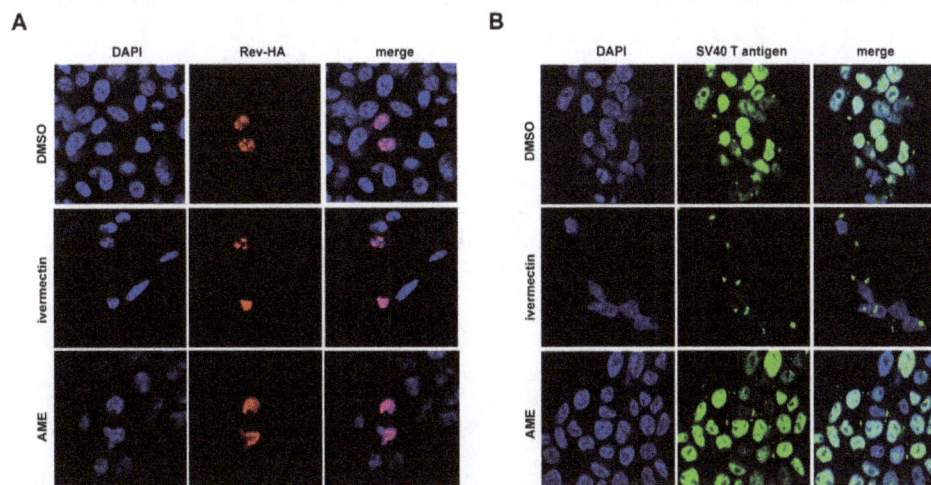

Figure 5. AME does not disturb the classical nuclear transport pathway. (**A**) HeLa cells were transfected with Rev-HA and treated with DMSO, ivermectin or AME. One day post-infection, cells were fixed and stained with mouse anti-HA antibody followed by TRITC-conjugated goat anti-mouse secondary antibody. The nuclei were stained with 4'6'-diamidino-2-phenylindole (DAPI). Cells were analyzed by confocal microscopy (with a 60× objective lens); (**B**) HEK293T cells were treated with DMSO, ivermectin, or AME. One day post-transfection, cells were fixed and stained with mouse anti-SV40 T antigen antibody followed by FITC-conjugated goat anti-mouse secondary antibody, and the nuclei were stained with 4'6'-diamidino-2-phenylindole (DAPI). Cells were examined by confocal microscopy. More than 25 cells per condition spanning three independent experiments were included for each panel.

3.4. AME Inhibits the Infection of Different HIV-1 Strains

Next, we further evaluated the anti-HIV-1 activity of AME in different cell lines. HEK293T, HeLa, Jurkat T, MT-2, and MT-4 cells were infected with HIV-Luc reporter viruses and then were treated with DMSO, RAL, or AME. AME inhibited HIV-1 infection by approximately 70%, 57%, 88%, 82%, and 80% in HEK293T, HeLa, Jurkat T, MT-2, and MT-4 cells, respectively (Figure 6A). Secondly, we tested the sensitivity of different HIV-1 strains to AME. To this end, we constructed two subtype C chimera viruses, ZM247Fv1Luc (founder/transmitted reporter virus) and Indie-C1-Luc (a primary isolate), and used these viruses to infect SupT1 cells in the presence of DMSO, RAL, or AME. The results showed that AME profoundly inhibited both viruses, with the inhibition rate of approximately 70% and 52% for Indie-C1-Luc and ZM247Fv1Luc, respectively (Figure 6B). Collectively, these data demonstrated that AME inhibits HIV of different origins in a variety of cell lines. Thirdly, we tested the activity of AME against different HIV-1 strains bearing the RAL-resistant mutations including V151L, G140SQ148H, Q148H, and Q148S. As expected, V151L, G140SQ148H, Q148H, and Q148S mutation conferred 5, 134, 23, and 26-fold resistance to RAL, respectively, yet these viruses were inhibited by AME as much as the wild type virus (Figure 6C). With respect to G140SQ148H and Q148S, they exhibit even more susceptibility to AME treatment than wild type.

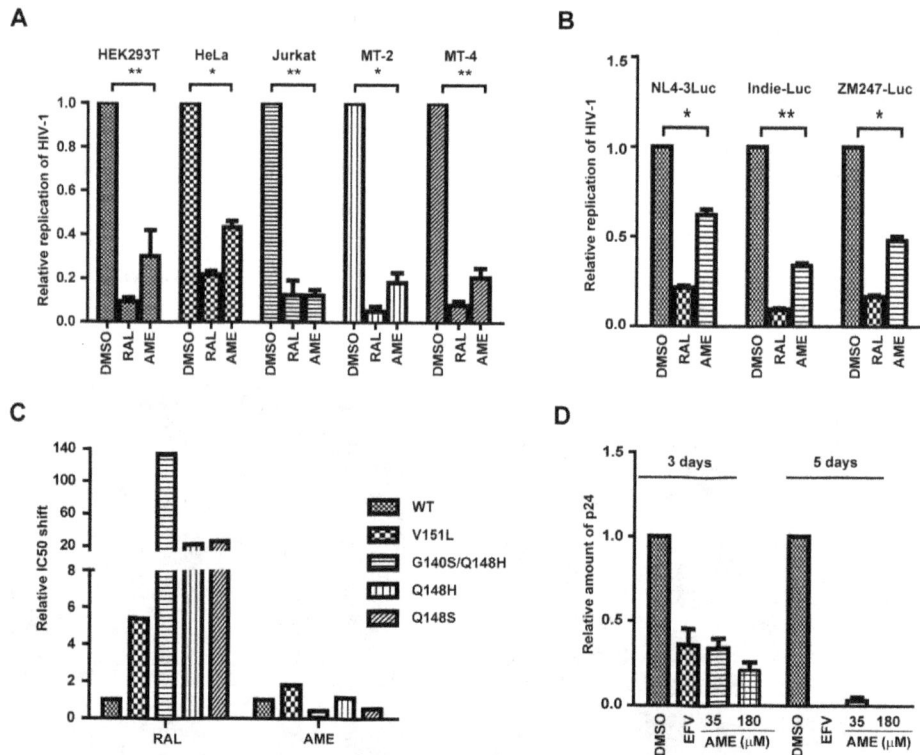

Figure 6. AME inhibits the infection of different HIV-1 strains in different cell lines. (**A**) $5 \times 10^5/\text{mL}$ Jurkat T, MT-2, MT-4, HeLa, and HEK293T cells were infected with VSV-G pseudotyped NL4-3Luc.R-E- viruses in the presence of DMSO, 25 nM RAL, and 60 µM AME. Forty-eight h post-infection, cells were harvested to measure luciferase activity; (**B**) SupT1 cells were infected with VSV-G pseudotyped subtype B NL4-3.luc.R-E- viruses, subtype C ZM247Fv1-Luc, or Indie-C1-Luc viruses treated with 12.5 nM RAL, 35 µM AME, or DMSO as the vehicle control. Two days post-infection, cells were harvested to measure luciferase activity; (**C**) SupT1 cells were infected with VSV-G pseudotyped NL4-3Luc.R-E- viruses or indicated RAL-resistant strains treated with AME, RAL, or DMSO at different concentrations. Cells were harvested to measure luciferase reporter activity. The IC_{50} for WT virus and RAL-resistant mutant virus was calculated; (**D**) Peripheral blood mononuclear cells (PBMCs) were infected with wild type NL4-3 HIV viruses treated with 2 nM EFV, 35 µM or 180 µM AME, or DMSO. Supernatant were harvested to measure the quantity of p24 antigen by ELISA at day 3 and day 5. Data represent the mean ± SD of three independent experiments. $* p < 0.05$, $** p < 0.01$.

Suppression of WT HIV replication by AME was also assessed in PBMCs. AME inhibited HIV replication by 66% and 75% at the concentrations of 35 and 180 µM, respectively, at day 3. The anti-HIV activity was more remarkable at day 5, with the inhibition rate of 95% and 100% at concentrations of 35 and 180 µM, respectively (Figure 6D). These data collectively demonstrate that AME inhibits different HIV-1 strains in different cell types and that the RAL-resistant HIV-1 is subject to AME inhibition.

4. Discussion

In the present study, we reported for the first time that AME, a natural product from endophytic fungus *Colletotrichum* sp., inhibits HIV-1 infection by targeting the nuclear import of viral integrase. Further studies revealed AME specifically blocked nuclear import of PIC and IN, rather than exerted a general effect on the importin α/β and importin β-mediated nuclear import. Furthermore, AME inhibited both wild type HIV-1 and RAL-resistant strains. These data collectively demonstrate the potential of AME or its derivatives for future development into a novel class of anti-HIV drugs: IN nuclear import inhibitors (INNIIs).

AME significantly reduced the nuclear accumulation of IN but not the SV40 large T antigen and Rev in living cells, indicating that AME selectively inhibits the nuclear import of IN, rather than the importin α/β and importin β-mediated pathway in general. Furthermore, AME exhibited a similar inhibitory effect on the nuclear localization of IN that was either over-expressed alone or as a part of PIC formed in the HIV-1 infected cells. This strongly suggests that AME inhibits HIV-1 replication by targeting either IN and/or IN-specific pathways. Despite that, the detailed mechanism of AME action remains to be elucidated; one possibility is that AME may target the interaction of IN with host factors that are required for nuclear transport. Several lines of evidence indicate that importin α/β are utilized in HIV PIC nuclear transport. To date, two members of importin α, Impα1 and Imp3, and two members of importin β, Imp7 and TRN-SR2, have been shown to interact with HIV-1 IN, while their precise roles in IN nuclear import remain unclear. It is conceivable that AME may bind to either HIV-1 IN or one of these adaptors or receptors and interrupts the interaction of IN with these key host factors. Future studies on the action of AME will provide insight into the detailed mechanism underlying nuclear import of HIV-1 PIC.

Early IN I drug design and discovery has mainly focused on the direct inhibition of enzyme catalytic activities, leading to the development of the first and second generation INSTIs. However, cross-resistance between INSTIs represents a major obstacle. HIV-1 PIC nuclear import inhibitors, such as AME shown in this study, do inhibit INSTI-resistant HIV-1, and, therefore, represent an important class of new anti-HIV drug with a novel mechanism of action. Furthermore, AME effectively inhibits the replication of different HIV-1 strains in several cell lines, suggesting a relatively broad anti-HIV activity of AME. More importantly, AME similarly inhibits both wild-type HIV-1 and RAL-resistant HIV-1, supporting the possible use of IN nuclear import inhibitors to control the infection of INSTI-resistant viruses. Taken together, these data demonstrate the potential of AME for the future development into a novel class of anti-HIV drug.

In summary, our data demonstrate that AME restricts HIV infection through directly targeting the nuclear import of HIV integrase. This inhibitory effect is highly specific and does not affect the canonical nuclear transport pathway in the host. Our study provides a novel strategy for developing new anti-HIV drugs.

Acknowledgments: This work was supported in part by 973 program (2012CB911102 CS), National Mega-project for Innovative Drugs (2012ZX09301002-001-015 and 2012ZX09301002-005-002), National Mega-Project for Infectious Disease (2013ZX10004601-002), and the Xiehe Scholar (S.C.). We thank the National Infrastructure of Microbial Resources (NIMR-2014-3) for providing valuable reagents.

Author Contributions: C.S. and D.J.W. conceived and designed the experiments; D.J.W., Z.J.Y., Y.Z.J., M.L., W.Y.B. and Y.H.S. performed the experiments; M.Z.Y., Z.J.M. and L.X.Y. analyzed the data; G.J.M. and G.Y. contributed RAL-resistant HIV-1 strains; Z.L.G. provided WT HIV-1 strain; G.M. provided AME; P.Z.G. contributed in vitro integrase assay; W.T. contributed analysis software; C.S. and D.J.W. wrote the paper.

References

1.   Zhao, G.; Wang, C.; Liu, C.; Lou, H. New developments in diketo-containing inhibitors of HIV-1 integrase. *Mini Rev. Med. Chem.* **2007**, *7*, 707–725. [CrossRef] [PubMed]

2.   Marchand, C.; Zhang, X.; Pais, G.C.; Cowansage, K.; Neamati, N.; Burke, T.R., Jr.; Pommier, Y. Structural determinants for HIV-1 integrase inhibition by beta-diketo acids. *J. Biol. Chem.* **2002**, *277*, 12596–12603. [CrossRef] [PubMed]

3.   McColl, D.J.; Chen, X. Strand transfer inhibitors of HIV-1 integrase: Bringing IN a new era of antiretroviral therapy. *Antivir. Res.* **2010**, *85*, 101–118. [CrossRef] [PubMed]

4.   Geretti, A.M.; Armenia, D.; Ceccherini-Silberstein, F. Emerging patterns and implications of HIV-1 integrase inhibitor resistance. *Curr. Opin. Infect. Dis.* **2012**, *25*, 677–686. [CrossRef] [PubMed]

5.   Wainberg, M.A.; Mesplede, T.; Quashie, P.K. The development of novel HIV integrase inhibitors and the problem of drug resistance. *Curr. Opin. Virol.* **2012**, *2*, 656–662. [CrossRef] [PubMed]

6.   Engelman, A.; Kessl, J.J.; Kvaratskhelia, M. Allosteric inhibition of HIV-1 integrase activity. *Curr. Opin. Chem. Biol.* **2013**, *17*, 339–345. [CrossRef] [PubMed]

7.   Depienne, C.; Mousnier, A.; Leh, H.; Le Rouzic, E.; Dormont, D.; Benichou, S.; Dargemont, C. Characterization of the nuclear import pathway for HIV-1 integrase. *J. Biol. Chem.* **2001**, *276*, 18102–18107. [CrossRef] [PubMed]

8.   Sherman, M.P.; Greene, W.C. Slipping through the door: HIV entry into the nucleus. *Microbes Infect.* **2002**, *4*, 67–73. [CrossRef]

9.   Raghavendra, N.K.; Shkriabai, N.; Graham, R.; Hess, S.; Kvaratskhelia, M.; Wu, L. Identification of host proteins associated with HIV-1 preintegration complexes isolated from infected CD4+ cells. *Retrovirology* **2010**, *7*, 66. [CrossRef] [PubMed]

10.   Luban, J. HIV-1 infection: Going nuclear with TNPO3/Transportin-SR2 and integrase. *Curr. Biol.* **2008**, *18*, R710–R713. [CrossRef] [PubMed]

11.   Levin, A.; Hayouka, Z.; Friedler, A.; Loyter, A. Transportin 3 and importin alpha are required for effective nuclear import of HIV-1 integrase in virus-infected cells. *Nucleus* **2010**, *1*, 422–431. [CrossRef] [PubMed]

12.   Hearps, A.C.; Jans, D.A. HIV-1 integrase is capable of targeting DNA to the nucleus via an importin alpha/beta-dependent mechanism. *Biochem. J.* **2006**, *398*, 475–484. [CrossRef] [PubMed]

13.   Ao, Z.; Danappa Jayappa, K.; Wang, B.; Zheng, Y.; Kung, S.; Rassart, E.; Depping, R.; Kohler, M.; Cohen, E.A.; Yao, X. Importin alpha3 interacts with HIV-1 integrase and contributes to HIV-1 nuclear import and replication. *J. Virol.* **2010**, *84*, 8650–8663. [CrossRef] [PubMed]

14.   Bouyac-Bertoia, M.; Dvorin, J.D.; Fouchier, R.A.; Jenkins, Y.; Meyer, B.E.; Wu, L.I.; Emerman, M.; Malim, M.H. HIV-1 infection requires a functional integrase NLS. *Mol. Cell.* **2001**, *7*, 1025–1035. [CrossRef]

15.   Levin, A.; Armon-Omer, A.; Rosenbluh, J.; Melamed-Book, N.; Graessmann, A.; Waigmann, E.; Loyter, A. Inhibition of HIV-1 integrase nuclear import and replication by a peptide bearing integrase putative nuclear localization signal. *Retrovirology* **2009**, *6*, 112. [CrossRef] [PubMed]

16.   Bukrinsky, M.I.; Haggerty, S.; Dempsey, M.P.; Sharova, N.; Adzhubei, A.; Spitz, L.; Lewis, P.; Goldfarb, D.; Emerman, M.; Stevenson, M. A nuclear localization signal within HIV-1 matrix protein that governs infection of non-dividing cells. *Nature* **1993**, *365*, 666–669. [CrossRef] [PubMed]

17.   Lee, K.; Ambrose, Z.; Martin, T.D.; Oztop, I.; Mulky, A.; Julias, J.G.; Vandegraaff, N.; Baumann, J.G.; Wang, R.; Yuen, W.; et al. Flexible use of nuclear import pathways by HIV-1. *Cell Host Microbe* **2010**, *7*, 221–233. [CrossRef] [PubMed]

18.   Haffar, O.K.; Popov, S.; Dubrovsky, L.; Agostini, I.; Tang, H.; Pushkarsky, T.; Nadler, S.G.; Bukrinsky, M. Two nuclear localization signals in the HIV-1 matrix protein regulate nuclear import of the HIV-1 pre-integration complex. *J. Mol. Biol.* **2000**, *299*, 359–368. [CrossRef] [PubMed]

19.   Matreyek, K.A.; Engelman, A. The requirement for nucleoporin NUP153 during human immunodeficiency virus type 1 infection is determined by the viral capsid. *J. Virol.* **2011**, *85*, 7818–7827. [CrossRef] [PubMed]

20.   Monette, A.; Pante, N.; Mouland, A.J. Examining the requirements for nucleoporins by HIV-1. *Future Microbiol.* **2011**, *6*, 1247–1250. [CrossRef] [PubMed]

21.   Christ, F.; Thys, W.; de Rijck, J.; Gijsbers, R.; Albanese, A.; Arosio, D.; Emiliani, S.; Rain, J.C.; Benarous, R.; Cereseto, A.; et al. Transportin-SR2 imports HIV into the nucleus. *Curr. Biol.* **2008**, *18*, 1192–1202. [CrossRef] [PubMed]

22.   Zaitseva, L.; Cherepanov, P.; Leyens, L.; Wilson, S.J.; Rasaiyaah, J.; Fassati, A. HIV-1 exploits importin 7 to maximize nuclear import of its DNA genome. *Retrovirology* **2009**, *6*, 11. [CrossRef] [PubMed]

23. Ao, Z.; Huang, G.; Yao, H.; Xu, Z.; Labine, M.; Cochrane, A.W.; Yao, X. Interaction of human immunodeficiency virus type 1 integrase with cellular nuclear import receptor importin 7 and its impact on viral replication. *J. Biol. Chem.* **2007**, *282*, 13456–13467. [CrossRef] [PubMed]

24. Mousnier, A.; Leh, H.; Mouscadet, J.F.; Dargemont, C. Nuclear import of HIV-1 integrase is inhibited in vitro by styrylquinoline derivatives. *Mol. Pharmacol.* **2004**, *66*, 783–788. [CrossRef] [PubMed]

25. Wagstaff, K.M.; Sivakumaran, H.; Heaton, S.M.; Harrich, D.; Jans, D.A. Ivermectin is a specific inhibitor of importin alpha/beta-mediated nuclear import able to inhibit replication of HIV-1 and dengue virus. *Biochem. J.* **2012**, *443*, 851–856. [CrossRef] [PubMed]

26. Wagstaff, K.M.; Rawlinson, S.M.; Hearps, A.C.; Jans, D.A. An AlphaScreen(R)-based assay for high-throughput screening for specific inhibitors of nuclear import. *J. Biomol. Screen* **2011**, *16*, 192–200. [CrossRef] [PubMed]

27. Marsden, M.D.; Avancena, P.; Kitchen, C.M.; Hubbard, T.; Zack, J.A. Single mutations in HIV integrase confer high-level resistance to raltegravir in primary human macrophages. *Antimicrob. Agents Chemother.* **2011**, *55*, 3696–3702. [CrossRef] [PubMed]

28. Mochizuki, N.; Otsuka, N.; Matsuo, K.; Shiino, T.; Kojima, A.; Kurata, T.; Sakai, K.; Yamamoto, N.; Isomura, S.; Dhole, T.N.; et al. An infectious DNA clone of HIV type 1 subtype C. *AIDS Res. Hum. Retrovir.* **1999**, *15*, 1321–1324. [CrossRef] [PubMed]

29. Salazar-Gonzalez, J.F.; Salazar, M.G.; Keele, B.F.; Learn, G.H.; Giorgi, E.E.; Li, H.; Decker, J.M.; Wang, S.; Baalwa, J.; Kraus, M.H.; et al. Genetic identity, biological phenotype, and evolutionary pathways of transmitted/founder viruses in acute and early HIV-1 infection. *J. Exp. Med.* **2009**, *206*, 1273–1289. [CrossRef] [PubMed]

30. Yang, Z.; Ding, J.; Ding, K.; Chen, D.; Cen, S.; Ge, M. Phomonaphthalenone A: A novel dihydronaphthalenone with anti-HIV activity from Phomopsis sp. HCCB04730. *Phytochem. Lett.* **2013**, *6*, 257–260. [CrossRef]

31. Desimmie, B.A.; Schrijvers, R.; Demeulemeester, J.; Borrenberghs, D.; Weydert, C.; Thys, W.; Vets, S.; van Remoortel, B.; Hofkens, J.; de Rijck, J.; et al. LEDGINs inhibit late stage HIV-1 replication by modulating integrase multimerization in the virions. *Retrovirology* **2013**, *10*, 57. [CrossRef] [PubMed]

8

A Conserved Residue, Tyrosine (Y) 84, in H5N1 Influenza A Virus NS1 Regulates IFN Signaling Responses to Enhance Viral Infection

Ben X. Wang [1,2], Lianhu Wei [1,3,4], Lakshmi P. Kotra [1,3,4], Earl G. Brown [5] and Eleanor N. Fish [1,2,*]

[1] Toronto General Hospital Research Institute, University Health Network, 67 College Street, Toronto, ON M5G 2M1, Canada; ben.wang@mail.utoronto.ca (B.X.W.); william.wei@utoronto.ca (L.W.); lkotra@uhnresearch.ca (L.P.K.)

[2] Department of Immunology, University of Toronto, 1 King's College Circle, Toronto, ON M5S 1A8, Canada

[3] Center for Molecular Design and Preformulations, University Health Network, 101 College Street, Toronto, ON M5G 1L7, Canada

[4] Department of Pharmaceutical Sciences, Leslie Dan Faculty of Pharmacy, University of Toronto, 144 College Street, Toronto, ON M5S 3M2, Canada

[5] Department of Biochemistry, Microbiology and Immunology, Faculty of Medicine, University of Ottawa, 451 Smyth Road, Ottawa, ON K1H 8M5, Canada; ebrown@uottawa.ca

* Correspondence: en.fish@utoronto.ca

Academic Editor: Andrew Mehle

Abstract: The non-structural protein, NS1, is a virulence factor encoded by influenza A viruses (IAVs). In this report, we provide evidence that the conserved residue, tyrosine (Y) 84, in a conserved putative SH2-binding domain in A/Duck/Hubei/2004/L-1 [H5N1] NS1 is critical for limiting an interferon (IFN) response to infection. A phenylalanine (F) substitution of this Y84 residue abolishes NS1-mediated downregulation of IFN-inducible STAT phosphorylation, and surface IFNAR1 expression. Recombinant IAV (rIAV) [H1N1] expressing A/Grey Heron/Hong Kong/837/2004 [H5N1] NS1-Y84F (rWSN-GH-NS1-Y84F) replicates to lower titers in human lung epithelial cells and is more susceptible to the antiviral effects of IFN-β treatment compared with rIAV expressing the intact H5N1 NS1 (rWSN-GH-NS1-wt). Cells infected with rWSN-GH-NS1-Y84F express higher levels of IFN stimulated genes (ISGs) associated with an antiviral response compared with cells infected with rWSN-GH-NS1-wt. In mice, intranasal infection with rWSN-GH-NS1-Y84F resulted in a delay in onset of weight loss, reduced lung pathology, lower lung viral titers and higher ISG expression, compared with mice infected with rWSN-GH-NS1-wt. IFN-β treatment of mice infected with rWSN-GH-NS1-Y84F reduced lung viral titers and increased lung ISG expression, but did not alter viral titers and ISG expression in mice infected with rWSN-GH-NS1-wt. Viewed altogether, these data suggest that the virulence associated with this conserved Y84 residue in NS1 is, in part, due to its role in regulating the host IFN response.

Keywords: influenza A viruses; non-structural protein 1; interferon-β; interferon signaling; interferon-stimulated genes

1. Introduction

H5N1 avian influenza A viruses (IAVs) that infect poultry and migratory birds pose a significant threat to global health and, since 2003, there have been 858 confirmed cases of H5N1 IAV infection in humans with a mortality rate of 53% [1]. While annual vaccines are effective in preventing seasonal IAV infections, they have limited use in the event of an outbreak of a newly emergent strain. Currently, the neuraminidase inhibitors oseltamivir (Tamiflu) and zanamivir (Relenza) are antivirals available to treat IAV infections. However, drug-resistant strains of IAVs, including pandemic H1N1 and H5N1, have been isolated [2–4].

Given the direct antiviral and immunomodulatory effects of interferons (IFNs)-α/β [5] and the importance of the innate immune response for limiting viral infection and spread [6], IFNs present as candidate broad-spectrum antivirals with the potential to act as a first-line treatment for existing and newly emergent IAV infections [7,8]. In a previous report, we provided evidence for the antiviral effects of IFN-α in limiting H5N1 and pandemic H1N1 2009 IAV replication in primary human lung cells [9]. Moreover, mice lacking a functional type I IFN receptor, IFNAR, exhibit significantly more weight loss and a more rapid time to death when infected with various IAVs including H5N1 and H1N1 subtypes, compared with mice with an intact IFN system [10,11]. Not surprisingly, IAVs have evolved mechanisms to evade and disrupt host IFN production, IFN signaling and IFN-inducible antiviral effector functions [9,12].

The non-structural protein 1 (NS1) is a virulence factor encoded by IAVs and is expressed in the nucleus and cytoplasm of host cells during the earliest stages of infection [13,14]. Functional as a dimer, NS1 is comprised of an N-terminal dsRNA-binding domain and a C-terminal protein-binding effector domain [15–17]. In the context of limiting an IFN response to infection, NS1 inhibits IFN-β production by preventing the activation of retinoic acid-inducible gene 1 (RIG-I) products [18,19]. In addition, NS1 can prevent the maturation of host mRNAs, including IFN-α/β mRNAs, by binding to and inhibiting cleavage and polyadenylation specific factor 4, 30 kDa subunit (CPSF4), and poly(A)-binding protein II (PABPII) [20,21]. Consequently, IAVs lacking NS1, or expressing truncated forms of NS1, induce higher levels of IFN-α/β mRNA expression and IFN production, and have been proposed as live-attenuated vaccines [22–24]. In addition to inhibiting the production of IFNs-α/β, in an earlier publication we provided indirect evidence that IAVs may also limit IFN signaling, mediated by NS1 disrupting IFN-inducible phosphorylation of signal transducer and activator of transcription (STAT) 1 and STAT2 [9].

The IAV NS1 N-terminal effector domain contains a Src homology (SH)3 and a putative SH2-binding motif, that are important for direct binding with p85β, the inhibitory subunit of phosphatidylinositol-3-kinase (PI3K) [25–27]. Binding of NS1 to the internal SH2 (i-SH2) domain of p85β leads to the activation of the PI3K-protein kinase B (AKT) pathway, to enhance viral replication. A tyrosine (Y) to phenylalanine (F) substitution at the strictly conserved residue 89 (Y89F) in the H1N1 NS1 putative SH2-binding domain prevented binding of NS1 to p85β, thus abrogating NS1-mediated AKT phosphorylation [25,26]. Additionally, this Y89F in the NS1 of IAV PR8 reduced virulence in infected mice [28].

SH2 domains are well-conserved motifs found in many intracellular signaling proteins, such as those responsible for initiating IFN-α/β signaling pathways [29,30] and may present as targets for NS1–host protein interactions that affect the host innate immune response to IAV infection. In this study, we used site-directed mutagenesis to alter the conserved Y84 residue within H5N1 NS1 in order to characterize its role in limiting the host IFN signaling response. In the context of IAV infection, we used reverse genetics to generate recombinant IAVs (rIAVs) [H1N1] encoding either a wildtype or mutant H5N1 NS1 and confirmed the importance of this putative SH2-binding domain for virus replication, providing evidence for its contribution to evasion of the host IFN response.

2. Materials and Methods

2.1. Cells and Reagents

Human cervical carcinoma HeLa cells, lung adenocarcinoma epithelial A549 cells, embryonic kidney HEK293T cells, Madin-Darby canine kidney (MDCK) cells, and mouse embryonic fibroblasts (MEFs) were purchased from ATCC (Manassas, VA, USA). STAT1$^{+/+}$ and STAT1$^{-/-}$ MEFs were provided by Dr. Leonidas C. Platanias (Robert H. Lurie Comprehensive Cancer Center, Chicago, IL, USA). All cells were maintained in Dulbecco's modified Eagle's medium (DMEM) supplemented with 10% fetal calf serum (FCS), 100 U/mL penicillin, and 100 µg/mL streptomycin (Invitrogen, Waltham, MA, USA) at 37 °C and 5% CO_2.

Human IFN-β-1a (Avonex, specific activity 1.2×10^7 U/mL), murine IFN-β1 (specific activity 3.6×10^7 U/mL), and an anti-human IFNAR1 antibody (unconjugated, clone AA3) were provided by Darren P. Baker (BiogenIdec, Cambridge, MA, USA). An anti-human IFNAR2 antibody (unconjugated, clone MMHAR-2) was purchased from PBL Assay Science (Piscataway, NJ, USA). An anti-mouse IgG (Alexa Fluor 647, H+L) was purchased from Invitrogen as a secondary antibody. Antibodies specific for human phospho (p)-AKT (Ser473), AKT, p-STAT1 (Tyr701), STAT1, p-STAT2 (Tyr690), STAT2, and HA-Tag (6E2) were purchased from Cell Signaling Technology (Danvers, MA, USA). An antibody specific for human α-tubulin was purchased from Sigma-Aldrich (St. Louis, MO, USA) and horseradish peroxidase (HRP)-conjugated anti-rabbit IgG and anti-mouse IgG secondary antibodies were purchased from GE Healthcare Life Sciences (Marlborough, MA, USA). Antibodies specific for mouse CD11b (BV421, clone M1/70) and CD45 (BV605, clone 30-F11) were purchased from BioLegend (San Diego, CA, USA). Antibody specific for mouse Ly6G was purchased from eBioscience (San Diego, CA, USA). Respective isotype control antibodies were purchased from BioLegend and eBioscience.

2.2. Mice

Male C57BL/6 mice, aged 6–8 weeks, were purchased from Taconic (Hudson, NY, USA) and housed in a pathogen-free environment. All experiments were approved by the Animal Care Committee of the Toronto General Hospital Research Institute.

2.3. In Silico Modeling

The crystallized structure of A/Puerto Rico/8/1934 [H1N1] NS1 and p85β i-SH2 domain complex (RCSB Protein Data Bank: 3L4Q) [31] was used to construct a model of avian A/Vietnam/1203/2004 [H5N1] NS1 (RCSB Protein Data Bank: 3F5T) [15] and p85β i-SH2 domain complex using SYBYL-X (Certara, Princeton, NJ, USA). The NS1 subunit from 3L4Q was removed and replaced with the NS1 subunit from 3F5T. Molecular interactions between 3F5T and the p85β i-SH2 domain subunit of 3L4Q were visualized.

2.4. Plasmids and Site-Directed Mutagenesis

Plasmid pBudCE4.1 (Invitrogen) co-expressing A/Duck/Hubei/L-1/2004 [H5N1] NS1 complementary DNA (cDNA; HA-tagged) and green fluorescent protein (GFP) was generated as previously described [9]. Plasmid encoding A/Grey Heron/Hong Kong/837/2004 [H5N1] NS gene was provided by Dr. Leo L.M. Poon (University of Hong Kong, Hong Kong). Plasmids (pLLB) [32] encoding the eight A/WSN/33 [H1N1] gene segments (HA, NA, NP, NS, PA, PB1, PB2, M) were provided by Dr. Earl G. Brown (University of Ottawa, Ottawa, ON, Canada). The A/Grey Heron/Hong Kong/837/2004 [H5N1] NS gene was cloned into the pLLB plasmid using homologous recombination as described previously [32]. Site-directed mutagenesis was performed to introduce a Y84F mutation in pBudCE4.1-NS1-HA-GFP and pLLB-A/Grey Heron/Hong Kong/837/2004 [H5N1]-NS using the QuikChange Site-Directed Mutagenesis Kit and XL1-Blue supercompetent cells purchased from Agilent Technologies (Santa Clara, CA, USA) following the manufacturer's protocol. Complimentary oligonucleotide primers (forward 5'GCCGGCTTCACGCTTCCTAACTGACATGAC3', reverse 5'GTCATGTCAGTTAGGAAGCG TGAAGCCGGC3') containing the desired Y84F mutation were synthesized by ACGT Corporation (Toronto, ON, Canada). The resulting pBudCE4.1-NS1-Y84F-HA-GFP plasmid and pLLB-A/Grey Heron/Hong Kong/837/2004 [H5N1] NS-Y84F gene were sequenced by ACGT Corporation to confirm the Y84F mutation.

2.5. Transfections

HeLa cells were seeded in 6-well plates at 2×10^5 cells/well in 2 mL 10% FCS DMEM and incubated at 37 °C in 5% CO_2 for 24 hours (h). Cells were transfected with 1.25 µg/well of pBudCE4.1-GFP (vector), pBudCE4.1-NS1-HA-GFP (NS1-wt), or pBudCE4.1-NS1-Y84F-HA-GFP (NS1-Y84F) using Lipofectamine™ LTX Reagent (Invitrogen) following the manufacturer's protocol and as previously described [9].

2.6. Western Immunoblots

Transfected HeLa cells were either left untreated or treated with 1×10^3 U/mL IFN-β-1a for 15 minutes (min) at 37 °C. Cells were lysed on ice using lysis buffer containing 1% Triton X-100, 0.5% NP-40, 150 mM NaCl, 10 mM Tris [pH 7.4], 1 mM EDTA, 1 mM EGTA, 0.2 mM Na_3VO_4, 0.2 mM PMSF, 10 µg/mL Aprotinin, 2 µg/mL Pepstatin A, and 1 mM $Na_4P_2O_7$. An amount of 25 µg of each sample lysate was used for Western immunoblots. Sample lysates were denatured in $5\times$ sample reducing buffer and resolved by SDS-PAGE. Proteins were transferred onto a nitrocellulose membrane and blocked with 5% BSA TBS-0.1% Tween-20 (TBS-T) for 1 h at room temperature. Membranes were probed with primary antibodies at a 1:1000 dilution in TBS-T overnight at 4 °C and secondary antibodies at a 1:10,000 dilution in TBS-T for 1 h at room temperature. Immunoblots were developed and proteins were visualized using SuperSignal West Pico Chemiluminescent Substrate Kit (Thermo Scientific, Waltham, MA, USA) following the manufacturer's protocol. Band intensities were quantitated by densitometry using ImageJ software (National Institutes of Health, Bethesda, MD, USA).

2.7. Reverse Genetics

The 5×10^5 HEK293T cells were transfected using Lipofectamine 2000 (Invitrogen) following the manufacturer's protocol. Twenty-four hours before transfection, HEK293T cells were seeded in 6-well plates coated with poly-D-lysine (Sigma-Aldrich). An amount of 1 µg of each pLLB plasmid encoding one of the A/WSN/33 [H1N1] gene segments (*HA, NA, NP, NS, PA, PB1, PB2, M*) was transfected into the HEK293T cells to generate wildtype rA/WSN/33 virus as previously described [33]. pLLB-A/Grey Heron/Hong Kong/837/2004 [H5N1] *NS* and pLLB-A/Grey Heron/Hong Kong/837/2004 [H5N1] *NS*-Y84F were used in place of pLLB-A/WSN/33 [H1N1] *NS* to generate rWSN-GH-NS1-wt and rWSN-GH-NS1-Y84F respectively. Sixteen hours post-transfection, medium was replaced with 0% FCS DMEM containing 1 µg/mL tosyl phenylalanyl chloromethyl ketone (TPCK)-treated trypsin (Sigma-Aldrich). Forty-eight hours post-transfection, medium containing viral progeny was overlayed onto a monolayer of MDCK cells for 72 h. Viral yield was determined by plaque assay.

2.8. Virus Infection

2.8.1. In Vitro

The 2×10^5 A549 cells, STAT1$^{+/+}$ and STAT1$^{-/-}$ MEFs were seeded in 24-well plates for 24 h and then washed twice with phosphate buffer solution (PBS) and infected in triplicate with each of the rA/WSN/33 viruses at a multiplicity of infection (MOI) of 0.01 in the presence of 0.5 µg/mL (MEFs) or 1 µg/mL (A549) TPCK-treated trypsin. Medium was collected at the indicated times post-infection and viral titers were determined by plaque assay in MDCK cells.

2.8.2. In Vivo

C57BL/6 mice 8–10 weeks of age were anesthetized by intraperitoneal injection with ketamine (Ketalean, Bimeda, Cambridge, ON, Canada) and xylazine (Rompun, Bayer, Mississauga, ON, Canada), and infected intranasally with 1×10^5 plaque-forming units (PFU) of rA/WSN/33-GH-NS1-wt, or rA/WSN/33-GH-NS1-Y84F diluted in 50 µL of PBS. Infected mice were monitored daily for weight-loss and sacrificed by cervical dislocation on days 1 and 3 post-infection. Lungs from infected mice ($n = 5$) were harvested, weighed, and stored at −80 °C. The lungs were then thawed and mechanically homogenized on ice in 500 µL of serum-free DMEM containing 1 µg/mL TPCK-treated trypsin. The homogenized lung tissues were centrifuged at $12,000 \times g$ and the supernatants were used to determine lung viral titers by plaque assay.

Lungs were also harvested for flow cytometry analysis of neutrophil infiltration ($n = 5$). Lungs were perfused by slowly injecting 10 mL of PBS into the right ventricle of the heart. The lungs were mashed and incubated at 37 °C for 30 min in the presence of 1 mM $CaCl_2$, 1.8 mM $MgCl_2$, 1 mg/mL collagenase D (Roche, Penzberg, Germany) and 1 mg/mL DNase I (Thermo Scientific). Isolated cells were passed through a 70 μm cell strainer to obtain a single-cell suspension and red blood cells were lysed using ammonium-chloride-potassium (ACK) lysing buffer (150 mM NH_4Cl, 10 mM $KHCO_3$, and 0.1 mM Na_2EDTA) for 5 min on ice. Cells were counted using a hemocytometer. Additional lungs were harvested on days 1 (wt, $n = 5$; Y84F, $n = 4$) and 3 (wt, $n = 5$; Y84F, $n = 3$) post-infection for histology. Lungs were placed into embedding cassettes and fixed using 4% formalin-PBS (Sigma-Aldrich) and stored at 4 °C.

2.8.3. IFN-β Treatment In Vivo

Infected C57BL/6 mice were treated with $1\times$ PBS or 1×10^5 U of murine IFN-β1 diluted in $1\times$ PBS by intraperitoneal injection at 8 h post-infection.

2.9. Plaque Assay

The 5×10^5 MDCK cells were seeded in 6-well plates for 24 h until they formed an 80% confluent monolayer. Samples containing rIAVs were serially diluted in serum-free DMEM containing 1 μg/mL TPCK-trypsin. MDCK cells were washed twice with PBS and infected with 800 μL of the serially diluted rIAVs. Infected MDCK cells were incubated at 37 °C for 1 h to allow virus adsorption. An amount of 2 mL of 0.65% agarose diluted in serum-free DMEM in the presence of 1 μg/mL TPCK-trypsin was then overlaid onto the infected MDCK cells. MDCK cells were incubated at 37 °C for 72 h, then fixed using a 3:1 methanol:acetic acid solution. Plaques were enumerated to determine the viral titer, recorded as the number of PFU/mL of medium or PFU/g of lung tissue.

2.10. RNA Extraction and cDNA Synthesis

RNA was extracted and purified from infected A549 cells and the homogenized lung tissues of infected mice using the RNeasy Mini Kit (Qiagen, Venlo, The Netherlands), according to the manufacturer's protocol. cDNAs were synthesized using 0.5 μg/sample of RNA, random primers, and M-MLV reverse transcriptase (Invitrogen), following the manufacturer's protocol. cDNAs were also synthesized from 1 μg of RNA purified from uninfected A549 cells and MEFs treated for 16 h with 1×10^3 U/mL of human IFN-β-1a and 1×10^3 U/mL of murine IFN-β1, respectively.

2.11. qPCR

Quantitative polymerase chain reaction (qPCR) was performed using the LightCycler FastStart DNA Master SYBR Green PLUS I kit (Roche) and a LightCycler (Roche), following the manufacturer's protocol as described previously [34]. Primers for target IFN stimulated genes (ISGs; Table 1) were synthesized by ACGT Corporation. Standard curves for each gene were generated using cDNAs from uninfected A549 cells and MEFs treated with 1×10^3 U/mL of human IFN-β-1a and 1×10^3 U/mL of murine IFN-β1, respectively. qPCR data were analyzed using LightCycler Data Analysis Software (Roche).

Table 1. List of human (*h*), mouse (*m*) and influenza A virus (*IAV*) qPCR primers used in this study.

Gene	Forward Primer (5′-3′)	Reverse Primer (5′-3′)
(*h*) HPRT1	TCCTCCTCTGCTCCGCCACC	TCACTAATCACGACGCCAGGGCT
(*h*) MxA	GATGATCAAAGGGATGTGGC	AGCTCGGCAACAGACTCTTC
(*h*) EIF2AK2	ACTTGGCCAAATCCACCTG	CCCAGATTTGACCTTCCTGA
(*m*) HPRT1	CATAACCTGGTTCATCATCGC	TCCTCCTCAGACCGCTTTT
(*m*) ISG15	CCCCAGCATCTTCACCTTTA	TGACTGTGAGAGCAAGCAGC
(*m*) OAS1	AGTTCTCCTCCACCTGCTCA	GGCTGTGGTACCCATGTTTT
(*m*) EIF2AK2	CTGTTGCAAGGCCAAAGTCT	GAACAAATCGTGACCGGAGT
(*m*) IFNA4	TATGTCCTCACAGCCAGCAG	TTCTGCAATGACCTCCATCA
(*m*) IFNB1	CCCAGTGCTGGAGAAATTGT	CCCTATGGAGATGACGGAGA
(*m*) CXCL1	TCTCCGTTCCTTGGGGACAC	CCACACTCAAGAATGGTCGC
(*m*) CXCL2	TCCAGGTCAGTTAGCCTTGC	CGGTCAAAAAGTTTGCCTTG
(*IAV*) M	AGATGAGTCTTCTAACCGAGGTCG	TGCAAAAACATCTTCAAGTCTCTG

qPCR: quantitative polymerase chain reaction; CXCL: C-X-C motif chemokine ligand; EIF2AK2: eukaryotic translation initiation factor 2 alpha kinase 2; HPRT: hypoxanthine-guanine phosphoribosyltransferase; IFN: interferon; ISG: IFN stimulated gene; M: matrix; MxA: myxovirus resistance 1; OAS: 2′-5′-oligoadenylate synthetase.

2.12. IFN-β ELISA

IFN-β production in the lungs of rIAV-infected C57BL/6 mice and by rIAV-infected A549 cells was quantified using the Legend Max ELISA kit (BioLegend) and the Verikine IFN-β enzyme-linked immunosorbent assay (ELISA) kit (PBL Assay Science), respectively, following the manufacturers' protocols. Culture medium and homogenized lung supernatants—diluted with 500 μL DMEM—containing viral progeny, were stored at −80 °C prior to use.

2.13. FACS Analysis of IFNAR1 and IFNAR2 Expression

Twenty-four hours post-transfection, HeLa cells were harvested using Versene (Gibco, Waltham, MA, USA). Cells were washed with fluorescence-activated cell sorting (FACS) buffer (2% FCS in PBS) and resuspended in 200 μL of FACS buffer containing anti-human IFNAR1 or anti-human IFNAR2 at a 1:100 dilution for 45 min on ice. Cells were then washed three times and resuspended in 200 μL of FACS buffer containing anti-mouse IgG (Alexa Fluor 647) at a 1:100 dilution for 30 min on ice. Untransfected and transfected HeLa cells incubated with anti-mouse IgG (Alexa Fluor 647) alone were used as controls. Flow cytometry was performed using a FACSCalibur (BD Biosciences, San Jose, CA, USA) and data were analyzed using FlowJo software (FlowJo, Ashland, OR, USA). Cells were gated based on GFP expression.

2.14. Histology and Identification of Lung Neutrophils

Harvested lungs were embedded in paraffin and 5 μm thin sections containing multiple lobes were mounted onto slides and stained with hematoxylin and eosin (H&E). Sections were scanned using an Aperio ScanScope XT slide scanner (Leica Biosystems, Wetzlar, Germany) at 20× magnification and images were analyzed using Aperio ImageScope software (Leica Biosystems).

Single cell suspensions were prepared from lung aspirates and cells were blocked with mouse serum (Sigma-Aldrich) for 15 min on ice prior to staining. The 5×10^5 cells/sample were stained with antibodies specific for mouse CD45, CD11b and Ly6G, or the appropriate isotype control antibodies for 45 min on ice. Compensations were conducted using anti-rat/hamster Ig, κ beads (BD Biosciences) and isotype control antibodies. Flow cytometry was performed using a LSR II (BD Biosciences) and data were analyzed using FlowJo software (FlowJo).

2.15. Statistical Analyses

An unpaired Student's *t*-test was used to analyze differences among groups. A paired Student's *t*-test was used to analyze differences among groups where *n* represents the same treatment from three independent experiments. p-values < 0.05 were considered statistically significant (* $p < 0.05$, ** $p < 0.01$, and *** $p < 0.001$).

3. Results

3.1. A Y84F Mutation in the H5N1 NS1 Conserved Putative SH2-Binding Domain Affects the Ability for NS1 to Upregulate AKT Phosphorylation

In an earlier publication, we provided evidence that cells expressing the A/Duck/Hubei/2004/L-1 [H5N1] NS1 are less responsive to the antiviral effects of IFN, exhibiting reduced IFN-inducible STAT1, STAT2 and STAT3 phosphorylation, thereby affecting the downstream events associated with STAT activation [9]. We have extended our studies to investigate the mechanism(s) whereby NS1 invokes these effects. Phosphorylation-independent binding of H1N1 NS1 to the p85β subunit of PI3K results in the phosphorylation of AKT, mediated by the catalytic activity of the p110 subunit, thereby enhancing viral replication [25,26]. This NS1-p85β binding has been ascribed to an SH2-binding domain in NS1, since a tyrosine to phenylalanine mutation at residue 89 (Y89F), within this domain, abrogated NS1 binding to host cell p85β and reduced IAV replication [25,26,28]. Notably, a number of IFN-inducible signaling effectors have SH2 domains, including STATs, from which we infer that a similar mechanism of NS1 binding to host transcription factors or signaling effectors may reduce IFN-inducible responses.

The A/Duck/Hubei/L-1/2004 (H5N1) NS1 is evolutionarily distinct from both A/Puerto Rico/8/34 [H1N1] and A/WSN/33 [H1N1] NS1 proteins, and contains a five amino acid deletion at residues 80–84. Due to these differences in the NS1 amino acid sequences, we generated an in silico model of the H5N1 NS1-p85β i-SH2 interaction, using published crystallized structures of an H1N1 NS1 and p85β i-SH2 complex (PDB: 3L4Q, green) [31] and an H5N1 NS1 (A/Vietnam/1203/2004) containing the same five amino acid deletion (PDB: 3F5T, red) [15] (Figure 1A). This in silico model shows that residue Y84 in the H5N1 NS1 putative SH2-binding domain may interact via hydrogen bonding with residue D569 in the p85β i-SH2 domain and that a Y84F substitution eliminates this interaction (Figure 1B).

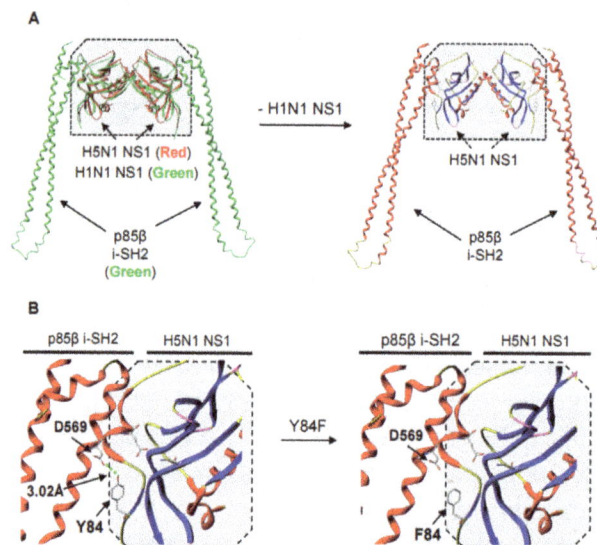

Figure 1. In silico modeling of the Y84F (tyrosine to phenylalanine) mutation in the putative Src homology 2 (SH2)-binding domain of A/Duck/Hubei/L-1/2004 [H5N1] non-structural protein 1 (NS1). (**A**) Ribbon diagrams of the A/Vietnam/1203/2004 [H5N1] H5N1 NS1 (PDB: 3F5T) and p85β complex, based on a crystallized structure of the H1N1 NS1 and p85β internal SH2 (i-SH2) domain complex (PDB: 3L4Q). (**B**) Ribbon diagrams showing the effect of the Y84F mutation on NS1-p85β binding.

Accordingly, we used site-directed mutagenesis to introduce the Y84F mutation within the conserved H5N1 NS1 putative SH2-binding domain, to examine its contribution to NS1-mediated down-regulation of IFN-inducible STAT phosphorylation and IAV virulence. In a first series of experiments, we examined the effects of expression of the wildtype NS1 (NS1-wt) or Y84F mutation (NS1-Y84F) in HeLa cells on AKT phosphorylation, a signaling effector downstream of PI3K. The objective was to demonstrate that, in contrast to NS1-wt, which is known to induce AKT phosphorylation, NS1-Y84F would fail to increase AKT phosphorylation. As anticipated, the results in Figure 2 reveal that 24 h post-transfection, cells expressing NS1-wt exhibit a 1.5-fold increase (significant $p < 0.01$) in AKT phosphorylation compared with cells expressing NS1-Y84F.

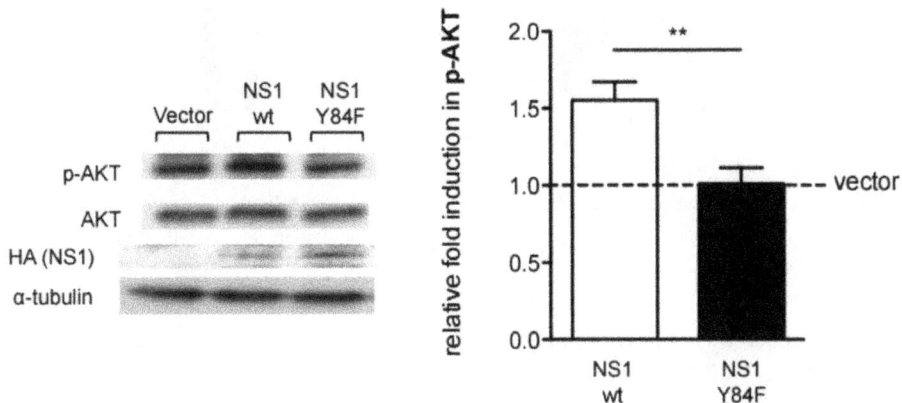

Figure 2. H5N1 NS1-Y84F is unable to upregulate protein kinase B (AKT) phosphorylation. HeLa cells were transfected with vector alone, vector carrying the NS1-wt complementary DNA (cDNA) (□), or NS1-Y84F cDNA (■). 24 hours (h) post-transfection, cells were lysed and lysates were resolved by SDS-PAGE and immunoblotted with an anti-phospho (p)-AKT (Ser473) antibody. The blot was then stripped and re-probed with an antibody against AKT. A separate aliquot of the same cell lysate was resolved by SDS-PAGE and immunoblotted with antibodies against HA (NS1) and α-tubulin. Band intensities were quantitated and the relative induction in p-AKT was determined, normalizing to AKT. Data are presented as the mean +/− standard error (SE) and are representative of three independent experiments. ** $p < 0.01$.

3.2. A Y84F Mutation Abrogates NS1-Mediated Inhibition of Type I IFN Signaling

Next, we performed a series of experiments to examine the effects of the Y84F mutation on the ability of NS1 to regulate the type I IFN signaling response. As mentioned, we have shown that H5N1 NS1 expression in HeLa cells inhibits IFN-inducible STAT1 and STAT2 phosphorylation [9]. Here, we show that the levels of IFN-inducible STAT1 and STAT2 phosphorylation are unaffected in cells expressing the mutant NS1-Y84F, compared with a reduction in cells expressing NS1-wt (Figure 3). Having demonstrated that expression of NS1-wt reduces cell surface IFNAR1 expression [9], we likewise examined whether the NS1-Y84F mutant would affect IFNAR1 cell surface expression. The data in Figure 4 reveal that in contrast to NS1-wt expression, which reduces IFNAR1 but not IFNAR2 expression, NS1-Y84F expression has no effect on IFNAR1 or IFNAR2 expression. The reduction in surface IFNAR1 expression in cells transfected with NS1-wt is similar in magnitude to the reduction observed in cells which have been treated with IFN-β (data not shown).

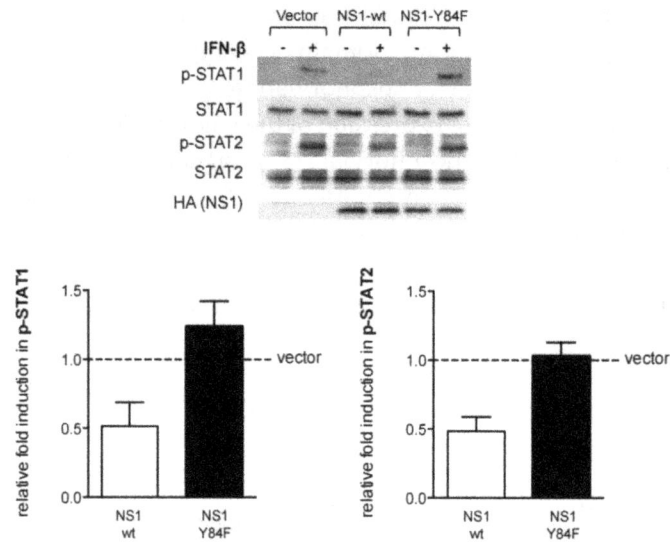

Figure 3. The Y84F mutation abrogates H5N1 NS1-mediated inhibition of interferon (IFN)-inducible signal transducer and activator of transcription (STAT) phosphorylation. HeLa cells were transfected with vector alone, or vector carrying the NS1-wt cDNA (□), or NS1-Y84F cDNA (■). 24 h post-transfection, cells were either left untreated or treated with 1000 U/mL of IFN-β for 15 minutes (min). Cells were lysed, lysates resolved by SDS-PAGE and immunoblotted with antibodies against p-STAT (Tyr701), p-STAT2 (Tyr690), or HA (NS1). Blots were then stripped and re-probed with antibodies against STAT1 or STAT2. Band intensities were quantitated and the relative induction in p-STAT1 and p-STAT2 was determined, normalizing to STAT1 and STAT2, respectively. Data are presented as the mean +/− SE and are representative of three independent experiments.

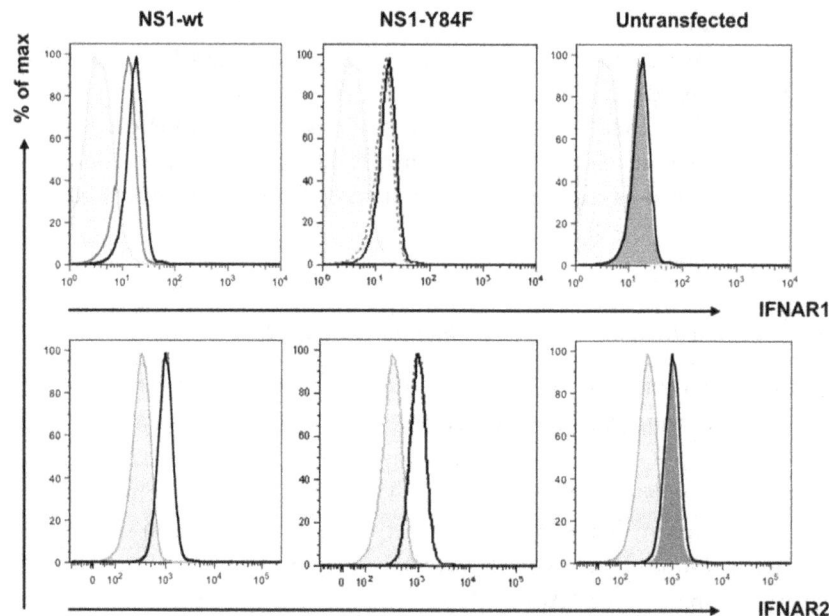

Figure 4. NS1-Y84F does not affect IFN-α/β receptor subunit (IFNAR) 1 expression. HeLa cells were transfected with vector alone (—, black), vector carrying the NS1-wt cDNA (—, grey), or NS1-Y84F cDNA (---). 24 h later, cells were stained with antibodies against IFNAR1 or IFNAR2. Transfected cells stained with the Alexa Fluor 647 secondary antibody alone served as the isotype control (—, light grey fill). Untransfected HeLa cells (no line, dark gray fill) were also stained. Cells were analyzed using a FACSCalibur, gating on green fluorescent protein (GFP)+ cells, then analyzing for IFNAR1 or IFNAR2 expression. Data are representative of two independent experiments.

3.3. Effects of the Conserved Putative SH2-Binding Domain in NS1 on Virus Replication

To further examine the importance of this Y84 residue within the putative SH2-binding domain in H5N1 NS1, we generated rIAVs expressing either the H5N1 NS1-wt or NS1-Y84F: rWSN-GH-NS1-wt and rWSN-GH-NS1-Y84F, respectively. Time course studies in A549 human lung epithelial cells revealed that rWSN-GH-NS1-wt grows to approximately 100-fold higher titers than rWSN-GH-NS1-Y84F (Figure 5A). For both rIAVs, viral titers increase up to 36 h post-infection, then decline at 48 h. These data support earlier published data [28] that the conserved putative SH2-binding domain of H1N1 NS1 is important for IAV replication in vitro. In addition, A549 cells infected with rWSN-GH-NS1-Y84F produced approximately 1.7-fold and 2.6-fold more IFN-β than cells infected with rWSN-GH-NS1-wt.

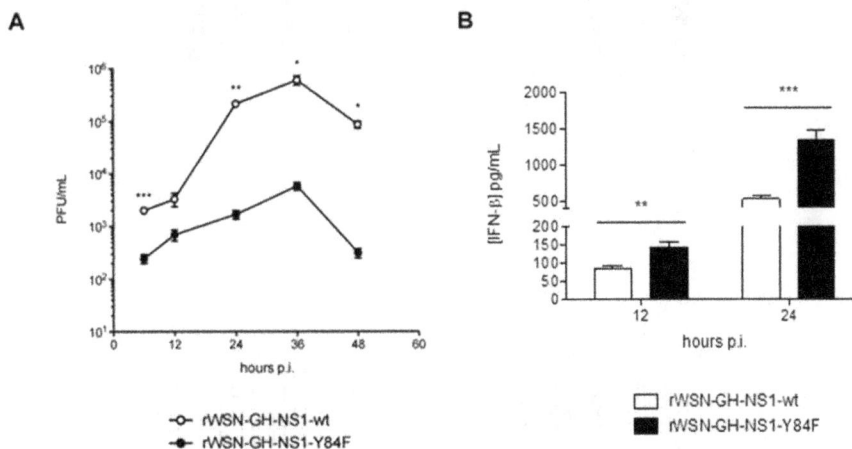

Figure 5. The Y84F mutation inhibits recombinant IAV (rIAV) [H1N1] replication and enhances IFN-β production in human A549 lung epithelial cells. (**A**) A549 cells were infected with rWSN-GH-NS1-wt (○) or rWSN-GH-NS1-Y84F (●) at a multiplicity of infection (MOI) of 0.01. Culture supernatants from rWSN-GH-NS1-wt (□) or rWSN-GH-NS1-Y84F (■) infected cells were collected at 6, 12, 24, 36, and 48 h post-infection and viral titers were determined by plaque assay in Madin-Darby canine kidney (MDCK) cells. (**B**) IFN-β levels were measured in the culture supernatants collected at 12 and 24 h post-infection by enzyme-linked immunosorbent assay (ELISA). Data are presented as the mean +/− SE and are representative of three (titration) and two (ELISA) independent experiments. * $p < 0.05$, ** $p < 0.01$, and *** $p < 0.001$.

Next, we conducted a series of experiments to investigate whether, as we had observed for HeLa cells expressing NS1-wt or NS1-Y84F, an intact putative SH2-binding domain influences the response to IFN-β treatment. Specifically, A549 cells were infected with rWSN-GH-NS1-wt or rWSN-GH-NS1-Y84F at a MOI of 0.01 for 12 h, and then treated with varying doses of IFN-β. This 12 h time point post-infection was selected to allow for NS1-wt or NS1-Y84F to be expressed in infected cells, yet early enough in the infection to preclude profound differences in viral titers. IFN-β treatment of uninfected A549 cells resulted in the expected increases in gene expression for the ISGs *EIF2AK2* (Figure 6A) and *MxA* (Figure 6B), both implicated in mediating antiviral activity. At 12 h post-IFN-β-treatment, *EIF2AK2* (Figure 6A) and *MxA* (Figure 6B) were induced in cells infected with rWSN-GH-NS1-Y84F, whereas cells infected with rWSN-GH-NS1-wt exhibited no ISG induction.

In subsequent experiments, we examined viral replication in IFN-β treated and rIAV infected A549 cells 36 h after IFN-β treatment, i.e., 48 h post-infection. This time point was chosen to best represent the outcome of ISG induction on IAV replication. Following IFN-β treatment, we observed a greater reduction in *M* gene expression in A549 cells infected with rWSN-GH-NS1-Y84F, compared to cells infected with rWSN-GH-NS1-wt (Figure 6C). Our measurements of virus in culture supernatants revealed that IFN-β treatment for 36 h reduced viral titers in a dose-dependent manner, albeit to a greater extent in the cells infected with rWSN-GH-NS1-Y84F: 1000 U/mL of IFN-β reduced the viral titer of rWSN-GH-NS1-wt by 1-log (17-fold), in comparison to 50 U/mL of IFN-β that reduced the titer of rWSN-GH-NS1-Y84F by 1-log (40-fold; Figure 6D). Notably, IFN-β doses of 100 and 1000 U/mL reduced viral titers in rWSN-GH-NS1-Y84F infected A549s to <10 PFU/mL.

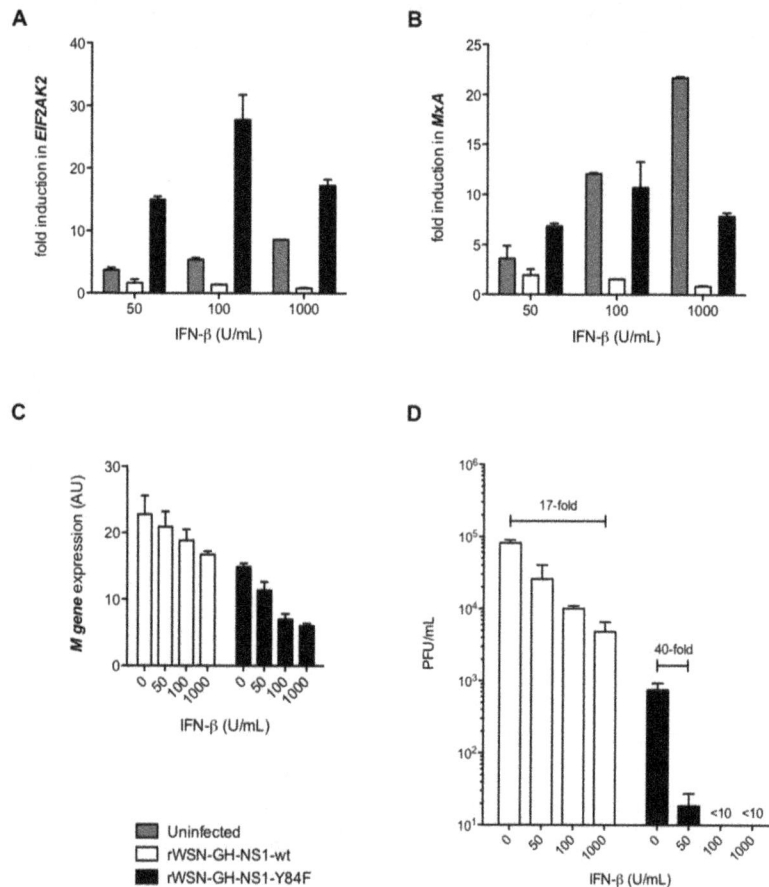

Figure 6. The Y84F mutation confers greater sensitivity to the antiviral effects of IFN-β. A549 cells were left uninfected (■) or infected with rWSN-GH-NS1-wt (□) or rWSN-GH-NS1-Y84F (■) at a MOI of 0.01. At 12 h post-infection, cells were either left untreated, or treated with 50, 100, or 1000 U/mL of IFN-β. RNA was purified from cells at 24 h post-infection (i.e., 12 h post-IFN-β treatment) and cDNA was synthesized. Quantitative polymerase chain reaction (qPCR) was performed to determine the relative expression of IFN-inducible (**A**) *EIF2AK2* and (**B**) *MxA*, normalized to *HPRT1*. Data are presented as the fold-induction of each IFN stimulated gene (ISG) relative to infected or uninfected A549 cells that were not treated with IFN-β. (**C**) IAV *M* gene expression was determined by qPCR at 48 h post-infection (36 h post-IFN-β treatment). (**D**) Viral titers in culture supernatants collected at 48 h post-infection (36 h post-IFN-β treatment) were determined by plaque assay in MDCK cells. Data are presented as the mean +/− SE and are representative of two independent experiments.

To determine whether pre-treatment of cells with IFN-β would override the inhibitory effects of NS1-wt, we treated A549s cells with increasing doses of IFN-β 16 h prior to infection with the rIAVs (i.e., to eliminate the inhibitory effects of NS1 by generating IFN-inducible antiviral responses that would precede infection). The results in Figure 7 show a greater than 1-log fold reduction in viral titers at 24 (Figure 7A) and 48 (Figure 7B) h post-infection, when cells are pre-treated with different doses of IFN-β, for both rIAVs, in comparison to untreated cells that are more susceptible to viral replication following infection with the rIAV expressing NS1-wt. Notably, pre-treatment of A549 cells with 10 U/mL of IFN-β prior to infection with rWSN-GH-NS1-wt was sufficient to reduce the viral titers at 24 and 48 h by 1-log fold.

A

B

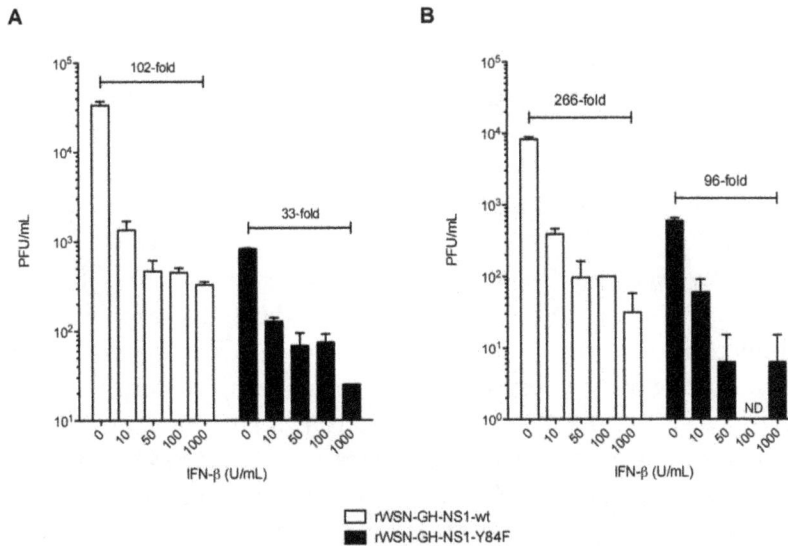

Figure 7. IFN-β pre-treatment inhibits rIAV replication. A549 cells were either left untreated or treated with 50, 100, or 1000 U/mL of IFN-β for 16 h and then infected with rWSN-GH-NS1-wt (□) or rWSN-GH-NS1-Y84F (■) at a MOI of 0.01. Culture supernatants were collected at (**A**) 24 and (**B**) 48 h post-infection (40 and 64 h post-IFN-β treatment, respectively) and viral titers were determined by plaque assay in MDCK cells. Data are presented as the mean +/− SE and are representative of two independent experiments. ND ~not detected.

In subsequent experiments, we infected STAT1$^{+/+}$ and STAT1$^{-/-}$ MEFs with the rIAVs to further determine whether the differences in viral replication that we observed between rWSN-GH-NS1-wt and rWSN-GH-NS1-Y84F were due primarily to effects of NS1 on the IFN-α/β response. STAT1$^{-/-}$ mice and STAT1$^{-/-}$ MEFs are unable to respond to IFN-α/β [35,36]. In STAT1$^{+/+}$ MEFs, viral titers increased by approximately 2.6-fold for rWSN-GH-NS1-wt between 6 and 36 h post-infection, whereas a marginal 1.5-fold increase in viral titers was observed for rWSN-GH-NS1-Y84F. In contrast, both rIAVs replicated in STAT1$^{-/-}$ MEFs with rWSN-GH-NS1-wt and rWSN-GH-NS1-Y84F viral titers increasing by approximately 7.6- and 10.9-fold, respectively (Figure 8).

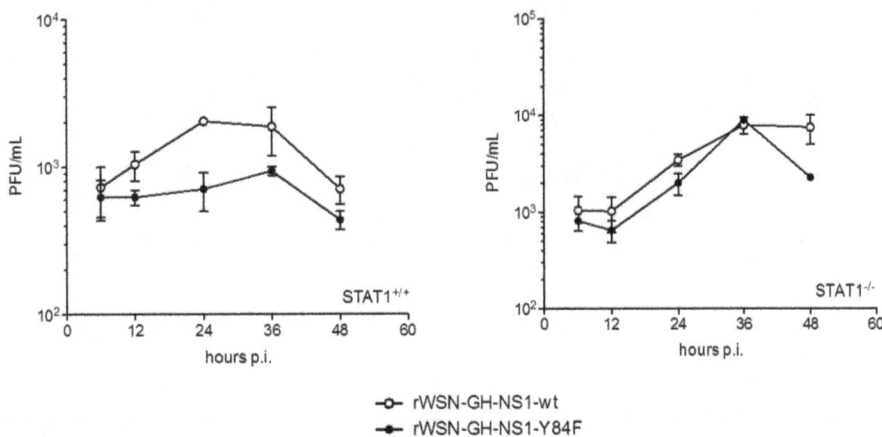

Figure 8. rWSN-GH-NS1-wt and rWSN-GH-NS1-Y84F replicate to higher titers in STAT1$^{-/-}$ mouse embryonic fibroblasts (MEFs) in comparison to STAT1$^{+/+}$ MEFs. STAT1$^{+/+}$ and STAT1$^{-/-}$ MEFs were infected with rWSN-GH-NS1-wt (○) or rWSN-GH-NS1-Y84F (●) at a MOI of 0.01. Culture supernatants were collected at 6, 12, 24, 36 and 48 h post-infection and viral titers were determined by plaque assay in MDCK cells. Data are presented as the mean +/− SE and are representative of two independent experiments.

3.4. The Conserved Putative SH2-Binding Domain in NS1 Contributes to IAV Virulence In Vivo, Affecting the IFN Response

In a final series of experiments, we examined the effects of the conserved putative SH2-binding domain in NS1 on IAV virulence in vivo, in the context of the IFN response. Studies were conducted in C57BL/6 mice to compare the infectivity of rWSN-GH-NS1-wt with rWSN-GH-NS1-Y84F, following intranasal inoculation. Initial readouts for infectivity were weight loss and lung viral loads on days 1 and 3 post-infection.

C57BL/6 mice received an intranasal inoculation of 1×10^5 PFU of either the rIAV expressing the NS1-wt or the rIAV expressing the NS1-Y84F. Weight loss and lung ISG expression were recorded on days 1 and 3 post-infection (Figure 9). Mice infected with virus expressing the mutant NS1-Y84F lost less than 5% of their starting body weight by day 3 post-infection, whereas mice infected with virus expressing the NS1-wt lost greater than 10% of their starting body weight over the same time period (Figure 9A). Furthermore, mice infected with virus expressing NS1-wt had higher lung viral titers than mice infected with virus expressing NS1-Y84F on days 1 and 3 post-infection (Figure 9B).

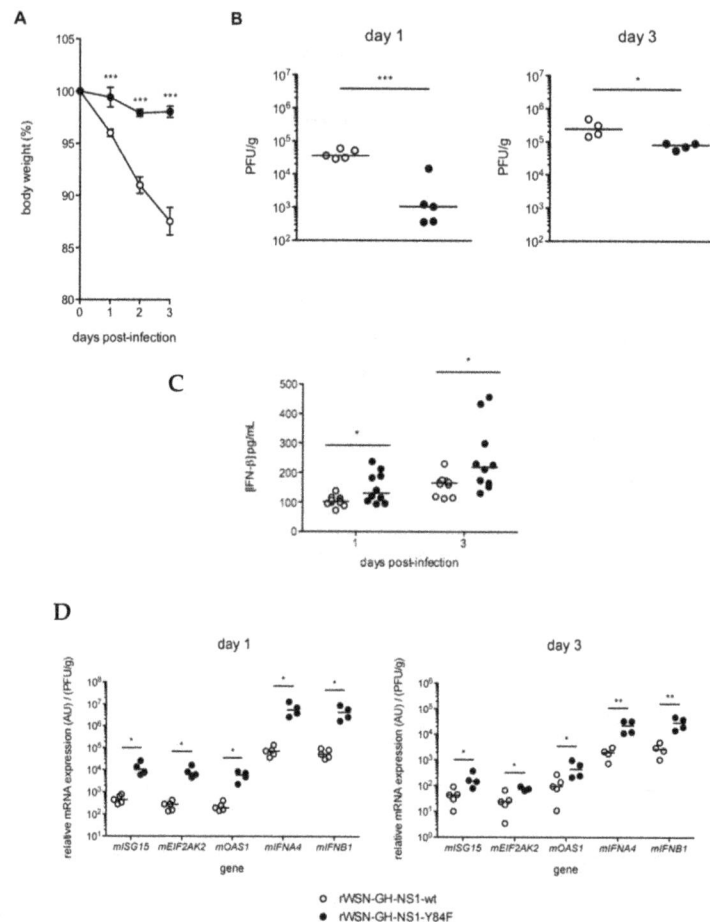

Figure 9. Mice infected with rIAV expressing the Y84F mutant NS1 experience delayed weight-loss and lower lung viral titers on day 1 and day 3 post-infection. C57BL/6 mice were infected with 1×10^5 plaque-forming units (PFU) of rWSN-GH-NS1-wt (○) or rWSN-GH-NS1-Y84F (●) by intranasal inoculation. (**A**) Weight-loss was monitored up to day 3 post-infection. Mice were euthanized on days 1 and 3 post-infection and (**B**) lung viral titers were determined by plaque assay in MDCK cells and (**C**) IFN-β production in the lungs was determined by ELISA. (**D**) RNA was purified from the lung tissues of rIAV-infected mice on days 1 and 3 post-infection and cDNA was synthesized. qPCR was performed to determine the relative expression of murine *ISG15*, *EIF2AK2*, *OAS1*, *IFNA4*, and *IFNB1*, normalized to *HPRT1* and lung viral titer. Data are representative of three independent experiments. * $p < 0.05$, ** $p < 0.01$, and *** $p < 0.001$.

Viral titers increased for both rIAVs expressing NS1-wt and NS1-Y84F between days 1 and 3 post-infection. To investigate whether the IFN response contributed to these differences in lung viral titers, IFN-β production and ISG expression were examined in the lungs of mice on days 1 and 3 post-infection. In comparison to rWSN-GH-NS1-wt infected mice, we show that IFN-β production is elevated in the lungs of rWSN-GH-NS1-Y84F infected mice on both days 1 and 3 post-infection (Figure 9C). Lung ISG expression, which included expression levels for IFN-α4 and IFN-β, was considered a measure of the IFN response to infection. We observed approximately 0.5 to 1-log fold greater expression of *ISG15*, *EIF2AK2*, *OAS1*, *IFNA4*, and *IFNB1* in the lungs of rWSN-GH-NS1-Y84F infected mice on days 1 and 3 post-infection compared with the lungs of rWSN-GH-NS1-wt infected mice (Figure 9D).

Lung histology and H&E staining revealed more cell infiltrates in the lungs of rWSN-GH-NS1-wt infected mice compared with mice infected with rWSN-GH-NS1-Y84F (Figure 10A). Flow cytometry analysis of lung infiltrates on day 1 post-infection showed that there were of the order of 1-log fold more neutrophils (CD45+, CD11b+, Ly6G+) in the lungs of rWSN-GH-NS1-wt infected mice compared with the lungs from mice infected with rWSN-GH-NS1-Y84F (Figure 10B). Consistent with greater neutrophil numbers, mice infected with virus expressing NS1-wt exhibited 7.5-fold and 15-fold greater *CXCL1* and *CXCL2* gene expression, respectively, in their lungs compared with mice infected with virus expressing NS1-Y84F on day 1 post-infection (Figure 10C). CXCL1 and CXCL2 are the major chemoattractants responsible for recruiting neutrophils.

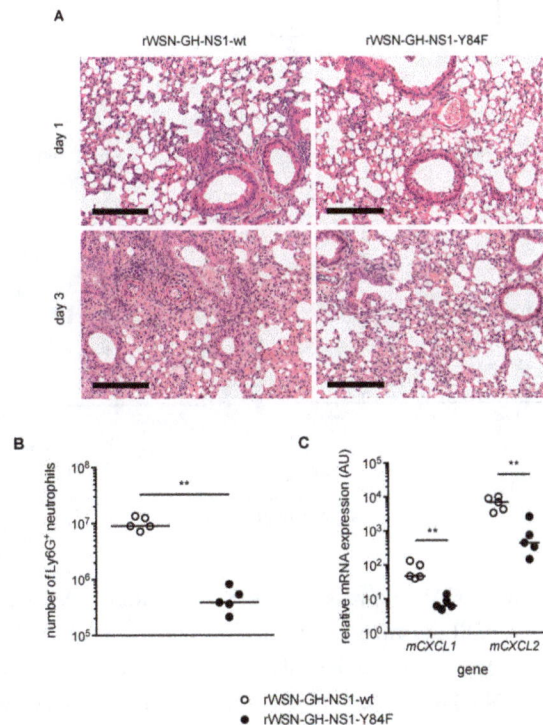

Figure 10. Mice infected with rIAV expressing the Y84F mutant NS1 experience reduced lung pathology and neutrophil infiltration early in infection. C57BL/6 mice were infected with 1×10^5 PFU of rWSN-GH-NS1-wt (○) or rWSN-GH-NS1-Y84F (●) by intranasal inoculation. (**A**) Lungs were harvested on days 1 and 3 post-infection and processed to prepare thin tissue sections (5 μm) for H&E staining. Black bar = 200 μm. (**B**) Lungs were harvested on day 1 post-infection and perfused with PBS. Lung tissues were mashed mechanically and ammonium-chloride-potassium (ACK) lysis was performed to remove red blood cells. Neutrophils were quantified by flow cytometry with antibodies targeting CD45, CD11b, and Ly6G. (**C**) Lungs were harvested on day 1 post-infection, RNA was extracted, and cDNA synthesized. qPCR was performed to determine the relative expression of murine *CXCL1* and *CXCL2*, normalized to *HPRT1*. Data are representative of two independent experiments. ** $p < 0.01$.

To examine the potential differential effects of IFN treatment on viral replication following infection with the different recombinant viruses, mice were infected with virus expressing the NS1-wt or virus expressing the NS1-Y84F mutant and at 8 h post-infection mice were treated with a single dose of 1×10^5 U of murine IFN-β1. While IFN-β treatment did not alter the lung viral titers of mice infected with rWSN-GH-NS1-wt, we observed a marginal reduction in the lung viral titers of mice infected with rWSN-GH-NS1-Y84F on days 1 and 3 post-infection (Figure 11A). Additionally, we observed increases in the expression of *EIF2AK2* (24-fold), *OAS1* (10-fold), *IFNA4* (10-fold) and *IFNB1* (10-fold) in the lungs of rWSN-GH-NS1-Y84F infected mice treated with IFN-β on day 1 post-infection compared with the lungs of infected, but untreated mice (Figure 11B). In contrast, there was no difference in the relative expression of these ISGs in the lungs of mice infected with rWSN-GH-NS1-wt, with or without IFN-β treatment.

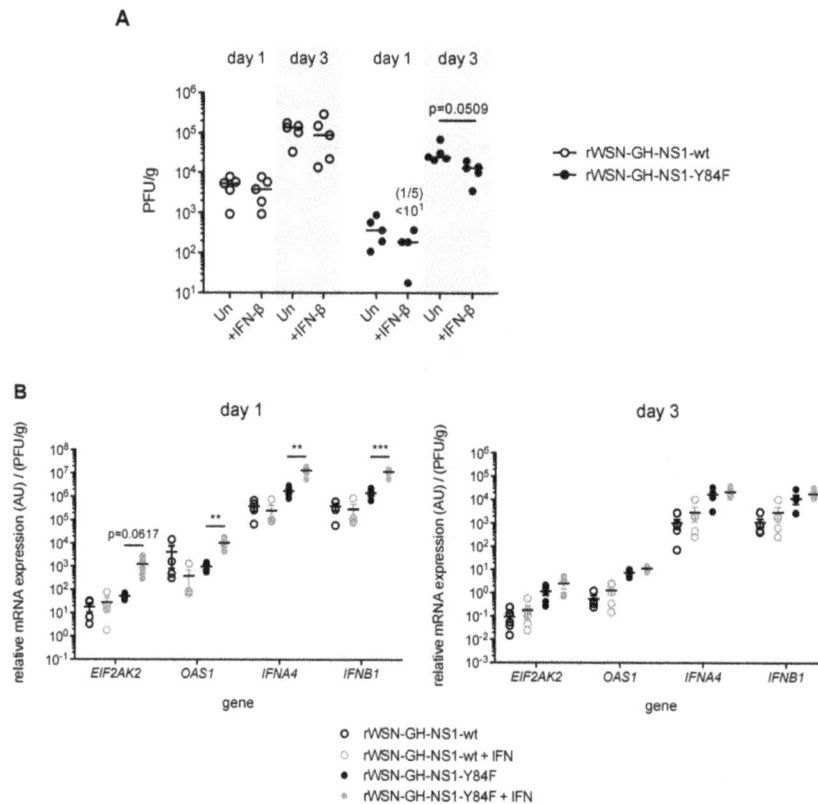

Figure 11. Mice infected with rIAV expressing the Y84F mutant NS1 are more sensitive to the antiviral effects of IFN-β. C57BL/6 mice were infected with 1×10^5 PFU of rWSN-GH-NS1-wt (○) or rWSN-GH-NS1-Y84F (●) by intranasal inoculation and injected by the intraperitoneal route with PBS or 1×10^5 U of murine IFN-β1 at 8 h post-infection. (**A**) Lung viral titers were determined on day 1 and 3 post-infection by plaque assay in MDCK cells. On day 1 post-infection, one of the five mice in the Y84F-infected and IFN-treated cohort, did not have a detectable lung viral titer (titer < 10^1 PFU/mL). (**B**) RNA was purified from the lung tissues of rIAV infected and PBS or murine IFN-β1 treated mice on days 1 and 3 post-infection. cDNA was synthesized and qPCR was performed to determine the relative expression of murine *EIF2AK2*, *OAS1*, *IFNA4*, and *IFNB1*, normalized to *HPRT1* and lung viral titer. Data are representative of two independent experiments. ** $p < 0.01$, and *** $p < 0.001$.

4. Discussion

Viruses have evolved to encode non-structural proteins in their genomes that interact with host factors to enable viral replication. Moreover, there is accumulating evidence for virus–host protein–protein interactions mediated by SH2 binding: binding of IAV NS1 to the i-SH2 domain of p85β to activate PI3K signaling to enhance viral replication [25,26]; the Nef protein of human immunodeficiency virus (HIV)-1 is critical for high titer viral replication and its function is dependent on interactions with the Src family kinase, Hck, stabilized by SH2 binding interactions [37]; the Epstein–Barr virus latency-associated membrane protein, LMP2A, interacts with the signaling scaffold, Shb, mediated by SH2 domain interactions to activate AKT [38]; in silico studies have suggested a molecular model for STAT3 and STAT6 SH2 interactions with the g2-Herpesvirus saimiri Tip protein [39].

Notably, many of the virulence factors encoded by viruses target an IFN response, specifically by binding to and inhibiting the activities of STATs, thereby preventing the induction of an antiviral state. Binding of the measles virus P protein to the SH2 domain of STAT1 limits an IFN response to infection [40]. Other examples include the Nipah virus V, P, and W proteins [41], Sendai virus C protein [42], and hepatitis C virus (HCV) core protein [43,44]. Furthermore, binding of these viral proteins to the host protein SH2-binding domain is often tyrosine phosphorylation-independent, e.g., HIV-1 Nef protein and hepatitis C virus core protein SH2-binding interactions. Our earlier studies revealed that expression of avian H5N1 NS1 in HeLa cells led to a block in IFN signaling [9]. H5N1 NS1 reduced IFN-inducible tyrosine phosphorylation of STAT1, STAT2 and STAT3 and inhibited the nuclear translocation of phospho-STAT2 and the formation of IFN-inducible STAT1:STAT1-, STAT1:STAT3- and STAT3:STAT3-DNA complexes. We attributed the inhibition of IFN-inducible STAT signaling by NS1 in HeLa cells to be a consequence, in part, of NS1-mediated inhibition of expression of the IFN receptor subunit, IFNAR1. In support of this NS1-mediated inhibition, we observed a reduction in expression of IFNAR1 in ex vivo human non-tumor lung tissues infected with H5N1 and H1N1 viruses. Indeed, studies by Zurney et al. [45] comparing cardiotropic reovirus infection of cardiac myocytes and cardiac fibroblasts identified a correlation between greater surface expression of IFNAR1 in cardiac fibroblasts and greater IFN-inducible STAT phosphorylation and induction of ISGs resulting in reduced infectivity [45].

Herein, we have extended our earlier studies to interrogate the mechanism of NS1-mediated inhibition of the IFN response we observed. Given the evidence for viral protein binding to SH2 domains in host proteins, and cognizant of the strictly conserved putative SH2-binding domain in NS1 contributing to IAV virulence, we focused our studies on this domain. Both A/Duck/Hubei/L-1/2004 and A/Grey Heron/Hong Kong/837/2004 H5N1 NS1 SH2-binding domains contain a conserved tyrosine residue at position 84. Highly pathogenic H5N1 IAVs that emerged after the year 2000 have a five amino acid deletion in the linker region between the NS1 dsRNA-binding domain and the protein-binding effector domain, where the conserved tyrosine residue has shifted from position 89 to 84 [15,46]. This five amino acid deletion has been shown to increase the virulence of H5N1 IAVs [46], and based on our current studies, residue Y84 in the H5N1 NS1 putative SH2-binding domain is still important for the activation of AKT, similar to H1N1 NS1 proteins, which contain the conserved tyrosine residue at position 89 [25,26,28]. Our results suggest that host AKT-activation mediated by the NS1 putative SH2-binding domain is well conserved, highlighting its critical role in IAV replication. Indeed, close scrutiny of the adaptive mutations in H5N1 NS1 associated with increased virulence or host switching to mouse reveals that the putative SH2-binding domain and, more specifically the conserved tyrosine residue within this domain, is intact [47,48].

Beyond AKT activation, we show that the conserved putative SH2-binding domain in H5N1 NS1 influences the virulence of IAVs, specifically by limiting an IFN-induced antiviral response. Our data suggest that the inhibitory effects of this putative SH2-binding domain are associated with inhibition of IFN-induced STAT phosphorylation and subsequent ISG expression. Specifically, we provide evidence that, in contrast to cells expressing the intact NS1, cells expressing NS1 with a single Y84F mutation in this conserved putative SH2-binding domain are able to respond fully to IFN treatment in terms of STAT phosphorylation. It is noteworthy that our measure of IFN responsiveness, phosphorylation of STATs, occurs rapidly, within 15 min of IFN treatment. Additionally, cells expressing the Y84F mutant NS1 revert to normal IFNAR1 cell surface expression. We infer from these expression studies that the Y84 residue within NS1 is important for modulating host STAT phosphorylation and IFNAR1 expression, but further studies are required to determine the mechanism and whether NS1 induced PI3K activation may be involved in the observed phenotype. Moreover, studies to determine if NS1-mediated inhibition of IFN signaling is conserved in seasonal IAV strains, which do not encode a deletion in the linker region, may also be warranted.

These in vitro data provided the basis for generating recombinant H1N1 viruses (A/WSN/33 [H1N1]) encoding either the intact (rWSN-GH-NS1-wt) or Y84F mutant H5N1 NS1 (rWSN-GH-NS1-Y84F). In vitro, we showed that the virus expressing the intact H5N1 NS1 replicates to a greater extent in human lung epithelial cells than the virus expressing the SH2-binding domain mutant of NS1, which induced higher levels of IFN-β production. Indeed, IFNs-α/β are induced by IFN signaling in a positive feedback loop. In addition, cells infected with the recombinant virus expressing the mutant NS1 responded to IFN-β treatment with greater levels of ISG expression than cells infected with the virus expressing intact NS1, yet to a lesser extent than uninfected and IFN-β treated cells, in support of virus inhibiting the IFN response and that this is partially mediated by residue Y84 within the putative SH2-binding domain of NS1. The ISGs selected were based on their known antiviral properties: *EIF2AK2* encodes protein kinase RNA-activated (PKR), a host antiviral effector that inhibits cellular translation [49,50]. *OAS1* encodes 2'-5'-oligoadenylate synthetase 1, which activates RNase L in the presence of IAV viral RNA, resulting in viral and cellular RNA degradation [51]. In line with previous IAV replication studies performed in STAT1$^{+/+}$ and STAT1$^{-/-}$ MEFs [36], both rWSN-GH-NS1-wt and rWSN-GH-NS1-Y84F replicated poorly in STAT1$^{+/+}$ MEFs in comparison to STAT1$^{-/-}$ MEFs, thereby further highlighting a potential role for residue Y84 of NS1 in regulating the type I IFN response, and the importance of the host IFN response for limiting viral replication.

Extending these infection studies in vivo, we confirmed that the recombinant H1N1 virus expressing the H5N1 NS1 i-SH2-binding mutant replicated to lower levels than the virus expressing the intact NS1, the mice exhibiting a less severe course of disease, in terms of lung pathology and extent of lung infiltrating neutrophils. Gene expression levels for *CXCL1* and *CXCL2* were higher in the lungs of mice infected with the virus expressing intact NS1, consistent with their role in recruiting neutrophils. In the context of an IAV infection, neutrophils have been shown to play a role in lung tissue damage and increasing the severity of infection [52,53]. In addition, the IFN response to infection, in terms of IFN-β production and transcriptional induction of ISGs—including genes for IFN-α and IFN-β—were more extensive in the lungs of the mice infected with the virus expressing the mutant NS1 than in the lungs of the mice expressing the intact NS1. There is evidence also that CXCL1 and CXCL2, are down-regulated by IFN [54,55]. We speculate that the stronger IFN response in the mice infected with the virus expressing the mutant NS1 contributed to the reduced lung viral replication and less severe lung pathology. This is supported by our findings when mice were treated with IFN-β. Mice infected with virus expressing the mutant NS1 responded to IFN treatment with a more robust transcriptional induction of ISGs than mice infected with the virus expressing an intact NS1 on day 1 post-infection. This was reflected in the modest reduction in lung viral titers in IFN-β-treated mice infected with virus encoding the Y84F mutant NS1, which was not observed in mice infected with virus expressing an intact NS1.

These studies highlight the importance of an IFN response to the control of IAV infections and the potential for IFN treatment to be considered when there are outbreaks of highly virulent IAV strains. Additionally, these studies suggest that targeting the conserved tyrosine residue in the putative SH2-binding domain in IAV NS1 may be a therapeutic strategy in the absence of available vaccines. In ongoing studies, we are identifying additional host protein-binding partners that interact with and/or that are affected by Y84 within the NS1 putative SH2-binding domain in the context of an IFN response.

5. Conclusions

Our studies support a role for the strictly conserved residue, Y84, in a putative SH2-binding domain in H5N1 NS1 in inhibiting the host IFN antiviral response. Specifically, using rIAVs expressing a wildtype NS1 or NS1 encoding a Y84F mutation, we show in vitro and in vivo that targeting the conserved tyrosine residue inhibits virus replication and confers greater sensitivity to the antiviral effects of IFN.

Acknowledgments: We would like to thank Darren P. Baker (formerly of BiogenIdec, Cambridge, MA, USA) for providing reagents and antibodies and Leo L.M. Poon (University of Hong Kong, Hong Kong) for providing the A/Grey Heron/Hong Kong/837/2004 [H5N1] NS gene. E.N.F. is a Tier 1 Canada Research Chair.

Author Contributions: B.X.W. and E.N.F. conceived and designed the experiments; B.X.W. performed the experiments. L.W. and L.P.K. performed the in silico modeling; E.G.B. provided the reverse genetics system used to generate the rIAVs used in this study; B.X.W. and E.N.F. analyzed the data; B.X.W. and E.N.F. wrote the paper.

References

1. Cumulative Number of Confirmed Human Cases for Avian Influenza A (H5N1) Reported to WHO. Available online: http://www.who.int/influenza/human_animal_interface/H5N1_cumulative_table_archives/en/ (accessed on 16 March 2017).

2. Baz, M.; Abed, Y.; Simon, P.; Hamelin, M.E.; Boivin, G. Effect of the neuraminidase mutation H274Y conferring resistance to oseltamivir on the replicative capacity and virulence of old and recent human influenza A (H1N1) viruses. *J. Infect. Dis.* **2010**, *201*, 740–745. [CrossRef] [PubMed]

3. De Jong, M.D.; Tran, T.T.; Truong, H.K.; Vo, M.H.; Smith, G.J.; Nguyen, V.C.; Bach, V.C.; Phan, T.Q.; Do, Q.H.; Guan, Y.; et al. Oseltamivir resistance during treatment of influenza A (H5N1) infection. *N. Engl. J. Med.* **2005**, *353*, 2667–2672. [CrossRef] [PubMed]

4. Hurt, A.C.; Holien, J.K.; Parker, M.; Kelso, A.; Barr, I.G. Zanamivir-resistant influenza viruses with a novel neuraminidase mutation. *J. Virol.* **2009**, *83*, 10366–10373. [CrossRef] [PubMed]

5. González-Navajas, J.M.; Lee, J.; David, M.; Raz, E. Immunomodulatory functions of type I interferons. *Nat. Rev. Immunol.* **2012**, *12*, 125–135. [CrossRef] [PubMed]

6. Takeuchi, O.; Akira, S. Innate immunity to virus infection. *Immunol. Rev.* **2009**, *227*, 75–86. [CrossRef] [PubMed]

7. Diamond, M.S.; Farzan, M. The broad-spectrum antiviral functions of IFIT and IFITM proteins. *Nat. Rev. Immunol.* **2013**, *13*, 46–57. [CrossRef] [PubMed]

8. Wang, B.X.; Fish, E.N. The yin and yang of viruses and interferons. *Trends Immunol.* **2012**, *33*, 190–197. [CrossRef] [PubMed]

9. Jia, D.; Rahbar, R.; Chan, R.W.; Lee, S.M.; Chan, M.C.; Wang, B.X.; Baker, D.P.; Sun, B.; Peiris, J.S.; Nicholls, J.M.; et al. Influenza virus non-structural protein 1 (NS1) disrupts interferon signaling. *PLoS ONE* **2010**, *5*, e13927. [CrossRef] [PubMed]

10. Liedmann, S.; Hincius, E.R.; Anhlan, D.; McCullers, J.A.; Ludwig, S.; Ehhardt, C. New virulence determinants contribute to the enhanced immune response and reduced virulence of an influenza A virus A/PR8/34 variant. *J. Infect. Dis.* **2014**, *209*, 532–541. [CrossRef] [PubMed]

11. Szretter, K.J.; Gangappa, S.; Belser, J.A.; Zeng, H.; Chen, H.; Matsuoka, Y.; Sambhara, S.; Swayne, D.E.; Tumpey, T.M.; Katz, J.M. Early control of H5N1 influenza virus replication by the type I interferon response in mice. *J. Virol.* **2009**, *83*, 5825–5834. [CrossRef] [PubMed]

12. García-Sastre, A. Induction and evasion of type I interferon responses by influenza viruses. *Virus Res.* **2011**, *162*, 12–18. [CrossRef] [PubMed]

13. Quinlivan, M.; Zamarin, D.; García-Sastre, A.; Cullinane, A.; Chambers, T.; Palese, P. Attenuation of Equine Influenza Viruses though Truncations of the NS1 Protein. *J. Virol.* **2005**, *79*, 8431–8439. [CrossRef] [PubMed]

14. Salvatore, M.; Basler, C.F.; Parisien, J.P.; Horvath, C.M.; Bourmakina, S.; Zheng, H.; Muster, T.; Palese, P.; García-Sastre, A. Effects of influenza A virus NS1 protein on protein expression: the NS1 protein enhances translation and is not required for shutoff of host protein synthesis. *J. Virol.* **2002**, *76*, 1206–1212. [CrossRef] [PubMed]

15. Bornholdt, Z.A.; Prasad, B.V. X-ray structure of NS1 from a highly pathogenic H5N1 influenza virus. *Nature* **2008**, *456*, 985–988. [CrossRef] [PubMed]

16. Bornholdt, Z.A.; Prasad, B.V. X-ray structure of influenza virus NS1 effector domain. *Nat. Struct. Mol. Biol.* **2006**, *13*, 559–560. [CrossRef] [PubMed]

17. Liu, J.; Lynch, P.A.; Chien, C.Y.; Montelione, G.T.; Krug, R.M.; Berman, H.M. Crystal structure of the unique RNA-binding domain of the influenza virus NS1 protein. *Nat. Struct. Biol.* **1997**, *4*, 896–899. [CrossRef] [PubMed]

18. Guo, Z.; Chen, L.M.; Zeng, H.; Gomez, J.A.; Plowden, J.; Fujita, T.; Katz, J.M.; Donis, R.O.; Sambhara, S. NS1 protein of influenza A virus inhibits the function of intracytoplasmic pathogen sensor, RIG-I. *Am. J. Respir. Cell. Mol. Biol.* **2007**, *36*, 263–269. [CrossRef] [PubMed]

19. Mibayashi, M.; Martínez-Sobrido, L.; Loo, Y.M.; Cárdenas, W.B.; Gale, M., Jr.; García-Sastre, A. Inhibition of retinoic acid-inducible gene I-mediated induction of beta interferon by the NS1 protein of influenza A virus. *J. Virol.* **2007**, *81*, 514–524. [CrossRef] [PubMed]

20. Li, Y.; Chen, Z.Y.; Wang, W.; Baker, C.C.; Krug, R.M. The 3′-end-processing factor CPSF is required for the splicing of single-intron pre-mRNAs in vivo. *RNA* **2001**, *7*, 920–931. [CrossRef] [PubMed]

21. Nemeroff, M.E.; Barabino, S.M.; Li, Y.; Keller, W.; Krug, R.M. Influenza virus NS1 protein interacts with the cellular 30 kDa subunit of CPSF and inhibits 3′end formation of cellular pre-mRNAs. *Mol. Cell.* **1998**, *1*, 991–1000. [CrossRef]

22. Haye, K.; Burmakina, S.; Moran, T.; García-Sastre, A.; Fernandez-Sesma, A. The NS1 protein of a human influenza virus inhibits type I interferon production and the induction of antiviral responses in primary human dendritic and respiratory epithelial cells. *J. Virol.* **2009**, *83*, 6849–6862. [CrossRef] [PubMed]

23. Solórzano, A.; Webby, R.J.; Lager, K.M.; Janke, B.H.; García-Sastre, A.; Richt, J.A. Mutations in the NS1 protein of swine influenza virus impair anti-interferon activity and confer attenuation in pigs. *J. Virol.* **2005**, *79*, 7535–7543. [CrossRef] [PubMed]

24. Wacheck, V.; Egorov, A.; Groiss, F.; Pfeiffer, A.; Fuereder, T.; Hoeflmayer, D.; Kundi, M.; Popow-Kraupp, T.; Redlberger-Fritz, M.; Mueller, C.A.; et al. A novel type of influenza vaccine: Safety and immunogenicity of replication-deficient influenza virus created by deletion of the interferon antagonist NS1. *J. Infect. Dis.* **2010**, *201*, 354–362. [CrossRef] [PubMed]

25. Hale, B.G.; Batty, I.H.; Downes, C.P.; Randall, R.E. Binding of influenza A virus NS1 protein to the inter-SH2 domain of p85 suggests a novel mechanism for phosphoinositide 3-kinase activation. *J. Biol. Chem.* **2008**, *283*, 1372–1380. [CrossRef] [PubMed]

26. Hale, B.G.; Jackson, D.; Chen, Y.H.; Lamb, R.A.; Randall, R.E. Influenza A virus NS1 protein binds p85beta and activates phosphatidylinositol-3-kinase signaling. *Proc. Natl. Acad. Sci. USA* **2006**, *103*, 14194–14199. [CrossRef] [PubMed]

27. Shin, Y.K.; Li, Y.; Liu, Q.; Anderson, D.H.; Babiuk, L.A.; Zhou, Y. SH3 binding motif 1 in influenza A virus NS1 protein is essential for PI3K/Akt signaling pathway activation. *J. Virol.* **2007**, *81*, 12730–12739. [CrossRef] [PubMed]

28. Hincius, E.R.; Hennecke, A.K.; Gensler, L.; Nordhoff, C.; Anhlan, D.; Vogel, P.; McCullers, J.A.; Ludwig, S.; Ehhardt, C. A single point mutation (Y89F) within the non-structural protein 1 of influenza A viruses limits epithelial cell tropism and virulence in mice. *Am. J. Pathol.* **2012**, *180*, 2361–2374. [CrossRef] [PubMed]

29. Gupta, S.; Yan, H.; Wong, L.H.; Ralph, S.; Krolewski, J.; Schindler, C. The SH2 domains of Stat1 and Stat2 mediate multiple interactions in the transduction of IFN-alpha signals. *EMBO J.* **1996**, *15*, 1075–1084. [PubMed]

30. Wallweber, H.J.A.; Tam, C.; Franke, Y.; Starovasnik, M.A.; Lupardus, P.J. Structural basis of IFNα receptor recognition by TYK2. *Nat. Struct. Mol. Biol.* **2014**, *21*, 443–448. [CrossRef] [PubMed]

31. Hale, B.G.; Kerry, P.S.; Jackson, D.; Precious, B.L.; Gray, A.; Killip, M.J.; Randall, R.E.; Russell, R.J. Structural insights into phosphoinositide 3-kinase activation by the influenza A virus NS1 protein. *Proc. Natl. Acad. Sci. USA* **2010**, *107*, 1954–1959. [CrossRef] [PubMed]

32. Liu, Q.; Wang, S.; Ma, G.; Pu, J.; Forbes, N.E.; Brown, E.G.; Liu, J.H. Improved and simplified recombineering approach for influenza virus reverse genetics. *J. Mol. Genet. Med.* **2009**, *3*, 225–231. [PubMed]

33. Martínez-Sobrido, L.; García-Sastre, A. Generation of recombinant influenza virus from plasmid DNA. *J. Vis. Exp.* **2010**. [CrossRef] [PubMed]

34. Rogers, E.; Wang, B.X.; Cui, Z.; Rowley, D.R.; Ressler, S.J.; Vyakarnam, A.; Fish, E.N. WFDC1/ps20: A host factor that influences the neutrophil response to murine hepatitis virus (MHV) 1 infection. *Antiviral Res.* **2012**, *96*, 158–168. [CrossRef] [PubMed]

35. Durbin, J.E.; Hackenmiller, R.; Simon, M.C.; Levy, D.E. Targeted disruption of the mouse Stat1 gene results in compromised innate immunity to viral disease. *Cell* **1996**, *84*, 443–450. [CrossRef]

36. García-Sastre, A.; Durbin, R.K.; Zheng, H.; Palese, P.; Gertner, R.; Levy, D.E.; Durbin, J.E. The role of interferon in influenza virus tissue tropism. *J Virol.* **1998**, *72*, 8550–8558. [PubMed]

37. Alvarado, J.J.; Tarafdar, S.; Yeh, J.I.; Smithgall, T.E. Interaction with the Src homology (SH3-SH2) region of the Src-family kinase Hck structures the HIV-1 Nef dimer for kinase activation and effector recruitment. *J. Biol. Chem.* **2014**, *289*, 28539–28553. [CrossRef] [PubMed]

38. Matskova, L.V.; Helmstetter, C.; Ingham, R.J.; Gish, G.; Lindholm, C.K.; Ernberg, I.; Pawson, T.; Winberg, G. The Shb signalling scaffold binds to and regulates constitutive signals from the Epstein-Barr virus LMP2A membrane protein. *Oncogene* **2007**, *26*, 4908–4917. [CrossRef] [PubMed]

39. Mazumder, E.D.; Jardin, C.; Vogel, B.; Heck, E.; Scholz, B.; Lengenfelder, D.; Sticht, H.; Ensser, A. A molecular model for the differential activation of STAT3 and STAT6 by the herpesviral oncoprotein tip. *PLoS ONE* **2012**, *7*, e34306. [CrossRef] [PubMed]

40. Devaux, P.; Priniski, L.; Cattaneo, R. The measles virus phosphoprotein interacts with the linker domain of STAT1. *Virology* **2013**, *444*, 250–256. [CrossRef] [PubMed]

41. Shaw, M.L.; Garcia-Sastre, A.; Palese, P.; Basler, C.F. Nipah virus V and W proteins have a common STAT1-binding domain yet inhibit STAT1 activation from the cytoplasmic and nuclear compartments, respectively. *J. Virol.* **2004**, *78*, 5633–5641. [CrossRef] [PubMed]

42. Garcin, D.; Marq, J.B.; Strahle, L.; le Mercier, P.; Kolakofsky, D. All four Sendai Virus C proteins bind Stat1, but only the larger forms also induce its mono-ubiquitination and degradation. *Virology* **2002**, *295*, 256–265. [CrossRef] [PubMed]

43. Anjum, S.; Afzal, M.S.; Ahmad, T.; Aslam, B.; Waheed, Y.; Shafi, T.; Qadri, I. Mutations in the STAT1-interacting domain of the hepatitis C virus core protein modulate the response to antiviral therapy. *Mol. Med. Rep.* **2013**, *8*, 487–492. [CrossRef] [PubMed]

44. Lin, W.; Kim, S.S.; Yeung, E.; Kamegaya, Y.; Blackard, J.T.; Kim, K.A.; Holtzman, M.J.; Chung, R.T. Hepatitis C virus core protein blocks interferon signaling by interaction with the STAT1 SH2 domain. *J. Virol.* **2006**, *80*, 9226–9235. [CrossRef] [PubMed]

45. Zurney, J.; Howard, K.E.; Sherry, B. Basal expression levels of IFNAR and Jak-STAT components are determinants of cell-type-specific differences in cardiac antiviral responses. *J. Virol.* **2007**, *81*, 13668–13680. [CrossRef] [PubMed]

46. Long, J.X.; Peng, D.X.; Liu, Y.L.; Wu, Y.T.; Liu, X.F. Virulence of H5N1 avian influenza virus enhanced by a 15-nucleotide deletion in the viral nonstructural gene. *Virus Genes* **2008**, *36*, 471–478. [CrossRef] [PubMed]

47. Forbes, N.E.; Ping, J.; Dankar, S.K.; Jia, J.J.; Selman, M.; Keleta, L.; Zhou, Y.; Brown, E.G. Multifunctional adaptive NS1 mutations are selected upon human influenza virus evolution in the mouse. *PLoS ONE* **2012**, *7*, e31839. [CrossRef] [PubMed]

48. Dankar, S.K.; Wang, S.; Ping, J.; Forbes, N.E.; Keleta, L.; Li, Y.; Brown, E.G. Influenza A virus NS1 gene mutations F103L and M106I increase replication and virulence. *Virol. J.* **2011**, *8*, 13. [CrossRef] [PubMed]

49. Li, S.; Min, J.Y.; Krug, R.M.; Sen, G.C. Binding of the influenza A virus NS1 protein to PKR mediates the inhibition of its activation by either PACT or double-stranded RNA. *Virology* **2006**, *349*, 13–21. [CrossRef] [PubMed]

50. Bergmann, M.; Garcia-Sastre, A.; Carnero, E.; Pehamberger, H.; Wolff, K.; Palese, P.; Muster, T. Influenza virus NS1 protein counteracts PKR-mediated inhibition of replication. *J. Virol.* **2000**, *74*, 6203–6206. [CrossRef] [PubMed]

51. Silverman, R.H. Viral encounters with 2′,5′-oligoadenylate synthetase and RNase L during the interferon antiviral response. *J. Virol.* **2007**, *81*, 12720–12729. [CrossRef] [PubMed]

52. Narasaraju, T.; Yang, E.; Samy, R.P.; Ng, H.H.; Poh, W.P.; Liew, A.A.; Phoon, M.C.; van Rooijen, N.; Chow, V.T. Excessive neutrophils and neutrophil extracellular traps contribute to acute lung injury of influenza pneumonitis. *Am. J. Pathol.* **2011**, *179*, 199–210. [CrossRef] [PubMed]

53. Perrone, L.A.; Plowden, J.K.; García-Sastre, A.; Katz, J.M.; Tumpey, T.M. H5N1 and 1918 pandemic influenza virus infection results in early and excessive infiltration of macrophages and neutrophils in the lungs of mice. *PLoS Pathog.* **2008**, *4*, e1000115. [CrossRef] [PubMed]

54. Stock, A.T.; Smith, J.M.; Carbone, F.R. Type I IFN suppresses Cxcr2 driven neutrophil recruitment into the sensory ganglia during viral infection. *J. Exp. Med.* **2014**, *211*, 751–759. [CrossRef] [PubMed]

55. Seo, S.U.; Kwon, H.J.; Ko, H.J.; Byun, Y.H.; Seong, B.L.; Uematsu, S.; Akira, S.; Kweon, M.N. Type I interferon signaling regulates Ly6C(hi) monocytes and neutrophils during acute viral pneumonia in mice. *PLoS Pathog.* **2011**, *7*, e1001304. [CrossRef] [PubMed]

9

TMPRSS2 and MSPL Facilitate Trypsin-Independent Porcine Epidemic Diarrhea Virus Replication in Vero Cells

Wen Shi [1], Wenlu Fan [1], Jing Bai [1], Yandong Tang [2], Li Wang [1], Yanping Jiang [1], Lijie Tang [1], Min Liu [3], Wen Cui [1], Yigang Xu [1,*] and Yijing Li [1,*]

[1] College of Veterinary Medicine, Northeast Agricultural University, Harbin 150030, China; wenshi_china@163.com (W.S.); fanwenlu1230@163.com (W.F.); bj0815@126.com (J.B.); wanglicau@163.com (L.W.); jiangyanping2017@126.com (Y.J.); tanglijie@neau.edu.cn (L.T.); cuiwen_200@163.com (W.C.)

[2] Harbin Veterinary Research Institute, Chinese Academy of Agricultural Sciences, Harbin 150001, China; tangyandong2008@163.com

[3] College of Animal Science and Technology, Northeast Agricultural University, Harbin 150030, China; liumin-707@163.com

*   Correspondence: yigangxu_china@sohu.com (Y.X.); liyijing@neau.edu.cn (Y.L.)

Academic Editors: Linda Dixon and Simon Graham

Abstract: Type II transmembrane serine proteases (TTSPs) facilitate the spread and replication of viruses such as influenza and human coronaviruses, although it remains unclear whether TTSPs play a role in the progression of animal coronavirus infections, such as that by porcine epidemic diarrhea virus (PEDV). In this study, TTSPs including TMPRSS2, HAT, DESC1, and MSPL were tested for their ability to facilitate PEDV replication in Vero cells. Our results showed that TMPRSS2 and MSPL played significant roles in the stages of cell–cell fusion and virus–cell fusion, whereas HAT and DESC1 exhibited weaker effects. This activation may be involved in the interaction between TTSPs and the PEDV S protein, as the S protein extensively co-localized with TMPRSS2 and MSPL and could be cleaved by co-expression with TMPRSS2 or MSPL. Moreover, the use of Vero cells expressing TMPRSS2 and MSPL facilitated PEDV replication in the absence of exogenous trypsin. In sum, we identified two host proteases, TMPRSS2 and MSPL, which may provide insights and a novel method for enhancing viral titers, expanding virus production, and improving the adaptability of PEDV isolates in vitro.

Keywords: porcine epidemic diarrhea virus; type II transmembrane serine protease; TMPRSS2; MSPL; virus replication

1. Introduction

Porcine epidemic diarrhea (PED) is caused by porcine epidemic diarrhea virus (PEDV) and is an acute and highly contagious enteric viral disease in nursing pigs. It is characterized by vomiting and lethal, watery diarrhea, and is becoming a global problem [1–7]. PED was first reported in feeder pigs and fattening swine in the United Kingdom in 1971 [8]. Since then, the disease has emerged in many pig-producing countries in Europe and Asia, resulting in tremendous economic losses to the pig industry. PEDV mainly infects the villous epithelial cells of the small intestine, which are rich in proteases, and causes atrophy of the villi, resulting in dehydration and diarrhea. Currently, although PEDV can be propagated in Vero cells treated with trypsin, which mediates the activation of virions for membrane fusion by cleaving the spike (S) glycoprotein [9,10], propagation of PEDV in vitro in a more productive manner remains a continued challenge. Sometimes, PEDV that has been isolated from clinical samples gradually loses its infectivity during further passages in cell cultures [1]. Therefore, the development of novel strategies to control PEDV is urgently required.

PEDV is a group I coronavirus (CoV) consisting of an enveloped virus with a single-stranded, positive-sense RNA genome of approximately 30 kb [11]. The S glycoproteins of CoVs are class I fusion proteins that are generated in a locked conformation to prevent premature triggering of the fusion mechanism and are subsequently prepared for action by proteolytic processing in a step called priming [12,13]. The S protein can be cleaved by endogenous proteases, which is thought to be necessary for inducing cell–cell fusion and virus-cell fusion [12,14–17]. Some endogenous proteases present in the pig small intestine potentially facilitate the entry of PEDV virions into intestinal epithelial cells [18]. However, in vitro, PEDV-infected cells produce syncytia only after treatment with an exogenous protease such as trypsin. This exogenous protease cleavage event leads to cell–cell and virus–cell fusion [14,16,19–21]. Therefore, the proteases responsible for PEDV activation may be potential therapeutic targets.

Recently, a type of trypsin-like serine protease termed type II transmembrane serine proteases (TTSPs) was reported to cleave and activate influenza virus and coronavirus surface proteins, allowing multicycle replication in the absence of trypsin. As previously described, transmembrane protease serine 2 (TMPRSS2) and human airway trypsin-like protease (HAT) can facilitate the spread of human influenza viruses [22–25]. TMPRSS2 and TMPRSS4 play important roles in influenza virus replication, supporting the spread of influenza virus in the absence of trypsin [26]. Subsequent studies confirmed that TMPRSS2 also can activate the spike protein of human coronaviruses, such as severe acute respiratory syndrome coronavirus (SARS-CoV) [27–29] and Middle East respiratory syndrome coronavirus (MERS-CoV) [30,31]. Zmora et al. evaluated seven TTSPs previously reported to activate the surface proteins of influenza A viruses (FLUAVs), MERS-CoV, and SARS-CoV and found that mosaic serine protease large-form (MSPL) and differentially expressed squamous cell carcinoma gene 1 (DESC1) contributed to viral spread in the host [32]. Moreover, the role of TMPRSS2 in the release of PEDV from infected cells was clarified [33]. However, the effects of transmembrane serine proteases on host infection by animal coronaviruses, especially PEDV, have not been thoroughly studied thus far.

In this study, to explore the mechanism of PEDV infection, optimize culture methods, and improve the proliferation of PEDV in vitro, the TTSPs TMPRSS2, HAT, DESC1, and MSPL were assessed to determine their effects on PEDV replication in Vero cells. The results may provide a novel approach to propagating PEDV in vitro as well as potential therapeutic targets for controlling PEDV infection.

2. Materials and Methods

2.1. Plasmids and Primers

pDONR223 plasmids containing the TMPRSS2 gene (BC051839), HAT gene (BC125195), DESC1 gene (BC113412), and MSPL gene (BC114928) were kindly provided by Biogot Technology, Public Protein/Plasmid Library, Nanjing, China. Recombinant pCMV-Myc plasmids expressing human TMPRSS2, HAT, DESC1, or MSPL were constructed following gene amplification and digestion by EcoRI and BglII. All polymerase chain reaction (PCR) primers used in this study are listed in Table 1.

Table 1. Primers used in the study.

Primers	Primer Sequence (5′→3′)	Targets (ID)
Primers for the construction of TTSPs plasmids		
TMPRSS2-F	CCGGAATTCGGATGGCTTTGAACTCAGGG	TMPRSS2
TMPRSS2-R	GGAAGATCTTTAGCCGTCTGCCCTCAT	(BC051839)
HAT-F	CCGGAATTCGGATGTATAGGCCAGCACG	HAT
HAT-R	GGAAGATCTCTAGATCCCAGTTTGTTG	(BC125195)
DESC1-F	CCGGAATTCGGATGATGTATCGGCCAGATG	DESC1
DESC1-R	GGAAGATCTTTAGATACCAGTTTTTG	(BC113412)
MSPL-F	CCGGAATTCGGATGGAGAGGGACAGCC	MSPL
MSPL-R	GGAAGATCTTTAGGATTTTCTGAATCG	(BC114928)
Primers for identification of PEDV by real-time PCR		
PN-F	ACTGAGGGTGTTTTCTGGGTTGC	Nucleocapsid gene of PEDV
PN-R	GGTTCAACAATCTCAACTACACTGG	(DQ072726)
Beta-actin-F	AAGGATTCATATGTGGGCGATG	β-actin gene of Vero cells
Beta-actin-R	TCTCCATGTCGTCCCAGTTGGT	(AB004047)
Primers for identification of swine TTSPs mRNA by real-time PCR		
sw-TMPRSS2-F	CACCCGAACTATGACCCCAAGACC	Swine-TMPRSS2
sw-TMPRSS2-R	CATAGCGGCGTTCAGCACCTC	(XM_013982601)
sw-HAT-F	ACAACGCACAATAACTCCCTCTG	Swine-HAT
sw-HAT-R	GACATTGTTCTGTTGAAGGCTGG	(XM_013978756)
sw-DESC1-F	TGCTGCTGATTTTTAGATTTCGCTC	Swine-DESC1
sw-DESC1-R	AGGGGGTCCTACAGCATCTTG	(XM_013978755)
sw-MSPL-F	CCCATAAGTGGCTTCCCGTC	Swine-MSPL
sw-MSPL-R	TGTAGATGCTCTCCTGGATGGTG	(XM_013989517)
sw-GAPDH-F	AAGGTCGGAGTGAACGGATTTG	Swine-GAPDH
sw-GAPDH-R	GCCTTGACTGTGCCGTGGAAC	(XM_013991162)
Primers for identification of TTSPs in Vero cells		
m-TMPRSS2-F	ACCGCCAGGTGTTGGACCTTAC	m-TMPRSS2
m-TMPRSS2-R	GACACGCCATCGCACCAGTTAG	(XM_007968781)
m-HAT-F	AGTGTGTGTCTCCCAGCTGCTAC	m-HAT
m-HAT-R	TCGGTAGGTTGTCACTCGGGTAT	(XM_007998573)
m-DESC1-F	GGTGGAACAGAAGTAGAAGAGGG	m-DESC1
m-DESC1-R	CACATCACCTGGGTGAAACTC	(XM_007998564)
m-MSPL-F	TGACCCTGTCCGCTCACATCCAC	m-MSPL
m-MSPL-R	AAATCGCACCTCACTCTCCATCTTG	(XM_008021030)

2.2. Cell and Virus Culture

The swine intestinal epithelial cell (IEC) line [34–37] and Vero cells (ATCC, Manassas, VA, USA) were cultured in Dulbecco's modified Eagle medium (DMEM; Gibco, Grand Island, NY, USA) containing 10% fetal calf serum (FCS; Gibco). Cell-adapted PEDV strain LJB/03 from our laboratory [38–40] was propagated in Vero cells and IECs. Briefly, the confluent cell monolayer was washed once with sterile phosphate-buffered saline (PBS) and incubated at 37 °C for 1 h with PEDV LJB/03 supplemented with 21 µg/mL trypsin; then, the inoculum was removed, the cells were washed twice with PBS, and the maintenance medium (DMEM) was supplemented with 5 µg/mL trypsin. Cell cultures were harvested until the cytopathic effect (CPE) exceeded 80%. After freeze-thaw treatment, the supernatants were collected and stored at −80 °C until required.

2.3. Expression of TTSPs in Transfected Vero Cells

Vero cells were transfected with pCMV-Myc expressing TMPRSS2, HAT, DESC1, or MSPL, using Lipofectamine LTX & Plus Reagent (Invitrogen, Life Technologies, Carlsbad, CA, USA). Then, 3 µg/well of recombinant plasmid DNA was diluted into 500 µL of Opi-MEM I reduced-serum medium (Gibco) without serum and mixed with an equal volume of PLUS reagent gently. The mixture was incubated at room temperature (RT; 20–25 °C) for 5 min. Lipofectamine LTX was added, and the complexes were allowed to form by incubation for 30 min. The DNA-Lipofectamine LTX complexes were then added to each well containing cells and medium. In parallel, Vero cells transfected with the same concentration of empty plasmid were used as a control. Post-transfection cells were cultured in 6-well plates at a density of 1.5×10^5/well and cultivated for 48 h. For analysis of TTSP expression by immunofluorescence, cells were washed three times with PBS and fixed with 4% paraformaldehyde at RT for 15 min; then, the cells were permeabilized with 0.2% Triton X-100 in PBS at RT for 10 min and blocked in PBS with 0.3% bovine serum albumin at 37 °C for 30 min. Subsequently, the cells were treated with mouse anti-Myc antibody (Sigma, St. Louis, MO, USA) and rhodamine Red-X-coupled anti-mouse (ZSGB-BIO, Beijing, China) as primary and secondary antibodies, respectively, followed by counterstaining with 4′,6-diamidino-2-phenylindole (DAPI, Beyotime, Shanghai, China). Then, a fluorescence microscope (Leica, Wetzlar, Germany) was used to visualize staining. For analysis of TTSP expression by western blot, cells were washed with PBS, and detached with 200 µL of cell lysis buffer (Beyotime) containing 1 mM phenylmethanesulfonyl fluoride (PMSF; Beyotime). Cells were subjected to sonication, mixed with $5\times$ sodium dodecyl sulfate (SDS) loading buffer, and denatured in boiling water for 10 min. Following SDS-polyacrylamide gel electrophoresis (SDS-PAGE), proteins were transferred to a polyvinylidene fluoride membrane (Merck Millipore, Darmstadt, Germany), and immunoblots were developed using mouse anti-Myc antibody (Sigma) as the primary antibody and horseradish peroxidase (HRP)-conjugated goat anti-mouse antibody (Thermo, Waltham, MA, USA) as the secondary antibody. As a loading control, mouse anti-actin antibody (Sigma) was used. For analysis of TTSPs by flow cytometry, Vero cells transfected with TTSP-encoding plasmid were detached, washed with PBS, incubated with ice-cold ethanol for 10 min, and then stained with mouse anti-Myc primary antibodies (Sigma) followed by DyLight 647-coupled anti-mouse secondary antibodies (Dianova, BioLeaf Biotech, Shanghai, China). After three washings with PBS, cells were fixed with 2% paraformaldehyde and staining was analyzed with an Aria II flow cytometer (BD Biosciences, San Jose, CA, USA).

2.4. Quantitative Real-Time PCR Analysis

The viral RNA of PEDV propagated in Vero cells was extracted using E.Z.N.A. Total RNA Kit I (Omega Bio-Tek, Doraville, GA, USA) following the manufacturer's protocol. Complementary DNA (cDNA) was produced via reverse transcription, using oligo(dT)$_{15}$ (Takara, Tokyo, Japan) and the Superscript Reverse Transcriptase Reagent Kit (Takara) according to the manufacturer's instructions. Then, an ABI 7500 real-time PCR system (Applied Biosystems, Carlsbad, CA, USA) was used to determine viral mRNA transcript levels with SYBR Premix EX Taq II (Takara) according to the manufacturer's recommendations. The specific real-time PCR primers targeting the N gene of PEDV and the β-actin gene of Vero cells are described in Table 1. Real-time PCR was performed under the following conditions: 40 cycles of 30 s at 95 °C, 3 s at 95 °C, and 30 s at 60 °C. The average cycle threshold (Ct) for each individual assay was calculated from triplicate measurements using the instrument's software in "auto Ct" mode (ABI 7500 system software, version 2.3). Relative Ct values of three independent tests were calculated by the $2^{-\Delta\Delta Ct}$ method. Levels of N transcripts were normalized to those of β-actin transcripts in the same sample, and the $2^{-\Delta\Delta Ct}$ value of viral RNA in each sample was analyzed in parallel. There were no specific signals detected in any negative controls.

2.5. Determination of Viral Titer of PEDV Propagated in Vero Cells Expressing TTSPs

Prior to investigating the infectivity of PEDV LJB/03 propagated in Vero cells transiently expressing TTSPs, the viral titer was determined by plaque assay. In brief, after digestion, suspended Vero cells were transfected with 3 µg/well of pCMV-Myc plasmids expressing TMPRSS2, HAT, DESC1, or MSPL, with the empty pCMV-Myc plasmid used as a control. Then, the Vero cells were seeded into 6-well plates at 1.5×10^5/well, and after 24 h, the cells were infected at a multiplicity of infection (MOI) of 0.1 in an infection medium with 3 µg/mL trypsin or PBS. After 1 h of viral adsorption, the inoculum was removed, and the cells were washed twice with PBS and fixed with 3 mL of Minimum Essential Medium (MEM, Gibco) with 0.8% agarose. When CPEs appeared, cells were stained with MEM containing 0.01% Neutral Red Solution (Sigma), and syncytia were counted as plaque under a microscope. The viral titer is expressed as plaque-forming units (PFU)/mL.

2.6. Determination of Effects of TTSPs and TTSP Inhibitor on Viral Replication

To analyze the effects of TTSPs on viral replication, the replication kinetics of intracellular viral RNA were determined by quantitative real-time PCR. Vero cells were transfected with 1 μg/well of pCMV-Myc plasmids expressing a TTSP (TMPRSS2, HAT, DESC1, or MSPL) or empty pCMV-Myc plasmid (control) and seeded in 24-well plates. Then, the cells were infected with PEDV at a multiplicity of infection (MOI) of 0.01 and supplemented with 3 μg/mL trypsin or PBS. After viral adsorption, the cells were washed twice with PBS and cultured with DMEM. At different time points post-infection, the cells were collected and subjected to quantitative real-time PCR detection as described above. To examine the viral replication in Vero cells treated with a TTSP inhibitor, TTSP-transfected Vero cells were pretreated with 200 μM or 500 μM of the TTSP inhibitor AEBSF-HCl (Sigma) or PBS for 1 h, as previously published [41]. Then, the treated cells were infected with PEDV LJB/03 at an MOI of 0.01 for 1 h; at 12 h post-infection, levels of viral replication were determined by quantitative real-time PCR.

2.7. Analysis of PEDV and TTSP Co-Localization

To determine the cellular localization of the S protein of PEDV and the TTSPs, Vero cells were transfected with pCMV-Myc plasmids expressing TMPRSS2, HAT, DESC1, or MSPL, or with empty plasmid serving as a negative control. At 24 h post-transfection, the cells were washed with PBS and infected with PEDV LJB/03 at an MOI of 1. The pCMV-Myc-transfected cells were infected with PEDV in the absence or presence of 3 μg/mL trypsin. At 24 h post-infection, the cells were fixed with 4% paraformaldehyde, permeabilized with 0.2% Triton X-100, and blocked with 0.3% bovine serum albumin. Then, the cells were incubated with mouse anti-Myc antibody (Sigma) and rabbit anti-PEDV S protein polyclonal antibody (developed in our laboratory) at RT for 1 h. After washing with PBS three times, the cells were incubated with fluorescein isothiocyanate (FITC)-conjugated goat anti-rabbit IgG (ZSGB-BIO) and Alexa Fluor 647-labeled goat anti-mouse IgG (H + L) (ZSGB-BIO) secondary antibodies at RT for 1 h. After washing, the cells were treated with DAPI (Beyotime). The coverslips were mounted on glass microscope slides in mounting buffer and examined using a laser scanning microscope (Leica TCS SP2, Wetzlar, Germany). Further image analysis, including calculation of the Pearson correlation coefficient (PCC), was performed with Image J with Just Another Colocalization Plugin [32,42].

2.8. Cleavage of PEDV S Protein by TTSPs

To determine the cleavability of S protein by TTSPs, PEDV strain LJB/03 S protein was cloned into plasmid pCMV-HA. Then, 293T cells were seeded into six-well plates at a density of 2×10^5/well and cotransfected with 2 μg/well of plasmid encoding PEDV S with a N-terminal HA tag and 2 μg/well TTSPs-expressing plasmid or an empty plasmid by using Lipofectamine LTX & Plus Reagent (Invitrogen) according to the manufacturer's protocol. At 48 h posttransfection, the cells were harvested and subjected to sonication and denatured. The lysates were separated by SDS-PAGE and blotted onto polyvinylidene fluoride membrane (Merck Millipore). The PEDV S protein with a N-terminal HA antigenic tag was detected by staining with mouse monoclonal antibody specific for the HA tag (Sigma), followed by incubation with an HRP-conjugated goat anti-mouse antibody (Thermo). As a loading control, expression of β-actin was detected with an anti-β-actin antibody (Sigma).

2.9. Analysis of TTSP Activation of PEDV for Cell–Cell Fusion

To determine the effects of TTSPs on PEDV for cell-cell fusion, the CheckMate Mammalian Two-Hybrid System (Promega, Madison, WI, USA) was used. In brief, Vero target cells were transfected with either empty pCMV-Myc plasmid or pCMV-Myc plasmids encoding TTSPs in combination with the pG5-luc plasmid, which carries the firefly luciferase reporter gene under the control of a promoter containing five GAL4-binding sites. In parallel, Vero effector cells were transfected with the plasmids pACT (containing the herpes simplex virus VP16 activation domain upstream of a multiple cloning region) and pBind (expressing *Renilla reniformis* luciferase under the control of the SV40 promoter). After 24 h, the effector cells were detached, diluted in fresh medium, and added to the target cells. After 24 h of co-cultivation, the cells were washed with PBS and infected with PEDV LJB/03 at an MOI of 1 and supplemented with 1 μg/mL or 0.1 μg/mL trypsin or PBS. Cell–cell fusion was quantified by determining luciferase activity in cell lysates with a commercially available kit (Promega) after 48 h of co-cultivation.

2.10. Quantitative Analysis of TTSP Expression in the Normal and PEDV-Infected Piglet Small Intestine/IECs

Total RNA samples obtained from small intestine tissues of three normal and three PEDV-infected piglets were used to quantify gene expression levels of TTSPs. cDNA was produced using Superscript Reverse Transcriptase Reagent Kit (Takara) according to the manufacturer's instructions, and a quantitative real-time PCR assay was performed in triplicate with SYBR® Premix EX Taq II (Takara) using the *GAPDH* gene as a control. The primers used are shown in Table 1. Average Ct values calculated for *TMPRSS2, HAT, DESC1,* and *MSPL* were normalized by subtraction from the Ct values obtained for *GAPDH* as an internal control. Template-free cDNA reaction mixtures were analyzed in parallel, and no specific signal was detected in any of these experiments. The piglets were handled and maintained under strict ethical considerations according to international recommendations for animal welfare. In addition, the TTSP expression levels in the normal IECs and PEDV-infected IECs were also subjected to quantitative real-time PCR detection as described above.

2.11. Adaptation of PEDV Isolated from Clinical Samples to Vero Cells Transiently Expressing TTSPs

To analyze the adaptation of PEDV strains isolated from clinical samples to Vero cells transfected with pCMV-Myc plasmids expressing TTSPs, two PEDV-positive small intestine tissue samples (A and B) collected from outbreaks of severe acute diarrhea in suckling piglets in 2013 and 2014 in China were tested. Twenty-four hours after transfection with plasmids expressing TTSPs or empty plasmid (control), Vero cells were infected with processed viral samples supplemented with PBS or 3 μg/mL trypsin for 72 h. Following three serial passages, total RNA was extracted to assess the relative RNA levels of PEDV by qPCR.

2.12. Statistical Analysis

All experiments were repeated 3–5 times. Data were statistically analyzed by one-way ANOVA, using GraphPad Prism v5.0 software. $p < 0.05$ was considered statistically significant.

3. Results

3.1. Expression of TTSPs in Transfected Vero Cells

The genes encoding TMPRSS2, HAT, DESC1, and MSPL were cloned into pCMV-Myc plasmids and transfected to Vero cells individually. Expression of the four TTSPs in transfected Vero cells was detected via indirect immunofluorescence, western blot assay and fluorescence-activated cell sorting (FACS). As shown in Figure 1A,B, the four TTSPs were successfully expressed in Vero cells transiently transfected with pCMV-Myc plasmids expressing TMPRSS2, HAT, DESC1, or MSPL. TTSPs were expressed at the cellular plasma membrane, and the number of HAT-positive cells was lower than that for the other three TTSP-positive cells (Figure 1A). The expression of most proteases was readily detectable with the proper predicted size for each, as previously published [32], but the proteases were not expressed in Vero cells transfected with empty pCMV-Myc plasmid (Figure 1B). TTSPs are synthesized as inactive single-chain zymogens and undergo self-cleavage into active forms during or after transport to cell surfaces [43,44]. DESC1 and MSPL were found to be activated and to form bands presenting the cleaved catalytic domain of mature forms. Moreover, we analyzed TTSP expression levels by fluorescence-activated cell sorting (FACS) of stained cells. The most prominent signal was measured in TMPRSS2-expressing cells, followed by MSPL- and DESC1-expressing cells. The fluorescence signal obtained from HAT-expressing cells was the weakest (Figure 1C).

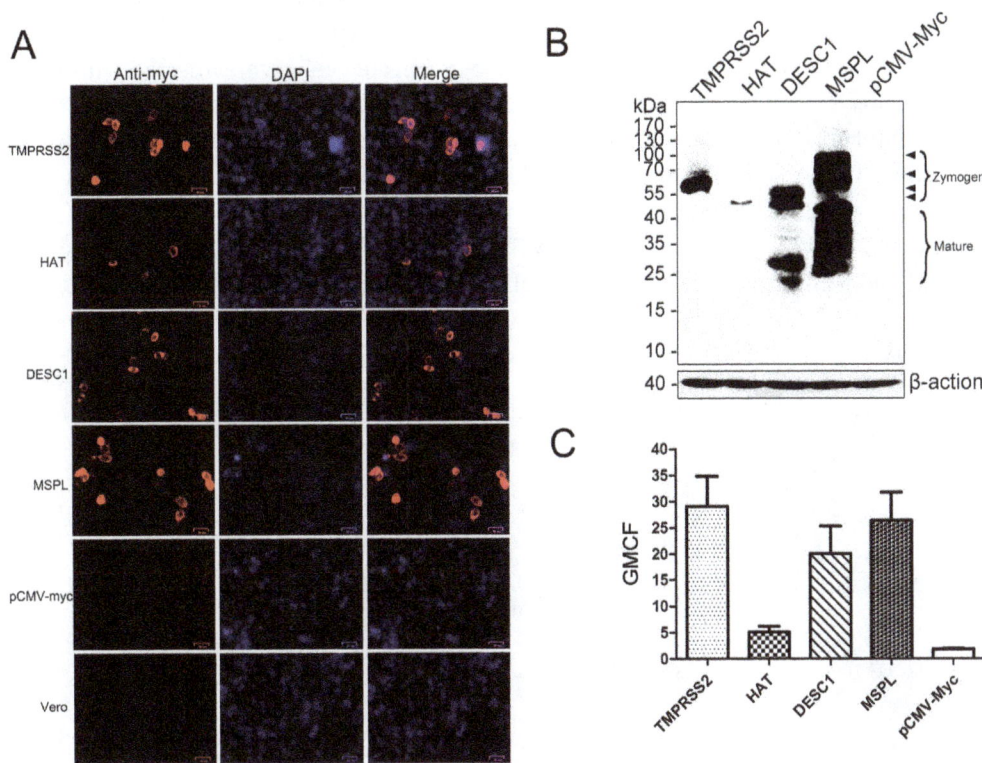

Figure 1. Expression of type II transmembrane serine proteases (TTSPs) in transfected Vero cells. (**A**) Post-transfection, the expression of TMPRSS2, HAT, DESC1, and MSPL in transfected Vero cells was detected via indirect immunofluorescence. Bar = 25 μm. Magnification, ×200; (**B**) TTSP expression in transfected Vero cells as determined by western blot. Zymogens and the mature form are indicated; (**C**) TTSPs expression was detected by FACS. The geometric mean channel fluorescence (GMCF) measured in a representative experiment performed with triplicate samples is shown. Error bars indicate standard deviations of three independent experiments.

3.2. Effects of TTSPs and TTSP Inhibitor on Viral Replication

Prior to this investigation, the presence of endogenously expressed TTSPs in Vero cells was analyzed by RT-PCR assay with primers targeting monkey-borne TMPRSS2 (XM_007968781), HAT (XM_007998573), DESC1 (XM_007998564), and MSPL (XM_008021030) genes (Table 1). No *TMPRSS2, HAT, DESC1,* or *MSPL* mRNA was detected in the Vero cells in this study. Next, the effects of these proteases on PEDV replication in Vero cells exogenously expressing TTSPs were examined. Following transfection, Vero cells were infected with PEDV in the presence or absence of trypsin. As shown in Figure 2A, in the absence of trypsin, the viral titers of PEDV propagated in Vero cells transfected with pCMV-Myc expressing TMPRSS2 and MSPL were clearly higher than those of PEDV propagated in Vero cells expressing HAT and DESC1. Among the TTSPs, the viral titers in Vero cells expressing TMPRSS2 and MSPL were almost $10^{2.5}$ to $10^{4.5}$ times higher than those in the empty-plasmid group and even 3- to 30-fold higher than those in Vero cells cultured with trypsin (3 μg/mL). Moreover, the viral RNA levels in each group were determined by qPCR at different time points post-infection. As shown in Figure 2B, at 72 h post-infection, viral RNA levels in Vero cells expressing MSPL were significantly higher than those in the other groups, and the viral mRNA relative quantity in trypsin-treated cells was slightly higher than that in TMPRSS2-transfected cells. However, the efficacy of TMPRSS2 in activating PEDV replication was almost the same as that of 3 μg/mL trypsin and was higher than that of HAT or

DESC1 at 84 h post-infection. These findings indicate that MSPL and TMPRSS2 play important roles in PEDV infection.

Furthermore, the TTSP inhibitor AEBSF-HCl was used to evaluate the effects of TTSPs on trypsin-independent PEDV entry. The cytotoxicity of AEBSF-HCl at the recommended concentrations was first tested to exclude cytotoxic effects. Then, TTSP-transfected Vero cells were treated with AEBSF-HCl and infected with PEDV. The pCMV-Myc-transfected cells were infected with PEDV in the absence of trypsin. At 12 h post-infection, viral RNA levels were determined by qPCR. As shown in Figure 2C, AEBSF-HCl induced strong inhibitory activity, resulting in dose-dependent decreases in the viral RNA levels. The viral RNA levels of PEDV in MSPL-transfected Vero cells treated with 500 µM AEBSF-HCl were significantly higher than those of cells transfected with other TTSPs. These results also indicate that TTSPs such as TMPRSS2 and MSPL play an important role in PEDV entry, suggesting that TMPRSS2 and MSPL promote PEDV replication better than trypsin.

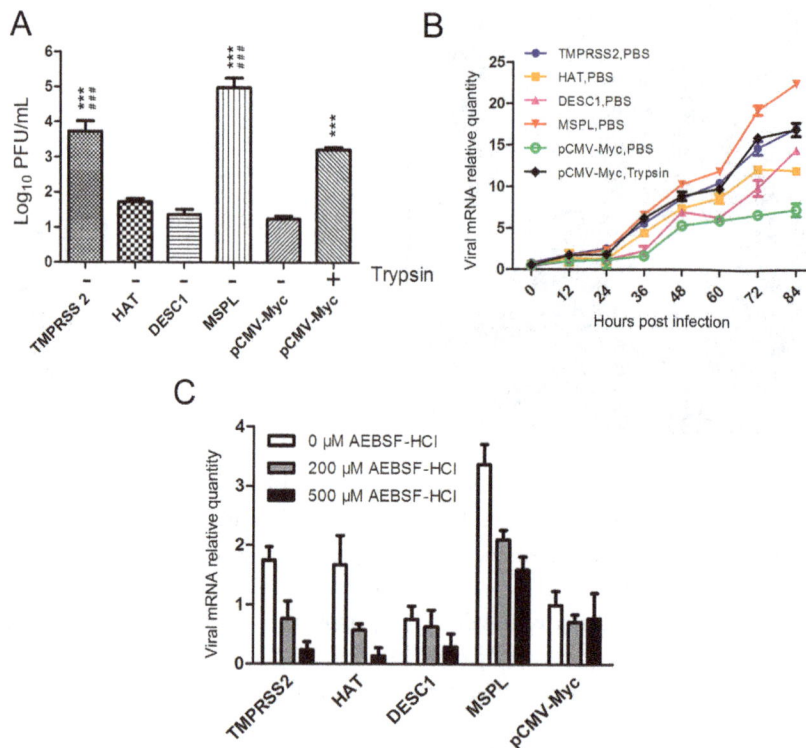

Figure 2. Effects of TTSPs and TTSP inhibitor on viral replication. (**A**) Porcine epidemic diarrhea virus (PEDV) titers following the expression of TTSPs in Vero cells. Viral titers were determined by plaque assay. *** $p < 0.001$ vs. empty pCMV-Myc plasmid; ### $p < 0.001$ vs. empty pCMV-Myc plasmid with 3 µg/mL trypsin; (**B**) Replication kinetics of intracellular viral RNA in Vero cells expressing TTSPs. Relative quantity of the empty pCMV-Myc plasmid with PBS at 0 h = 1; (**C**) Viral replication after TTSP inhibitor treatment. Error bars indicate the standard error of three independent experiments. The relative quantity of the empty pCMV-Myc plasmid with 0 µM AEBSF-HCl treatment = 1.

3.3. TTSP Activation of PEDV for Cell–Cell Fusion

We evaluated the impact of TTSP expression on PEDV-infected Vero cells, using a cell–cell fusion assay (Figure 3). Among the four TTSPs, the expression of MSPL and TMPRSS2 in Vero target cells significantly promoted fusion with Vero effector cells following PEDV infection; in particular, MSPL facilitated cell fusion better than 1 µg/mL trypsin treatment. In contrast, transfection with HAT and DESC1 did not promote cell–cell fusion that was observed with the empty plasmid control. These results indicate that MSPL and TMPRSS2 facilitate PEDV replication.

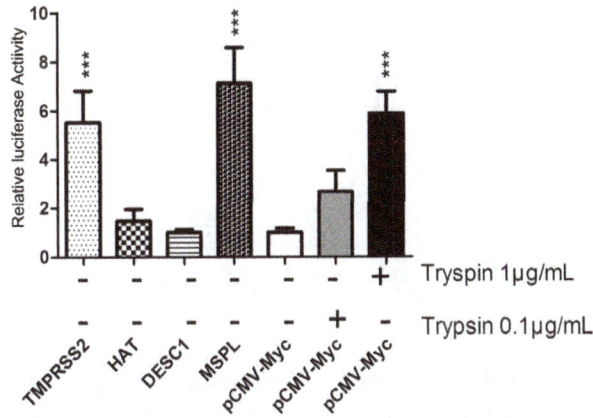

Figure 3. TMPRSS2 and MSPL activation of PEDV for cell–cell fusion. The results of a representative experiment performed with triplicate samples are shown; *** $p < 0.001$ vs. pCMV-Myc without trypsin. Relative quantity of pCMV-Myc without trypsin = 1. Error bars indicate standard error of the mean.

3.4. Co-Localization of TTSPs and PEDV

In this study, the co-localization of the four TTSPs with the PEDV S protein was investigated in infected Vero cells to determine the mechanism of PEDV activation by TTSPs. As shown in Figure 4A, immunofluorescence staining of TTSP-transfected Vero cells infected with PEDV revealed that the PEDV S protein was extensively co-localized with MSPL and TMPRSS2 but not with HAT or DESC1. This assessment was confirmed upon determination of the PCC for TTSPs and S protein signals. The S signals correlated well with those of TMPRSS2 and MSPL, indicating extensive co-localization, whereas little correlation was measured for the S protein and HAT or DESC1 signals (Figure 4B). Thus, the cellular localizations of S protein and TMPRSS2 or MSPL overlap extensively, indicating that MSPL and TMPRSS2 may interact with S protein, activating PEDV replication in Vero cells.

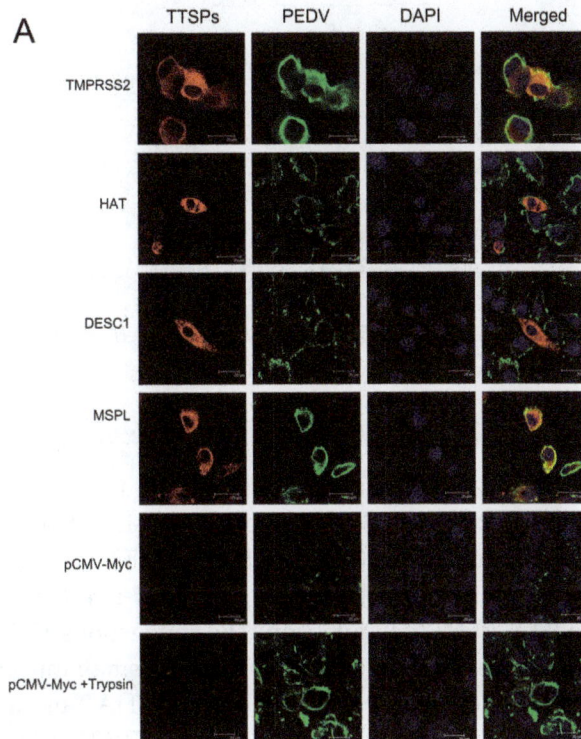

Figure 4. *Cont.*

B

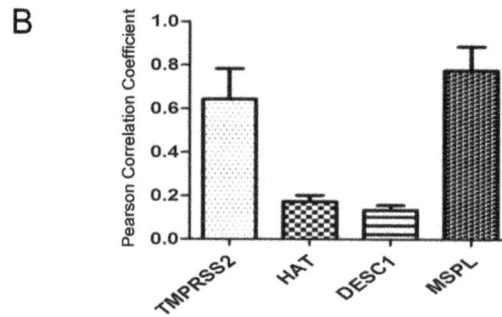

Figure 4. Analysis of TTSP and PEDV S protein co-localization. (**A**) Analysis of TTSP and PEDV S protein co-localization using a laser scanning microscope. Bar = 20–25 μm. Magnification, ×400; (**B**) The co-localization of TTSPs and S protein was determined by calculation of Pearson correlation coefficient (PCC). The average PCC measured for three to five cells from separate experiments is shown; error bars indicate the standard errors of the means.

3.5. Effects of TTSPs on PEDV S Protein Cleavage

We further assessed if the TTSPs studied were able to cleave the S protein of PEDV. As shown in Figure 5, the full-length PEDV S proteins migrating at 200 kDa were detected using anti-HA antibody reacting with the N-terminal of the PEDV S protein. Cleavage of PEDV S was detected upon the coexpression of TMPRSS2 and MSPL. The size of cleavage fragments were the same, approximately 35 kDa. In contrast, coexpressing of HAT or DESC1 did not facilitate PEDV S cleavage. Shirato et al. found that PEDV S protein could be cleaved by co-expression with TMPRSS2, the cleavage C-terminal fraction of S protein detected was 160 kDa [33]. Therefore, our study further confirmed the roles of TMPRSS2 and MSPL in the PEDV S protein activation. The effects of TMPRSS2 and MSPL on PEDV S protein cleavage may be responsible for facilitating the replication of PEDV.

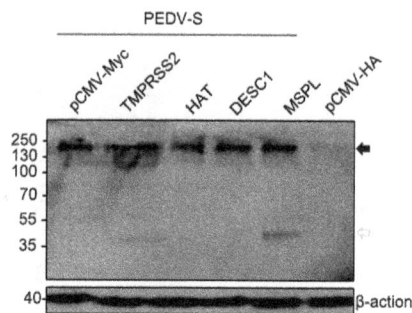

Figure 5. TMPRSS2 and MSPL cleave the PEDV S protein. Black-filled arrowheads, uncleaved S protein; white-filled arrowheads, N-terminal cleavage fragments.

3.6. Determination of TTSP Expression in the Normal and PEDV-Infected Piglet Small Intestine/IECs

We performed real-time RT-PCR analysis of the mRNA levels of TMPRSS2, HAT, DESC1, and MSPL in the small intestine tissues of normal and PEDV-infected piglets. The expression of DESC1 in the normal piglets was the highest, followed by HAT, TMPRSS2, and MSPL; moreover, the mRNA levels of all TTSPs increased in the small intestine of PEDV-infected piglets (Figure 6A). We also detected the TTSP level in IECs after PEDV infection. The endogenous TTSP level is up-regulated in IECs after PEDV infection, which was similar to that in piglet small intestine tissues infected with PEDV (Figure 6B). These results suggest that TMPRSS2, MSPL, HAT and to a higher degree, DESC1 are expressed in piglet small intestine, and that the endogenous TTSP level is up-regulated after PEDV infection. However, whether the endogenous presence of TTSPs in the small intestines of piglets contributes to viral spread in infected piglets remains to be determined.

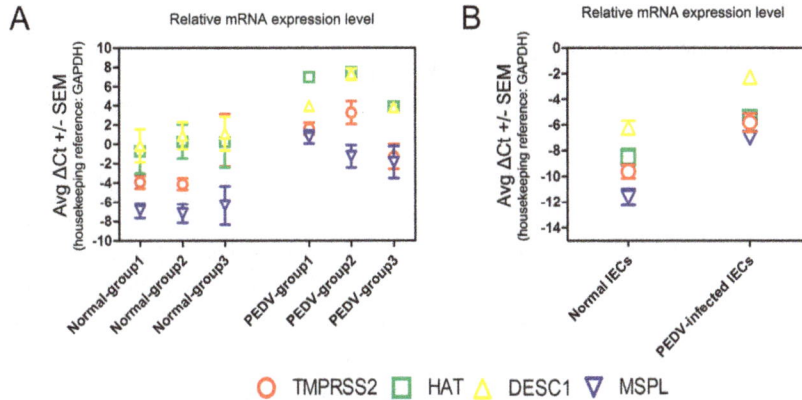

Figure 6. Expression of TTSPs in the normal and PEDV-infected porcine small intestine/intestinal epithelial cells (IECs). (**A**) Expression of TTSPs in the normal and PEDV-infected porcine small intestine; (**B**) Expression of TTSPs in the normal and PEDV-infected IECs.

3.7. TTSPs Facilitate Propagation of PEDV Isolates in Vero Cells

Two PEDV-positive piglet small intestine samples were used to test the abilities of TMPRSS2, HAT, DESC1, and MSPL to facilitate PEDV. TTSP-transfected Vero cells were infected with processed viral samples, and the relative RNA levels of PEDV propagated in Vero cells expressing TMPRSS2, MSPL, HAT, or DESC1 were determined for three serial passages. As shown in Figure 7A, TMPRSS2 and MSPL facilitated strain A replication to an almost 20-fold higher extent than the TTSP control (pCMV-Myc group) after three serial passages. Additionally, HAT could also promote culture of strain A in Vero cells. We speculate that there may be an active site of HAT on the S protein of strain A. The effect of trypsin on the isolation of strain B was not significant. However, the effect of strain B cultured in Vero cells expressing TMPRSS2 and MSPL was better than trypsin treatment; the viral mRNA level in Vero cells expressing MSPL was two times higher than that of the trypsin group after three serial passages (Figure 7B). TMPRSS2 and MSPL facilitated the propagation of the two PEDV isolates (strains A and B) in Vero cells efficiently and steadily, suggesting a promising approach for PEDV propagation of clinical samples in the absence of trypsin treatment.

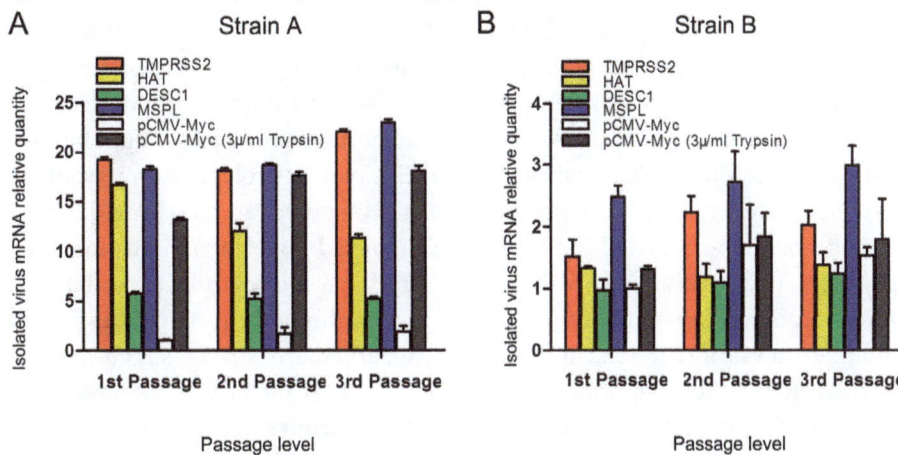

Figure 7. Culture of PEDV isolated from pig intestine in Vero cells transiently expressing TTSPs in three serial passages. (**A**) Isolation of PEDV strain A; (**B**) Isolation of PEDV strain B. The relative quantity of pCMV-Myc without trypsin at the 1st passage = 1. Error bars indicate standard error of the mean. Results shown are those of a representative experiment performed with triplicate samples.

4. Discussion

During the 1970s and 1990s, PEDV caused widespread epidemics in multiple swine-producing countries in Europe [5,45–47]. Since then, severe outbreaks have emerged in a number of Asian countries, including Japan [48], China [49], South Korea [50], and Thailand [51]. Recently, PEDV has been spreading rapidly among swine farms in the United States, resulting in high piglet mortality in more than 32 states [52,53], and similar outbreaks have also been reported in Canada and Mexico [1,3,4,6]. Currently, severe PED is one of the most important diseases affecting pig farming in China [54]. However, the effectiveness of the CV777-based vaccine has been questioned because PED outbreaks have also occurred in vaccinated herds [55]. Therefore, there is an urgent need to improve the protective efficacy of vaccines and to develop new vaccines. However, PEDV isolation in vitro remains challenging, as the isolated virus may gradually lose infectivity upon continued passaging in cell cultures supplemented with trypsin [1]. Recently, we attempted to isolate PEDV from clinical samples, using porcine intestinal epithelial cells (IECs) and found that PEDV isolates were better adapted to growth in IECs than in Vero cells [34], indicating that some trypsin-like proteases present in the IECs facilitated the propagation of PEDV. Moreover, previous research has suggested that several TTSPs located in the mucosal epithelium play critical roles in viral infectivity through the activation of viral surface proteins [23,27,31,32]. At the beginning of this study, we designed primers according to the predicted sequences of swine TTSPs from national center for biotechnology information (NCBI), and attempted to amplify the full-length TTSP genes from the trachea, bronchus, lung, and small intestine tissues of piglet and IECs. However, we failed to obtain the porcine TTSP genes. We speculate that differences may exist between the actual sequences and predicted sequences. Although TTSPs are the host proteases of respiratory and digestive tract mucosa, the TTSP expression levels may be low or limited in some conditions or over a period of time, which results in difficulty in obtaining actual porcine TTSP genes. Therefore, we studied human TTSPs (TMPRSS2, HAT, DESC1, and MSPL) to explore their effects on PEDV replication.

The TTSP family is composed of more than 20 members and divided into four subfamilies: the HAT/DESC subfamily, hepsin/TMPRSS/enteropeptidase subfamily (including TMPRSS2 and MSPL), matriptase subfamily, and corin subfamily. TMPRSS2 and MSPL are predominantly expressed in the fetal liver and kidney [56] and on the brush-border of the duodenum [57]. HAT is predominantly expressed in the trachea [58,59], whereas DESC1 is restricted to the epithelia of the skin and oral cavity [60,61]. In this study, we confirmed the presence of TMPRSS2, HAT, DESC1, and MSPL in the small intestines of normal piglets and IECs, and we also found that the mRNA levels of these TTSPs increased in PEDV-infected small intestine tissues and IECs. Whether or not the endogenous presence of TTSPs in the small intestines of piglets contributes to viral spread in infected piglets remains to be determined, and knock-out mice, as well as specific protease inhibitors, might be useful tools for these endeavors. Previous studies demonstrated that TTSPs play key roles in hormone or growth factor activation, epithelial differentiation, and the initiation of proteolytic cascades [62,63]. The mechanism underlying the effects of up-regulated TTSPs in PEDV-infected piglet small intestine tissues requires further investigation. The inhibitors of TMPRSS2 and MSPL may be potential candidates for treatment of PEDV. Moreover, according to a comparison of the promotion effects of these TTSPs on viral replication and titers in vitro, TMPRSS2 and MSPL were particularly strong, suggesting that members of the hepsin/TMPRSS/enteropeptidase subfamily may activate PEDV emergence due to their specific structure. Additional research is underway to determine whether other members of the hepsin/TMPRSS/enteropeptidase subfamily are able to activate PEDV.

The role of TTSPs in the release of PEDV from infected cells has been reported previously [33], although the mechanism by which TTSPs promote the propagation of animal coronaviruses remains unclear. In this study, to explore this mechanism, we first focused on whether TTSPs (TMPRSS2, HAT, DESC1, and MSPL) activated viral transmission via cell–cell fusion assay, and our results demonstrated that the activating effects of MSPL and TMPRSS2 were more robust than those of the other TTSPs. MSPL exhibited the strongest effect followed by TMPRSS2. It has been suggested that the addition of trypsin mediates cell–cell fusion in PEDV-infected cells [64], thus demonstrating that TMPRSS2 and MSPL exhibit trypsin-like characteristics that facilitate cell–cell fusion. However, it should be noted that the cell–cell fusion assay allows the interaction of cell surfaces on which large amounts of receptors and proteases may be expressed, and therefore it might not fully mirror virus–cell fusion.

Although serine proteases are reportedly involved in PEDV entry, it was previously unclear which of them are most effective [41]. Thus, we used the previously published TTSP inhibitor AEBSF-HCl [41] to assess whether TTSPs (TMPRSS2, HAT, DESC1, and MSPL) activate PEDV entry into cells in the absence of trypsin. We found that viral RNA levels were decreased in a dose-dependent manner following AEBSF-HCl treatment. Additionally, we found that the level of viral mRNA increased in a dose-dependent manner in Vero cells expressing TMPRSS2 and MSPL, but not HAT and DESC1 at the stage of virus entry [65]. These results suggest that the activating effect of TMPRSS2 and, in particular, MSPL, on PEDV entry into cells was greater than that of HAT and DESC1. Although Liu et al. found that TMPRSS2 did not increase the entry of PEDV pseudoviruses into Huh-7 cells [66], several studies suggested that some candidate cellular enzymes, such as TTSPs could activate PEDV replication [41,64], and that human TMPRSS2 has been shown to enhance the multicycle replication of PEDV [33].

To explore the key role of TTSPs in facilitating PEDV replication, we speculated that the S protein of PEDV might have interacted with TTSPs located on the cell surface during viral infection. Thus, an assessment of the co-localization of TTSPs with the PEDV S protein was performed, and our results showed that the PEDV S protein co-localized extensively with MSPL and TMPRSS2, indicating that these TTSPs might interact with the PEDV S protein to promote viral entry into cells. It is worth noting that PEDV-activating TTSPs co-localized with S protein, whereas inactive TTSPs did not, despite robust expression (such as DESC1) in the cellular system analyzed. It is therefore conceivable that the cellular localization of a TTSP, apart from its substrate specificity, might determine whether the protease can activate S and other viral glycoproteins; this possibility deserves further investigation. It has been confirmed that TTSPs cleave and activate the SARS-CoV and MERS-CoV S proteins, and the cleaved fragments of the S protein may induce subtle conformational changes that increase its sensitivity for binding to its receptor [15]. We also attempted to verify the cleavage of S protein by TTSPs, and we found TMPRSS2 and MSPL could cleave PEDV S protein with the same size of cleavage fragments, but HAT and DESC1 could not. Therefore, our study further confirmed the roles of TMPRSS2 and MSPL in the PEDV S protein activation. However, the mechanisms of S protein activation by TTSPs for PEDV entry have not been clearly demonstrated, and additional research is under way to further investigate these mechanisms.

Currently, the propagation of PEDV in vitro remains a continuing challenge, as viral infectivity gradually declines during serial passages in cell cultures. In this study, we confirmed that TMPRSS2 and MSPL effectively facilitate the isolation of PEDV in vitro in the absence of trypsin. Viral adaptation and growth in Vero cells expressing TMPRSS2 and MSPL were higher than those in control cells transfected with empty plasmid control and in cells treated with trypsin. These results indicate that TMPRSS2 and MSPL might be more conducive to PEDV isolation in vitro than exogenous proteases like trypsin, suggesting a promising approach for PEDV isolation in vitro, using Vero cell lines continuously expressing TMPRSS2 or MSPL. The establishment of Vero cell lines stably expressing TMPRSS2/MSPL may facilitate the use of attenuated cell-culture-adapted PEDV strains cultured in the absence of trypsin for vaccine development, which can reduce the cost and simplify the process in the PEDV vaccine production.

In conclusion, we first demonstrated that TMPRSS2 and MSPL facilitate the replication of the animal coronavirus PEDV and play a significant role in viral infection by promoting cell–cell fusion and virus–cell fusion. Whether or not the endogenous presence of TTSPs in the small intestines of piglets contributes to viral spread in infected piglets should be determined further. This study provides insights and a novel method for enhancing viral titers, expanding virus production, and improving the adaptability of PEDV isolates in vitro.

Acknowledgments: This work was supported by the National Natural Science Foundation of China (grant 31472226), the National Key Research and Development Program of China (grant 2016YFD0500100) and the National Key Technology R&D Program of China (grant 2015BAD12B02-7). We thank the Biogot Technology, Public Protein/Plasmid Library for providing the plasmids encoding full-length human transmembrane proteases. We also thank Yandong Tang for providing the valuable suggestions to this study.

Author Contributions: Y.L. and Y.X. conceived and designed the study. W.S., W.F., J.B., Y.T., L.W., W.C. and Y.J. performed the experiments. L.T. and M.L. analyzed and interpreted the data. W.S. wrote the paper. All authors read and approved the manuscript.

References

1.   Chen, Q.; Li, G.; Stasko, J.; Thomas, J.T.; Stensland, W.R.; Pillatzki, A.E.; Gauger, P.C.; Schwartz, K.J.; Madson, D.; Yoon, K.-J. Isolation and characterization of porcine epidemic diarrhea viruses associated with the 2013 disease outbreak among swine in the United States. *J. Clin. Microbiol.* **2014**, *52*, 234–243. [CrossRef] [PubMed]

2.   Debouck, P.; Pensaert, M. Experimental infection of pigs with a new porcine enteric coronavirus, CV 777. *Am. J. Vet. Res.* **1980**, *41*, 219–223. [PubMed]

3.   Kim, S.H.; Kim, I.J.; Pyo, H.M.; Tark, D.S.; Song, J.Y.; Hyun, B.H. Multiplex real-time RT-PCR for the simultaneous detection and quantification of transmissible gastroenteritis virus and porcine epidemic diarrhea virus. *J. Virol. Methods* **2007**, *146*, 172–177. [CrossRef] [PubMed]

4.   Ojkic, D.; Hazlett, M.; Fairles, J.; Marom, A.; Slavic, D.; Maxie, G.; Alexandersen, S.; Pasick, J.; Alsop, J.; Burlatschenko, S. The first case of porcine epidemic diarrhea in Canada. *Can. Vet. J.* **2015**, *56*, 149–152. [PubMed]

5.   Pensaert, M.B.; de Bouck, P. A new coronavirus-like particle associated with diarrhea in swine. *Arch. Virol.* **1978**, *58*, 243–247. [CrossRef] [PubMed]

6.   Trujillo-Ortega, M.E.; Beltran-Figueroa, R.; Garcia-Hernandez, M.E.; Juarez-Ramirez, M.; Sotomayor-Gonzalez, A.; Hernandez-Villegas, E.N.; Becerra-Hernandez, J.F.; Sarmiento-Silva, R.E. Isolation and characterization of porcine epidemic diarrhea virus associated with the 2014 disease outbreak in Mexico: Case report. *BMC Vet. Res.* **2016**, *12*, 132. [CrossRef] [PubMed]

7.   Wood, E.N. An apparently new syndrome of porcine epidemic diarrhoea. *Vet. Rec.* **1977**, *100*, 243–244. [CrossRef] [PubMed]

8.   Oldham, J. Letter to the editor. *Pig Farming* **1972**, *10*, 72–73. [CrossRef]

9.   Hofmann, M.; Wyler, R. Propagation of the virus of porcine epidemic diarrhea in cell culture. *J. Clin. Microbiol.* **1988**, *26*, 2235–2239. [PubMed]

10. Park, J.E.; Cruz, D.J.; Shin, H.J. Receptor-bound porcine epidemic diarrhea virus spike protein cleaved by trypsin induces membrane fusion. *Arch. Virol.* **2011**, *156*, 1749. [CrossRef] [PubMed]

11. Lai, M.M.; Cavanagh, D. The molecular biology of coronaviruses. *Adv. Virus Res.* **1997**, *48*, 1–100. [PubMed]

12. Bosch, B.J.; van der Zee, R.; de Haan, C.A.; Rottier, P.J. The coronavirus spike protein is a class I virus fusion protein: Structural and functional characterization of the fusion core complex. *J. Virol.* **2003**, *77*, 8801–8811. [CrossRef] [PubMed]

13. White, J.M.; Delos, S.E.; Brecher, M.; Schornberg, K. Structures and mechanisms of viral membrane fusion proteins: Multiple variations on a common theme. *Crit. Rev. Biochem. Mol.* **2008**, *43*, 189–219. [CrossRef] [PubMed]

14. Kawase, M.; Shirato, K.; Matsuyama, S.; Taguchi, F. Protease-mediated entry via the endosome of human coronavirus 229E. *J. Virol.* **2009**, *83*, 712–721. [CrossRef] [PubMed]

15. Matsuyama, S.; Taguchi, F. Two-step conformational changes in a coronavirus envelope glycoprotein mediated by receptor binding and proteolysis. *J. Virol.* **2009**, *83*, 11133–11141. [CrossRef] [PubMed]

16. Simmons, G.; Reeves, J.D.; Rennekamp, A.J.; Amberg, S.M.; Piefer, A.J.; Bates, P. Characterization of severe acute respiratory syndrome-associated coronavirus (SARS-CoV) spike glycoprotein-mediated viral entry. *Proc. Natl. Acad. Sci. USA* **2004**, *101*, 4240–4245. [CrossRef] [PubMed]

17. Spaan, W.; Cavanagh, D.; Horzinek, M.C. Coronaviruses: Structure and genome expression. *J. Gen. Virol.* **1988**, *69*, 2939–2952. [CrossRef] [PubMed]

18. Zamolodchikova, T.S. Serine proteases of small intestine mucos—Localization, functional properties, and physiological role. *Biochemistry* **2012**, *77*, 820–829. [PubMed]

19. Qiu, Z.; Hingley, S.T.; Simmons, G.; Yu, C.; Das Sarma, J.; Bates, P.; Weiss, S.R. Endosomal proteolysis by cathepsins is necessary for murine coronavirus mouse hepatitis virus type 2 spike-mediated entry. *J. Virol.* **2006**, *80*, 5768–5776. [CrossRef] [PubMed]

20. Yamada, Y.K.; Takimoto, K.; Yabe, M.; Taguchi, F. Acquired fusion activity of a murine coronavirus MHV-2 variant with mutations in the proteolytic cleavage site and the signal sequence of the S protein. *Virology* **1997**, *227*, 215–219. [CrossRef] [PubMed]

21. Yoshikura, H.; Tejima, S. Role of protease in mouse hepatitis virus-induced cell fusion. Studies with a cold-sensitive mutant isolated from a persistent infection. *Virology* **1981**, *113*, 503–511. [CrossRef]

22. Bottcher, E.; Matrosovich, T.; Beyerle, M.; Klenk, H.D.; Garten, W.; Matrosovich, M. Proteolytic activation of influenza viruses by serine proteases TMPRSS2 and HAT from human airway epithelium. *J. Virol.* **2006**, *80*, 9896–9898. [CrossRef] [PubMed]

23. Bottcher, E.; Freuer, C.; Steinmetzer, T.; Klenk, H.D.; Garten, W. MDCK cells that express proteases TMPRSS2 and HAT provide a cell system to propagate influenza viruses in the absence of trypsin and to study cleavage of HA and its inhibition. *Vaccine* **2009**, *27*, 6324–6329. [CrossRef] [PubMed]

24. Bottcher, E.; Freuer, C.; Sielaff, F.; Schmidt, S.; Eickmann, M.; Uhlendorff, J.; Steinmetzer, T.; Klenk, H.D.; Garten, W. Cleavage of influenza virus hemagglutinin by airway proteases TMPRSS2 and HAT differs in subcellular localization and susceptibility to protease inhibitors. *J. Virol.* **2010**, *84*, 5605–5614. [CrossRef] [PubMed]

25. Bertram, S.; Heurich, A.; Lavender, H.; Gierer, S.; Danisch, S.; Perin, P.; Lucas, J.M.; Nelson, P.S.; Pohlmann, S.; Soilleux, E.J. Influenza and SARS-coronavirus activating proteases TMPRSS2 and HAT are expressed at multiple sites in human respiratory and gastrointestinal tracts. *PLoS ONE* **2012**, *7*, e35876. [CrossRef] [PubMed]

26. Bertram, S.; Glowacka, I.; Blazejewska, P.; Soilleux, E.; Allen, P.; Danisch, S.; Steffen, I.; Choi, S.Y.; Park, Y.; Schneider, H.; et al. TMPRSS2 and TMPRSS4 facilitate trypsin-independent spread of influenza virus in Caco-2 cells. *J. Virol.* **2010**, *84*, 10016–10025. [CrossRef] [PubMed]

27. Glowacka, I.; Bertram, S.; Muller, M.A.; Allen, P.; Soilleux, E.; Pfefferle, S.; Steffen, I.; Tsegaye, T.S.; He, Y.; Gnirss, K.; et al. Evidence that TMPRSS2 activates the severe acute respiratory syndrome coronavirus spike protein for membrane fusion and reduces viral control by the humoral immune response. *J. Virol.* **2011**, *85*, 4122–4134. [CrossRef] [PubMed]

28. Matsuyama, S.; Nagata, N.; Shirato, K.; Kawase, M.; Takeda, M.; Taguchi, F. Efficient activation of the severe acute respiratory syndrome coronavirus spike protein by the transmembrane protease TMPRSS2. *J. Virol.* **2010**, *84*, 12658–12664. [CrossRef] [PubMed]

29. Shulla, A.; Heald-Sargent, T.; Subramanya, G.; Zhao, J.; Perlman, S.; Gallagher, T. A transmembrane serine protease is linked to the severe acute respiratory syndrome coronavirus receptor and activates virus entry. *J. Virol.* **2011**, *85*, 873–882. [CrossRef] [PubMed]

30. Gierer, S.; Bertram, S.; Kaup, F.; Wrensch, F.; Heurich, A.; Kramer-Kuhl, A.; Welsch, K.; Winkler, M.; Meyer, B.; Drosten, C.; et al. The spike protein of the emerging betacoronavirus EMC uses a novel coronavirus receptor for entry, can be activated by TMPRSS2, and is targeted by neutralizing antibodies. *J. Virol.* **2013**, *87*, 5502–5511. [CrossRef] [PubMed]

31. Shirato, K.; Kawase, M.; Matsuyama, S. Middle East respiratory syndrome coronavirus infection mediated by the transmembrane serine protease TMPRSS2. *J. Virol.* **2013**, *87*, 12552–12561. [CrossRef] [PubMed]

32. Zmora, P.; Blazejewska, P.; Moldenhauer, A.S.; Welsch, K.; Nehlmeier, I.; Wu, Q.; Schneider, H.; Pohlmann, S.; Bertram, S. DESC1 and MSPL activate influenza A viruses and emerging coronaviruses for host cell entry. *J. Virol.* **2014**, *88*, 12087–12097. [CrossRef] [PubMed]

33. Shirato, K.; Matsuyama, S.; Ujike, M.; Taguchi, F. Role of proteases in the release of porcine epidemic diarrhea virus from infected cells. *J. Virol.* **2011**, *85*, 7872–7880. [CrossRef] [PubMed]

34. Shi, W.; Jia, S.; Zhao, H.; Yin, J.; Wang, X.; Yu, M.; Ma, S.; Wu, Y.; Chen, Y.; Fan, W.; et al. Novel approach for isolation and identification of Porcine epidemic diarrhea virus (PEDV) strain NJ using porcine intestinal epithelial cells. *Viruses* **2017**, *9*, 19. [CrossRef] [PubMed]

35. Wang, J.; Hu, G.; Lin, Z.; He, L.; Xu, L.; Zhang, Y. Characteristic and functional analysis of a newly established porcine small intestinal epithelial cell line. *PLoS ONE* **2014**, *9*, e110916. [CrossRef] [PubMed]

36. Li, W.; Wang, G.; Liang, W.; Kang, K.; Guo, K.; Zhang, Y. Integrin beta3 is required in infection and proliferation of classical swine fever virus. *PLoS ONE* **2014**, *9*, e110911.

37. Xu, X.; Zhang, H.; Zhang, Q.; Dong, J.; Liang, Y.; Huang, Y.; Liu, H.J.; Tong, D. Porcine epidemic diarrhea virus E protein causes endoplasmic reticulum stress and up-regulates interleukin-8 expression. *Virol. J.* **2013**, *10*, 26. [CrossRef] [PubMed]

38. Junwei, G.; Baoxian, L.; Lijie, T.; Yijing, L. Cloning and sequence analysis of the N gene of porcine epidemic diarrhea virus LJB/03. *Virus Genes* **2006**, *33*, 215–219. [CrossRef] [PubMed]

39. Jinghui, F.; Yijing, L. Cloning and sequence analysis of the M gene of porcine epidemic diarrhea virus LJB/03. *Virus Genes* **2005**, *30*, 69–73. [CrossRef] [PubMed]

40. Mao, Y.Y.; Zhang, G.H.; Ge, J.W.; Jiang, Y.P.; Qiao, X.Y.; Cui, W.; Li, Y.J. Isolation and characteristics of virus culture of porcine epidemic diarrhea virus LJB/03. *Chin. J. Virol.* **2010**, *26*, 483–489.

41. Park, J.E.; Cruz, D.J.; Shin, H.J. Clathrin- and serine proteases-dependent uptake of porcine epidemic diarrhea virus into Vero cells. *Virus Res.* **2014**, *191*, 21–29. [CrossRef] [PubMed]

42. Bolte, S.; Cordelieres, F.P. A guided tour into subcellular colocalization analysis in light microscopy. *J. Microsc.-Oxf.* **2006**, *224*, 213–232. [CrossRef] [PubMed]

43. Afar, D.E.; Vivanco, I.; Hubert, R.S.; Kuo, J.; Chen, E.; Saffran, D.C.; Raitano, A.B.; Jakobovits, A. Catalytic cleavage of the androgen-regulated TMPRSS2 protease results in its secretion by prostate and prostate cancer epithelia. *Cancer Res.* **2001**, *61*, 1686–1692. [PubMed]

44. Miyake, Y.; Yasumoto, M.; Tsuzuki, S.; Fushiki, T.; Inouye, K. Activation of a membrane-bound serine protease matriptase on the cell surface. *J. Biochem.* **2009**, *146*, 273–282. [CrossRef] [PubMed]

45. Chasey, D.; Cartwright, S.F. Virus-like particles associated with porcine epidemic diarrhoea. *Res. Vet. Sci.* **1978**, *25*, 255–256. [PubMed]

46. Nagy, B.; Nagy, G.; Meder, M.; Mocsari, E. Enterotoxigenic *Escherichia coli*, rotavirus, porcine epidemic diarrhoea virus, adenovirus and calici-like virus in porcine postweaning diarrhoea in Hungary. *Acta Vet. Hung.* **1996**, *44*, 9–19. [PubMed]

47. Van Reeth, K.; Pensaert, M. Prevalence of infections with enzootic respiratory and enteric viruses in feeder pigs entering fattening herds. *Vet. Rec.* **1994**, *135*, 594–597. [PubMed]

48. Takahashi, K.; Okada, K.; Ohshima, K. An outbreak of swine diarrhea of a new-type associated with coronavirus-like particles in Japan. Nihon juigaku zasshi. *Jpn. J. Vet. Sci.* **1983**, *45*, 829–832. [CrossRef]

49. Xuan, H.; Xing, D.; Wang, D.; Zhu, W.; Zhao, F.; Gong, H. Study on the culture of porcine epidemic diarrhea virus adapted to fetal porcine intestine primary cell monolayer. *Chin. J. Vet. Sci.* **1984**, *4*, 202–208.

50. Kweon, C.H.; Kwon, B.J.; Jung, T.S.; Kee, Y.J.; Hur, D.H.; Hwang, E.K.; Rhee, J.C.; An, S.H. Isolation of porcine epidemic diarrhea virus (PEDV) in Korea. *Korean J. Vet. Res.* **1993**, *33*, 249–254.

51. Puranaveja, S.; Poolperm, P.; Lertwatcharasarakul, P.; Kesdaengsakonwut, S.; Boonsoongnern, A.; Urairong, K.; Kitikoon, P.; Choojai, P.; Kedkovid, R.; Teankum, K.; et al. Chinese-like strain of porcine epidemic diarrhea virus, Thailand. *Emerg. Infect. Dis.* **2009**, *15*, 1112–1115. [CrossRef] [PubMed]

52. Panel, E.A. Scientific Opinion on porcine epidemic diarrhoea and emerging pig deltacoronavirus. *EFSA J.* **2014**, *12*, 3877.

53. Song, D.; Moon, H.; Kang, B. Porcine epidemic diarrhea: A review of current epidemiology and available vaccines. *Clin. Exp. Vaccine Res.* **2015**, *4*, 166–176. [CrossRef] [PubMed]

54. Sun, R.Q.; Cai, R.J.; Chen, Y.Q.; Liang, P.S.; Chen, D.K.; Song, C.X. Outbreak of porcine epidemic diarrhea in suckling piglets, China. *Emerg. Infect. Dis.* **2012**, *18*, 161–163. [CrossRef] [PubMed]

55. Luo, Y.; Zhang, J.; Deng, X.; Ye, Y.; Liao, M.; Fan, H. Complete genome sequence of a highly prevalent isolate of porcine epidemic diarrhea virus in South China. *J. Virol.* **2012**, *86*, 9551. [CrossRef] [PubMed]

56. Dhanasekaran, S.M.; Barrette, T.R.; Ghosh, D.; Shah, R.; Varambally, S.; Kurachi, K.; Pienta, K.J.; Rubin, M.A.; Chinnaiyan, A.M. Delineation of prognostic biomarkers in prostate cancer. *Nature* **2001**, *412*, 822–826. [CrossRef] [PubMed]

57. Kitamoto, Y.; Yuan, X.; Wu, Q.; McCourt, D.W.; Sadler, J.E. Enterokinase, the initiator of intestinal digestion, is a mosaic protease composed of a distinctive assortment of domains. *Proc. Natl. Acad. Sci. USA* **1994**, *91*, 7588–7592. [CrossRef] [PubMed]

58. Yamaoka, K.; Masuda, K.I.; Ogawa, H.; Takagi, K.I.; Umemoto, N.; Yasuoka, S. Cloning and characterization of the cDNA for human airway trypsin-like protease. *J. Biol. Chem.* **1998**, *273*, 11895–11901. [CrossRef] [PubMed]

59.   Yasuoka, S.; Ohnishi, T.; Kawano, S.; Tsuchihashi, S.; Ogawara, M.; Masuda, K.; Yamaoka, K.; Takahashi, M.; Sano, T. Purification, characterization, and localization of a novel trypsin-like protease found in the human airway. *Am. J. Respir. Cell Mol.* **1997**, *16*, 300–308. [CrossRef] [PubMed]

60.   Hobson, J.P.; Netzel-Arnett, S.; Szabo, R.; Rehault, S.M.; Church, F.C.; Strickland, D.K.; Lawrence, D.A.; Antalis, T.M.; Bugge, T.H. Mouse DESC1 is located within a cluster of seven DESC1-like genes and encodes a type II transmembrane serine protease that forms serpin inhibitory complexes. *J. Biol. Chem.* **2004**, *279*, 46981–46994. [CrossRef] [PubMed]

61.   Lang, J.C.; Schuller, D.E. Differential expression of a novel serine protease homologue in squamous cell carcinoma of the head and neck. *Br. J. Cancer* **2001**, *84*, 237–243. [CrossRef] [PubMed]

62.   Bugge, T.H.; Antalis, T.M.; Wu, Q. Type II transmembrane serine proteases. *J. Biol. Chem.* **2009**, *284*, 23177–23181. [CrossRef] [PubMed]

63.   Szabo, R.; Bugge, T.H. Membrane-anchored serine proteases in vertebrate cell and developmental biology. *Annu. Rev. Cell Dev. Biol.* **2011**, *27*, 213–235. [CrossRef] [PubMed]

64.   Wicht, O.; Li, W.; Willems, L.; Meuleman, T.J.; Wubbolts, R.W.; van Kuppeveld, F.J.; Rottier, P.J.; Bosch, B.J. Proteolytic activation of the porcine epidemic diarrhea coronavirus spike fusion protein by trypsin in cell culture. *J. Virol.* **2014**, *88*, 7952–7961. [CrossRef] [PubMed]

65.   Shi, W.; (Northeast Agricultural University, Harbin, China). TMPRSS2 and MSPL promote PEDV replication in a dose-dependent manner at the stage of virus entry. in preparations. 2017.

66.   Liu, C.; Ma, Y.; Yang, Y.; Zheng, Y.; Shang, J.; Zhou, Y.; Jiang, S.; Du, L.; Li, J.; Li, F. Cell entry of porcine epidemic diarrhea coronavirus is activated by lysosomal proteases. *J. Biol. Chem.* **2016**, *291*, 24779–24786. [CrossRef] [PubMed]

10

Identifying and Characterizing Interplay between Hepatitis B Virus X Protein and Smc5/6

Christine M. Livingston [1], Dhivya Ramakrishnan [1], Michel Strubin [2], Simon P. Fletcher [1] and Rudolf K. Beran [1,*]

[1] Gilead Sciences, Foster City, CA 94404, USA; christine.marie.livingston@gmail.com (C.M.L.); Dhivya.Ramakrishnan@gilead.com (D.R.); simon.fletcher@gilead.com (S.P.F.)

[2] Department of Microbiology and Molecular Medicine, University Medical Center (CMU), 1211 Geneva, Switzerland; Michel.Strubin@unige.ch

* Correspondence: rudolf.beran@gilead.com

Academic Editors: Ulrike Protzer and Michael Nassal

Abstract: Hepatitis B X protein (HBx) plays an essential role in the hepatitis B virus (HBV) replication cycle, but the function of HBx has been elusive until recently. It was recently shown that transcription from the HBV genome (covalently-closed circular DNA, cccDNA) is inhibited by the structural maintenance of chromosome 5/6 complex (Smc5/6), and that a key function of HBx is to redirect the DNA-damage binding protein 1 (DDB1) E3 ubiquitin ligase to target this complex for degradation. By doing so, HBx alleviates transcriptional repression by Smc5/6 and stimulates HBV gene expression. In this review, we discuss in detail how the interplay between HBx and Smc5/6 was identified and characterized. We also discuss what is known regarding the repression of cccDNA transcription by Smc5/6, the timing of HBx expression, and the potential role of HBx in promoting hepatocellular carcinoma (HCC).

Keywords: HBx; HBV; DDB1; Smc5/6; cccDNA

1. Introduction

It is estimated that 250 million individuals are chronically infected with hepatitis B virus (HBV) [1,2]. Chronic hepatitis B (CHB) can lead to the development of cirrhosis and hepatocellular carcinoma (HCC), and more than 650,000 people die each year due to HBV-associated liver diseases. Nucleos(t)ide analogs and interferon-α (IFN-α) are approved for the treatment of CHB, but these therapies rarely lead to cure [2,3]. Thus, there is an urgent need to develop novel antiviral therapies.

HBV is a member of the *Hepadnaviridae* virus family. The HBV virion consists of an enveloped icosahedral capsid containing a 3.2 kb partially double-stranded DNA genome known as relaxed-circular DNA (rcDNA). Following cell binding and entry, rcDNA is deposited within the nucleus and is repaired to form covalently-closed circular DNA (cccDNA). cccDNA serves as the template for HBV pre-genomic RNA (pgRNA)—the intermediate form of the HBV genome—and also as the template for the transcription of all viral messenger RNAs (mRNAs). The HBV RNAs are translated into various viral proteins: the large, medium, and small envelope proteins (collectively HBsAg), E antigen (HBeAg), core, polymerase, and hepatitis B X protein (HBx) [4]. HBx is a 17 kDa protein conserved among mammalian hepadnaviruses [5,6] that is essential for HBV replication both in vitro and in vivo [7,8]. HBx is the only regulatory protein produced by HBV, and its role in the HBV lifecycle has long remained enigmatic.

HBx interactions with host proteins have been extensively studied to attempt to functionally define its role in the viral replication cycle. HBx has previously been reported to interact with a large number of host proteins [5,9–14]. However, the interaction with DNA-damage binding protein 1 (DDB1, also known as UVDDB-p127) was of particular interest because mutations that prevent X protein interaction with DDB1 inhibit hepadnavirus infection [5,13,15,16]. In addition, the structure of DDB1 complexed with a central peptide fragment of HBx has been solved [13]. DDB1 binds Cullin4 (Cul4) as part of an E3 ubiquitin ligase complex [17]. Various viruses hijack the DDB1–E3 ubiquitin ligase to promote the degradation of host proteins that would otherwise restrict viral replication. For example, the V protein of SV5 (a paramyxovirus) redirects the DDB1–E3 ubiquitin ligase to promote the degradation of Stat1 to prevent interferon signaling [18]. HIV Vpx also hijacks the DDB1–E3 ubiquitin ligase, but instead promotes the degradation of the antiviral factor SAMHD1 (SAM domain and HD domain-containing protein 1) [19]. It was therefore hypothesized that HBx binding to DDB1 could lead to proteasomal degradation of a specific cellular restriction factor [20,21].

In addition to binding DDB1, HBx has long been known to activate the transcription of a wide variety of genes encoded by episomal templates (i.e., closed-circular DNA molecules independent of cellular chromosomes), including cccDNA [22–28]. It was determined that HBx does so regardless of promoter or enhancer sequence, and thus acts as a non-specific transcriptional activator (transactivator) of episomal DNA. In contrast, HBx does not transactivate chromosomal genes [20,28]. Moreover, the transactivation of episomal DNA by HBx was shown to require an interaction of HBx with the DDB1–Cul4 ubiquitin ligase machinery [28]. Taken together, these observations suggest that HBx transactivation activity is dependent upon DDB1-mediated degradation of a cellular restriction factor.

2. HBx Promotes the Degradation of the Structural Maintenance of Chromosome 5/6 Complex, a Host Restriction Factor

In a recent study, we sought to identify the cellular factor(s) targeted for proteasomal degradation by HBx [20]. To do so, we expressed two tagged HBx-DDB1 fusion constructs in HepG2 cells: (1) wild-type HBx-DDB1, which binds Cul4 [13,29]; and (2) HBx-DDB1m4, which encodes a DDB1 mutant that cannot bind Cul4 [13]. Only wild-type HBx-DDB1 would be expected to bind the Cul4 E3 ubiquitin ligase as well as the cellular factor(s), and to target the cellular factor(s) for proteasomal degradation. In contrast, because HBx-DDB1m4 cannot bind the Cul4 E3 ubiquitin ligase, HBx-DDB1m4 would be expected to bind the cellular factor(s), but not target it for destruction. We then performed tandem affinity purification and identified the cellular proteins that bind these "baits" by mass spectrometry. As expected, wild-type HBx-DDB1 pulled down Cul4 and components of the E3 ubiquitin ligase. However, the mutant HBx-DDB1 pulled down the subunits of the structural maintenance of chromosome 5/6 (Smc5/6) complex (Smc5, Smc6, Nse1, Nse2, Nse3, and Nse4) (Figure 1). Consistent with this complex being targeted for proteosomal degradation by HBx, we found that Smc6 levels were lower in cells expressing HBx and in HBV-infected human hepatocytes in vitro and in vivo. Moreover, we observed that HBx selectively stimulatedgene expression from episomal DNA (including cccDNA) by targeting Smc5/6 for degradation. Chromatin immunoprecipitation (ChIP) experiments revealed that the Nse4 subunit (and presumably the entire Smc5/6 complex) directly bound episomal DNA, including cccDNA. Collectively, our observations suggest that Smc5/6 binds cccDNA to silence transcription in the absence of functional HBx. However, in the presence of functional HBx, Smc5/6 is degraded and cccDNA is transcribed (Figure 2).

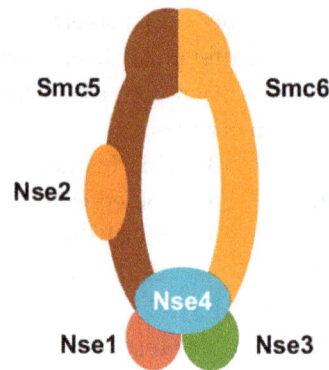

Figure 1. A cartoon representation of structural maintenance of chromosome 5/6 complex (Smc5/6). Smc5/6 is composed of Smc5, Smc6, Nse1, Nse2, Nse3, and Nse4.

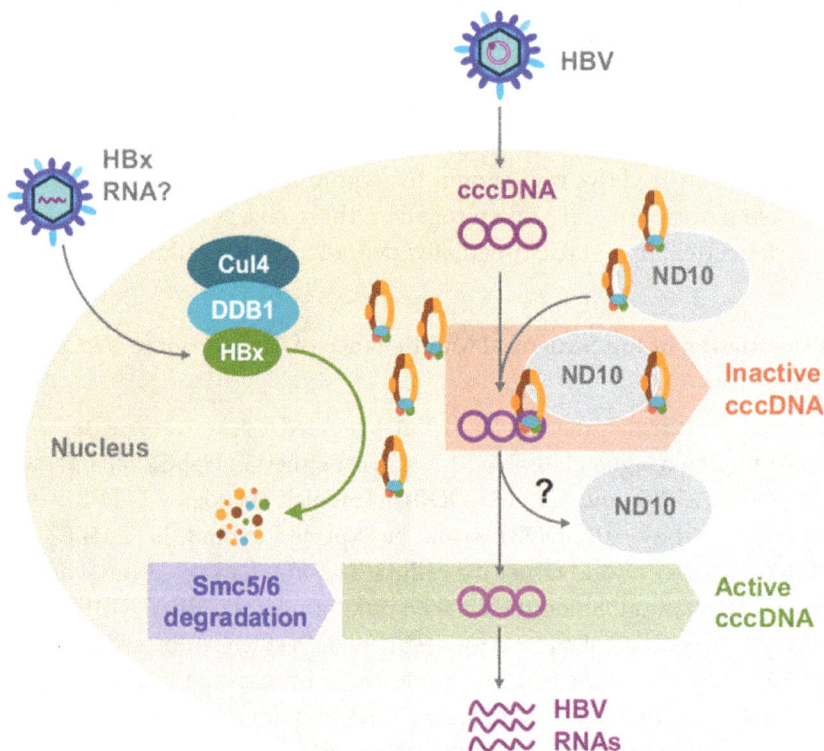

Figure 2. A model depicting the role of hepatitis B X protein (HBx) in hepatitis B virus (HBV) infection of a human hepatocyte. Relaxed-circular DNA (rcDNA) and possibly HBx RNA are deposited within the cell, and HBx protein may be translated from the HBx RNA. rcDNA is converted to covalently-closed circular DNA (cccDNA) and HBx binds Cullin4–DNA-damage binding protein1 (Cul4–DDB1). Structural maintenance of chromosomes 5/6 (Smc5/6) co-localizes with nuclear domain 10 (ND10) bodies. Cul4–DDB1–HBx targets Smc5/6 for ubiquitination. Smc5/6 is subsequently degraded by the proteasome, and cccDNA can now be transcribed.

The finding that HBx hijacks Cul4–DDB1 to promote the proteasomal degradation of Smc5/6 was subsequently confirmed by Murphy et al. [21] using a different method from Decorsière et al. [20]. Briefly, they expressed tagged HBx in HepG2 cells and treated with MLN4924 to inactivate E3-ubiquitin ligase activity. They subsequently purified HBx-interacting proteins through a tandem affinity purification strategy and analyzed tryptic peptides by liquid chromatography-tandem mass spectrometry. Using this method, they identified Smc5/6 subunits as putative HBx substrates.

Then, they also confirmed that HBV infection promotes the degradation of Smc5/6 in vitro and in vivo and that HBx activates episomal gene expression by promoting Smc5/6 degradation.

Besides demonstrating that HBx hijacks the DDB1–E3 ligase to target Smc5/6 for degradation, Murphy et al. [21] extended the work of Decorsière et al. [20]. Most notably, they determined that while Smc5/6 is targeted by HBx, the related chromosome maintenance complexes cohesin and condensin are not. Their findings also suggested that Smc5 and Smc6 may be directly polyubiquitinated in the presence of the Cul4–DDB1–HBx E3 ligase complex. These observations lend further support to the model that HBx selectively promotes degradation of Smc5/6 via an E3-ubiquitin ligase pathway.

While HBx is now known to promote Smc5/6 degradation, the details of how HBx modulates the interaction between the DDB1–E3 ligase and Smc5/6 remain to be determined. HBx contains a total of 154 residues, and residues 88–100 form a conserved alpha-helical motif called the H-box. The H-box is the minimal region required for DDB1 binding. Several point mutations in the HBx H-box region reduce HBx binding to DDB1 [30]. Apart from the H-box, plasmid-based HBV replication assays indicated that HBx residues 43–154 are essential for replication, while residues 1–42 are dispensable [31]. Point mutations or insertions at residues such as 58, 61, 68, 69, 119, 129, and 139 (all located outside of the H-box) also inhibited HBx binding to DDB1 [5,31]. These observations suggest that residues outside of the H-box influence DDB1 binding, consistent with studies of other DDB1-binding proteins [13,32,33]. Recently, a B-cell lymphoma 2 homology 3 (BH3)-like domain in HBx (residues 110–135) was identified. This BH3-like region also adopts an alpha-helical structure and binds very weakly to the BH3-binding groove of anti-apoptotic protein B-cell lymphoma 2 (Bcl-2) [34]. However, the biological relevance of an HBx interaction with Bcl-2 is unclear at this time, and it remains to be determined if the HBx BH3-like region is involved in Smc5/6 and/or DDB1 binding. Overall, the HBx mutagenesis studies suggest that the H-box—plus the region C-terminal to the H-box—are required for HBx function. Further studies will be needed to determine if HBx alone or both HBx and DDB1 interact with Smc5/6.

3. Smc5/6

Smc5/6 is a complex that directly binds DNA and is required for chromosome dynamics and stability [35,36]. Smc5/6 has been extensively studied in yeast (less so in mammals), and has been shown to play a role in homologous recombination as well as in resolving replication-induced DNA supercoiling [35–38]. In addition, a recent study demonstrated that Smc5/6 binds and topologically entraps plasmid DNA in an ATP-dependent manner [39].

Besides chromosome maintenance, our data suggests that Smc5/6 binds episomes (including cccDNA) and blocks episome transcription [20,28]. However, the mechanisms used by Smc5/6 to detect episomes and block their transcription are unclear. Smc5/6 recognition of episomes appears to be sequence-independent [20,28]. As for how Smc5/6 represses cccDNA transcription after detecting it, it is possible that Smc5/6 binding to cccDNA simply blocks RNA polymerase or other transcription factors from binding to cccDNA. A second possibility is that Smc5/6 recognizes the topological features unique to episomes undergoing transcription and topologically entraps episomes shortly after transcription initiates. This would prevent the proper movement of the transcription complex along the episome. In line with this second hypothesis, low levels of HBV transcription have been observed during the first few days after infection with HBx-negative HBV [7]. Further studies will be needed to elucidate a detailed understanding of Smc5/6 transcriptional repression on cccDNA and other episomes.

The spatial relationship between Smc5/6 and other nuclear components may be closely tied to the ability of Smc5/6 to recognize and inhibit cccDNA transcription. Recently, we reported that Smc5/6 co-localizes with promyelocytic leukemia protein (PML) and speckled protein of 100 kDa (Sp100) in primary human hepatocytes [40] (Figure 2). PML and Sp100 are major structural components of nuclear domain 10 bodies (ND10), which are dynamic nuclear protein aggregates [41,42]. ND10 have been shown to traffic to the incoming DNA genomes of various viruses and restrict viral transcription. Viruses targeted by ND10 include herpes simplex virus-1, Kaposi's sarcoma-associated herpesvirus, and human cytomegalovirus [43–47]. Along these lines, we recently showed that ND10 co-localization with Smc5/6 is required for repression of cccDNA transcription. Indeed, small interfering RNA (siRNA) knockdown of ND10 components dispersed Smc5/6 and rescued HBx-negative HBV transcription [40]. This suggests that cccDNA co-localizes with ND10 during its interaction with Smc5/6, though this has not yet been shown directly.

4. HBx RNA Is Present in Chronic HBV-Infected Patient Plasma

A major question that arises from the recent HBx studies is how is HBx expressed if it is required to alleviate transcriptional repression of cccDNA by Smc5/6? If Smc5/6 only recognizes episomal DNA undergoing transcription and low levels of cccDNA transcription occur before Smc5/6 repression [7], then some HBx mRNA might transcribe before Smc5/6 represses. An alternative possibility is that HBx protein or RNA is carried into the cell by HBV-like particles. We recently determined that Smc5/6 was degraded in the majority of HBV-infected human hepatocytes by the time cccDNA transcription could be detected [40], suggesting that HBx is expressed very early during infection. Indeed, using RNA sequencing (RNA-Seq) we detected RNA reads mapping to HBx during the first 24 h post-infection, whereas by 2 days post-infection the RNA reads mapped across the entire HBV genome with high abundance [40]. We also recently observed by RNA-Seq that HBx RNA is present in HBV preparations produced from cell culture as well as in chronic HBV-infected patient plasma [40]. At this time, it is unknown if the HBx RNA is packaged into HBV-like particles which can be secreted from infected cells, as has been reported for pgRNA [48]. Also, it is not known if this HBx RNA is translated into functional HBx. It is tempting to hypothesize that HBV counters Smc5/6 repression of cccDNA transcription very early during infection by delivering HBx RNA into the cell (Figure 2). However, much work remains to determine if this is the case.

5. The Potential Role of HBx Activity in Promoting Hepatocellular Carcinoma

The discovery that HBx potentially targets Smc5/6 for degradation has important implications for HBV pathogenesis. HBV infection has long been known to be associated with HCC development, though the mechanistic reasons are not fully understood [49]. Over-expression of HBx in non-dividing cells does not have deleterious effects [50]. This is consistent with the lack of a cellular stress response to HBV-infected hepatocytes [40]. In contrast, over-expression of HBx or depletion of Smc5/6 in dividing cells induces genomic instability [50,51]. Moreover, loss of Smc5/6 may predispose cells to genetic errors under conditions of DNA damage [52] (e.g., induced by necroinflammation in CHB), and reduced expression of the NSMCE2 (Nse2) subunit is associated with increased cancer incidence in mice [53]. Thus, it is plausible that HBx-promoted loss of Smc5/6 may be a contributing factor to the development of HBV-related HCC.

6. Conclusions

Our work and that of others has recently demonstrated that HBx-mediated degradation of Smc5/6 is necessary for cccDNA transcription. However, key questions remain concerning how HBx is expressed despite Smc5/6 repression of cccDNA transcription, how Smc5/6 detects episomal DNA and inhibits cccDNA transcription, how cccDNA transcription is activated once Smc5/6 suppression is relieved, how HBx physically interacts with both DDB1 and Smc5/6, and whether HBx contributes to HBV pathogenesis by promoting Smc5/6 degradation. These are important questions to resolve because there is a strong need to develop new antivirals that lead to a functional cure of CHB [54], and a HBx inhibitor may be a key component of a future curative regimen.

References

1. Schweitzer, A.; Horn, J.; Mikolajczyk, R.T.; Krause, G.; Ott, J.J. Estimations of worldwide prevalence of chronic hepatitis B virus infection: A systematic review of data published between 1965 and 2013. *Lancet* **2015**, *386*, 1546–1555. [CrossRef]

2. Lozano, R.; Naghavi, M.; Foreman, K.; Lim, S.; Shibuya, K.; Aboyans, V.; Abraham, J.; Adair, T.; Aggarwal, R.; Ahn, S.Y.; et al. Global and regional mortality from 235 causes of death for 20 age groups in 1990 and 2010: A systematic analysis for the global burden of disease study 2010. *Lancet* **2012**, *380*, 2095–2128. [CrossRef]

3. Kwon, H.; Lok, A.S. Hepatitis b therapy. *Nat. Rev. Gastroenterol. Hepatol.* **2011**, *8*, 275–284. [CrossRef] [PubMed]

4. Urban, S.; Schulze, A.; Dandri, M.; Petersen, J. The replication cycle of hepatitis B virus. *J. Hepatol.* **2010**, *52*, 282–284. [CrossRef] [PubMed]

5. Sitterlin, D.; Lee, T.H.; Prigent, S.; Tiollais, P.; Butel, J.S.; Transy, C. Interaction of the UV-damaged DNA-binding protein with hepatitis B virus X protein is conserved among mammalian hepadnaviruses and restricted to transactivation-proficient X-insertion mutants. *J. Virol.* **1997**, *71*, 6194–6199. [PubMed]

6. Van Hemert, F.J.; van de Klundert, M.A.; Lukashov, V.V.; Kootstra, N.A.; Berkhout, B.; Zaaijer, H.L. Protein X of hepatitis B virus: Origin and structure similarity with the central domain of DNA glycosylase. *PLoS ONE* **2011**, *6*, e23392. [CrossRef] [PubMed]

7. Lucifora, J.; Arzberger, S.; Durantel, D.; Belloni, L.; Strubin, M.; Levrero, M.; Zoulim, F.; Hantz, O.; Protzer, U. Hepatitis B virus X protein is essential to initiate and maintain virus replication after infection. *J. Hepatol.* **2011**, *55*, 996–1003. [CrossRef] [PubMed]

8. Tsuge, M.; Hiraga, N.; Akiyama, R.; Tanaka, S.; Matsushita, M.; Mitsui, F.; Abe, H.; Kitamura, S.; Hatakeyama, T.; Kimura, T.; et al. HBx protein is indispensable for development of viraemia in human hepatocyte chimeric mice. *J. Gen. Virol.* **2010**, *91*, 1854–1864. [CrossRef] [PubMed]

9. Wei, C.; Ni, C.; Song, T.; Liu, Y.; Yang, X.; Zheng, Z.; Jia, Y.; Yuan, Y.; Guan, K.; Xu, Y.; et al. The hepatitis B virus X protein disrupts innate immunity by downregulating mitochondrial antiviral signaling protein. *J. Immunol.* **2010**, *185*, 1158–1168. [CrossRef] [PubMed]

10. Zhang, S.M.; Sun, D.C.; Lou, S.; Bo, X.C.; Lu, Z.; Qian, X.H.; Wang, S.Q. HBx protein of hepatitis B virus (HBV) can form complex with mitochondrial hsp60 and hsp70. *Arch. Virol.* **2005**, *150*, 1579–1590. [CrossRef] [PubMed]

11. Melegari, M.; Scaglioni, P.P.; Wands, J.R. Cloning and characterization of a novel hepatitis B virus X binding protein that inhibits viral replication. *J. Virol.* **1998**, *72*, 1737–1743. [PubMed]

12. Benhenda, S.; Ducroux, A.; Riviere, L.; Sobhian, B.; Ward, M.D.; Dion, S.; Hantz, O.; Protzer, U.; Michel, M.L.; Benkirane, M.; et al. Methyltransferase PRMT1 is a binding partner of HBx and a negative regulator of hepatitis B virus transcription. *J. Virol.* **2013**, *87*, 4360–4371. [CrossRef] [PubMed]

13. Li, T.; Robert, E.I.; van Breugel, P.C.; Strubin, M.; Zheng, N. A promiscuous alpha-helical motif anchors viral hijackers and substrate receptors to the Cul4-DDB1 ubiquitin ligase machinery. *Nat. Struct. Mol. Biol.* **2010**, *17*, 105–111. [CrossRef] [PubMed]

14. Becker, S.A.; Lee, T.H.; Butel, J.S.; Slagle, B.L. Hepatitis B virus X protein interferes with cellular DNA repair. *J. Virol.* **1998**, *72*, 266–272. [PubMed]

15. Sitterlin, D.; Bergametti, F.; Tiollais, P.; Tennant, B.C.; Transy, C. Correct binding of viral X protein to UVDDB-p127 cellular protein is critical for efficient infection by hepatitis B viruses. *Oncogene* **2000**, *19*, 4427–4431. [CrossRef] [PubMed]

16. Sitterlin, D.; Bergametti, F.; Transy, C. UVDDB p127-binding modulates activities and intracellular distribution of hepatitis B virus X protein. *Oncogene* **2000**, *19*, 4417–4426. [CrossRef] [PubMed]

17. Shiyanov, P.; Nag, A.; Raychaudhuri, P. Cullin 4A associates with the UV-damaged DNA-binding protein ddb. *J. Biol. Chem.* **1999**, *274*, 35309–35312. [CrossRef] [PubMed]

18. Didcock, L.; Young, D.F.; Goodbourn, S.; Randall, R.E. The V protein of simian virus 5 inhibits interferon signalling by targeting STAT1 for proteasome-mediated degradation. *J. Virol.* **1999**, *73*, 9928–9933. [PubMed]

19. Hrecka, K.; Hao, C.; Gierszewska, M.; Swanson, S.K.; Kesik-Brodacka, M.; Srivastava, S.; Florens, L.; Washburn, M.P.; Skowronski, J. Vpx relieves inhibition of HIV-1 infection of macrophages mediated by the SAMHD1 protein. *Nature* **2011**, *474*, 658–661. [CrossRef] [PubMed]

20. Decorsiere, A.; Mueller, H.; van Breugel, P.C.; Abdul, F.; Gerossier, L.; Beran, R.K.; Livingston, C.M.; Niu, C.; Fletcher, S.P.; Hantz, O.; et al. Hepatitis B virus X protein identifies the Smc5/6 complex as a host restriction factor. *Nature* **2016**, *531*, 386–389. [CrossRef] [PubMed]

21. Murphy, C.M.; Xu, Y.; Li, F.; Nio, K.; Reszka-Blanco, N.; Li, X.; Wu, Y.; Yu, Y.; Xiong, Y.; Su, L. Hepatitis B virus X protein promotes degradation of Smc5/6 to enhance HBV replication. *Cell Rep.* **2016**, *16*, 2846–2854. [CrossRef] [PubMed]

22. Colgrove, R.; Simon, G.; Ganem, D. Transcriptional activation of homologous and heterologous genes by the hepatitis B virus X gene product in cells permissive for viral replication. *J. Virol.* **1989**, *63*, 4019–4026. [PubMed]

23. Twu, J.S.; Schloemer, R.H. Transcriptional trans-activating function of hepatitis B virus. *J. Virol.* **1987**, *61*, 3448–3453. [PubMed]

24. Spandau, D.F.; Lee, C.H. Trans-activation of viral enhancers by the hepatitis B virus X protein. *J. Virol.* **1988**, *62*, 427–434. [PubMed]

25. Seto, E.; Yen, T.S.; Peterlin, B.M.; Ou, J.H. Trans-activation of the human immunodeficiency virus long terminal repeat by the hepatitis B virus X protein. *Proc. Natl. Acad. Sci. USA* **1988**, *85*, 8286–8290. [CrossRef] [PubMed]

26. Cross, J.C.; Wen, P.; Rutter, W.J. Transactivation by hepatitis B virus X protein is promiscuous and dependent on mitogen-activated cellular serine/threonine kinases. *Proc. Natl. Acad. Sci. USA* **1993**, *90*, 8078–8082. [CrossRef] [PubMed]

27. Tang, H.; Delgermaa, L.; Huang, F.; Oishi, N.; Liu, L.; He, F.; Zhao, L.; Murakami, S. The transcriptional transactivation function of HBx protein is important for its augmentation role in hepatitis B virus replication. *J. Virol.* **2005**, *79*, 5548–5556. [CrossRef] [PubMed]

28. Van Breugel, P.C.; Robert, E.I.; Mueller, H.; Decorsiere, A.; Zoulim, F.; Hantz, O.; Strubin, M. Hepatitis B virus X protein stimulates gene expression selectively from extrachromosomal DNA templates. *Hepatology* **2012**, *56*, 2116–2124. [CrossRef] [PubMed]

29. Leupin, O.; Bontron, S.; Strubin, M. Hepatitis B virus X protein and simian virus 5 V protein exhibit similar UV-DDB1 binding properties to mediate distinct activities. *J. Virol.* **2003**, *77*, 6274–6283. [CrossRef] [PubMed]

30. Leupin, O.; Bontron, S.; Schaeffer, C.; Strubin, M. Hepatitis B virus X protein stimulates viral genome replication via a DDB1-dependent pathway distinct from that leading to cell death. *J. Virol.* **2005**, *79*, 4238–4245. [CrossRef] [PubMed]

31. Hodgson, A.J.; Hyser, J.M.; Keasler, V.V.; Cang, Y.; Slagle, B.L. Hepatitis B virus regulatory HBx protein binding to DDB1 is required but is not sufficient for maximal HBV replication. *Virology* **2012**, *426*, 73–82. [CrossRef] [PubMed]

32. Andrejeva, J.; Poole, E.; Young, D.F.; Goodbourn, S.; Randall, R.E. The p127 subunit (DDB1) of the UV-DNA damage repair binding protein is essential for the targeted degradation of STAT1 by the V protein of the Paramyxovirus simian virus 5. *J. Virol.* **2002**, *76*, 11379–11386. [CrossRef] [PubMed]

33. Angers, S.; Li, T.; Yi, X.; MacCoss, M.J.; Moon, R.T.; Zheng, N. Molecular architecture and assembly of the DDB1-CUL4A ubiquitin ligase machinery. *Nature* **2006**, *443*, 590–593. [CrossRef] [PubMed]

34. Jiang, T.; Liu, M.; Wu, J.; Shi, Y. Structural and biochemical analysis of Bcl-2 interaction with the hepatitis B virus protein HBx. *Proc. Natl. Acad. Sci. USA* **2016**, *113*, 2074–2079. [CrossRef] [PubMed]

35. Jeppsson, K.; Carlborg, K.K.; Nakato, R.; Berta, D.G.; Lilienthal, I.; Kanno, T.; Lindqvist, A.; Brink, M.C.; Dantuma, N.P.; Katou, Y.; et al. The chromosomal association of the Smc5/6 complex depends on cohesion and predicts the level of sister chromatid entanglement. *PLoS Genet.* **2014**, *10*, e1004680. [CrossRef] [PubMed]

36. Jeppsson, K.; Kanno, T.; Shirahige, K.; Sjogren, C. The maintenance of chromosome structure: Positioning and functioning of Smc complexes. *Nat. Rev. Mol. Cell Biol.* **2014**, *15*, 601–614. [CrossRef] [PubMed]

37. Potts, P.R. The yin and yang of the MMS21-SMC5/6 sumo ligase complex in homologous recombination. *DNA Repair (Amst)* **2009**, *8*, 499–506. [CrossRef] [PubMed]

38. De Piccoli, G.; Cortes-Ledesma, F.; Ira, G.; Torres-Rosell, J.; Uhle, S.; Farmer, S.; Hwang, J.Y.; Machin, F.; Ceschia, A.; McAleenan, A.; et al. Smc5-Smc6 mediate DNA double-strand-break repair by promoting sister-chromatid recombination. *Nat. Cell Biol.* **2006**, *8*, 1032–1034. [CrossRef] [PubMed]

39. Kanno, T.; Berta, D.G.; Sjogren, C. The Smc5/6 complex is an ATP-dependent intermolecular DNA linker. *Cell Rep.* **2015**, *12*, 1471–1482. [CrossRef] [PubMed]

40. Niu, C.; Livingston, C.M.; Li, L.; Beran, R.K.; Daffis, S.; Ramakrishnan, D.; Burdette, D.; Peiser, L.; Salas, E.; Ramos, H.; et al. The Smc5/6 complex restricts hbv when localized to ND10 without inducing an innate immune response and is counteracted by the HBV X protein shortly after infection. *PLoS ONE* **2017**, *12*, e0169648. [CrossRef] [PubMed]

41. Everett, R.D. Interactions between DNA viruses, ND10 and the DNA damage response. *Cell. Microbiol.* **2006**, *8*, 365–374. [CrossRef] [PubMed]

42. Everett, R.D.; Rechter, S.; Papior, P.; Tavalai, N.; Stamminger, T.; Orr, A. PML contributes to a cellular mechanism of repression of Herpes simplex virus type 1 infection that is inactivated by ICP0. *J. Virol.* **2006**, *80*, 7995–8005. [CrossRef] [PubMed]

43. Everett, R.D.; Maul, G.G. HSV-1 ie protein Vmw110 causes redistribution of PML. *EMBO J.* **1994**, *13*, 5062–5069. [PubMed]

44. Tavalai, N.; Papior, P.; Rechter, S.; Stamminger, T. Nuclear domain 10 components promyelocytic leukemia protein and hDaxx independently contribute to an intrinsic antiviral defense against human cytomegalovirus infection. *J. Virol.* **2008**, *82*, 126–137. [CrossRef] [PubMed]

45. Full, F.; Jungnickl, D.; Reuter, N.; Bogner, E.; Brulois, K.; Scholz, B.; Sturzl, M.; Myoung, J.; Jung, J.U.; Stamminger, T.; et al. Kaposi's sarcoma associated herpesvirus tegument protein ORF75 is essential for viral lytic replication and plays a critical role in the antagonization of ND10-instituted intrinsic immunity. *PLoS Pathog.* **2014**, *10*, e1003863. [CrossRef] [PubMed]

46. Adler, M.; Tavalai, N.; Muller, R.; Stamminger, T. Human cytomegalovirus immediate-early gene expression is restricted by the nuclear domain 10 component Sp100. *J. Gen. Virol.* **2011**, *92*, 1532–1538. [CrossRef] [PubMed]

47. Kelly, C.; Van Driel, R.; Wilkinson, G.W. Disruption of PML-associated nuclear bodies during human cytomegalovirus infection. *J. Gen. Virol.* **1995**, *76*, 2887–2893. [CrossRef] [PubMed]

48. Wang, J.; Shen, T.; Huang, X.; Kumar, G.R.; Chen, X.; Zeng, Z.; Zhang, R.; Chen, R.; Li, T.; Zhang, T.; et al. Serum hepatitis B virus RNA is encapsidated pregenome rna that may be associated with persistence of viral infection and rebound. *J. Hepatol.* **2016**, *65*, 700–710. [CrossRef] [PubMed]

49. Levrero, M.; Zucman-Rossi, J. Mechanisms of HBV-induced hepatocellular carcinoma. *J. Hepatol.* **2016**, *64*, S84–S101. [CrossRef] [PubMed]

50. Martin-Lluesma, S.; Schaeffer, C.; Robert, E.I.; van Breugel, P.C.; Leupin, O.; Hantz, O.; Strubin, M. Hepatitis B virus X protein affects s phase progression leading to chromosome segregation defects by binding to damaged DNA binding protein 1. *Hepatology* **2008**, *48*, 1467–1476. [CrossRef] [PubMed]

51. Gallego-Paez, L.M.; Tanaka, H.; Bando, M.; Takahashi, M.; Nozaki, N.; Nakato, R.; Shirahige, K.; Hirota, T. Smc5/6-mediated regulation of replication progression contributes to chromosome assembly during mitosis in human cells. *Mol. Biol. Cell* **2014**, *25*, 302–317. [CrossRef] [PubMed]

52. Wu, N.; Yu, H. The Smc complexes in DNA damage response. *Cell Biosci.* **2012**, *2*, 5. [CrossRef] [PubMed]

53. Jacome, A.; Gutierrez-Martinez, P.; Schiavoni, F.; Tenaglia, E.; Martinez, P.; Rodriguez-Acebes, S.; Lecona, E.; Murga, M.; Mendez, J.; Blasco, M.A.; et al. NSMCE2 suppresses cancer and aging in mice independently of its SUMO ligase activity. *EMBO J.* **2015**, *34*, 2604–2619. [CrossRef] [PubMed]

54. Fletcher, S.P.; Delaney, W.E.T. New therapeutic targets and drugs for the treatment of chronic hepatitis B. *Semin. Liver Dis.* **2013**, *33*, 130–137. [CrossRef] [PubMed]

11

MTase Domain of *Dendrolimus punctatus cypovirus* VP3 Mediates Virion Attachment and Interacts with Host ALP Protein

Lan Su [1,2], **Congrui Xu** [1,2], **Chuangang Cheng** [1], **Chengfeng Lei** [1] and **Xiulian Sun** [1,*]

[1] Wuhan Institute of Virology, Chinese Academy of Sciences, Wuhan 430071, China; sul@wh.iov.cn (L.S.); xucr@wh.iov.cn (C.X.); ccg083@126.com (C.C.); cflei@wh.iov.cn (C.L.)

[2] University of Chinese Academy of Sciences, Beijing 100049, China

* Correspondence: sunxl@wh.iov.cn

Academic Editor: Karyn Johnson

Abstract: *Dendrolimus punctatus cypovirus* (DpCPV) is an important pathogen of *D. punctatus*, but little is known about the mechanisms of DpCPV infection. Here, we investigated the effects of VP3, VP4 and VP5 structural proteins on the viral invasion. Both the C-terminal of VP3 (methyltransferase (MTase) domain) and VP4 (A-spike) bound to *Spodoptera exigua* midgut brush border membrane vesicles (BBMVs) in a dose-dependent manner, and the binding was inhibited by purified DpCPV virions. Importantly, anti-MTase and anti-VP4 antibodies inhibited viral binding to *S. exigua* BBMVs. Using far-Western blots, a 65 kDa protein in *Bombyx mori* BBMVs, identified as alkaline phosphatase protein (*Bm*ALP) by mass spectrometry, specifically interacted with DpCPV MTase. The interaction between MTase and *Bm*ALP was verified by co-immunoprecipitation in vitro. Pretreatment of *B. mori* BBMVs with an anti-ALP antibody or incubation of DpCPV virions with prokaryotically expressed *Bm*ALP reduced viral attachment. Additionally, *Bm*ALP inhibited DpCPV infection in *S. exigua* larvae. Our data provide evidence that the MTase domain and A-spike function as viral attachment proteins during the DpCPV infection process, and ALP is the ligand that interacts with DpCPV via the MTase domain. These results augment our understanding of the mechanisms used by cypoviruses to enter their hosts.

Keywords: DpCPV; attachment proteins; ALP; viral entry

1. Introduction

Cypovirus (CPV), a segmented double-stranded ribonucleic acid (dsRNA) virus within the genus *Cypovirus* (*Reoviridae* family), possesses a single capsid layer and is commonly embedded in the characteristic crystalline inclusion body called a polyhedron, which is formed in the cell cytoplasm of insects [1]. CPV infection occurs via the fecal-oral route and is mainly restricted to the larvae midgut. Accordingly, after the consumption of contaminated food by the larvae, the polyhedra dissolve in the highly alkaline environment of the midgut, releasing virions that adhere to and penetrate the plasma membrane of microvilli and settle in the cytoplasm of the columnar epithelial cells [2]. The recent study showed that CPV enter the cell by clathrin-mediated endocytosis employing integrin beta and receptor for activated protein kinase C (RACK1) [3]. Consequently, populations of progeny viruses are synthesized and embedded into a polyhedrin. As the disease progresses, the digestive and absorptive functions are severely affected and the larvae die one after another.

Dendrolimus punctatus cypovirus (DpCPV) is an important viral pathogen isolated from *D. punctatus*, with a relatively wide host range in *Lepidoptera*. In China, this virus species was developed in 2010 as a commercial insecticide to control the pine caterpillar, *D. punctatus*. Cryoelectron microscopy studies have documented the fine structures of the viral particles [4,5]. On the single capsid shell formed by VP1, three turret proteins exist: B-spikes formed by the C-terminal of VP3 (methyltransferase (MTase) domain), A-spikes formed by VP4 [6], and large protrusion proteins formed by VP5. The A-spike extends from the surface of the B-spike, but it is flexibly attached to the B-spike [5,7]. Examination of density slices indicates that the A-spike is directly bound to a polyhedron [5–7]. As the A-spike is the most projecting portion of CPV virions, and has a similar localization on the virion structure to the σ1 protein of mammalian orthoreovirus (MRV), it has been speculated that the CPV A-spike is probably associated with cell attachment and penetration [5,7,8]. Despite its precise functions in CPV infection, its interactions with host proteins have not been determined yet.

In general, viral attachment to one or more cellular receptors is the first step of viral invasion, and this process may determine the host range and cell tissue tropism of a virus [9,10]. It is believed that viral spike proteins perform essential functions in receptor binding and cell penetration, as seen with the VP7 protein of the bluetongue virus [9], the VP4 spikes of rotavirus [11] and the σ1 protein of MRV [12]. In *Reoviridae*, the entry details of reovirus and rotavirus have been well characterized. As the attachment protein of reovirus, an σ1 tail inserts into the virion via "turrets" formed by the λ2 protein, and its head projects away from the virion surface [13]. Reovirus initially tethers to sialylated glycans by the σ1 tail with low affinity, followed by higher affinity binding to junctional adhesion molecule A (JAM-A) by the σ1 head [14]. In the central nervous system, reovirus binds the Nogo receptor, NgR1, a leucine-rich repeat protein, to infect neurons [10]. The rotavirus cell entry is a multistep process. It is proposed that the rotavirus functional receptor is likely to be a complex of several cell components including gangliosides, integrins, and probably other proteins such as hsc70, which might be associated in lipid rafts and which need the lipid microdomain organization to efficiently mediate cell entry of rotaviruses [15].

In this study, we tested the binding characteristics of DpCPV turret proteins to midgut brush border membrane vesicles (BBMVs), followed by competition and neutralization assays. Furthermore, we showed that MTase and VP4 functioned as the viral attachment proteins in the DpCPV infection process and alkaline phosphatase protein (ALP) served as the MTase ligand.

2. Materials and Methods

2.1. Viruses, Larvae, and Antibodies

DpCPV was initially isolated from *D. punctatus* larvae in Macheng, Hubei, China, and propagated in *Spodoptera exigua* larvae. *S. exigua* and *Bombyx mori* larvae were obtained from the Core Facility and Technical Support, Wuhan Institute of Virology, Chinese Academy of Sciences. Polyclonal antibodies against the DpCPV virions, MTase domain of VP3 [16], VP4 [6] and VP5 [17] were prepared previously in our laboratory.

2.2. Recombinant Proteins

(i) Maltose binding protein (MBP)-fused DpCPV structural proteins. DpCPV VP3 was truncated to the galactosyltransferase (GTase) (1–366 aa) and MTase (406–1058 aa) domains based on its structure [5]. The gene segments encoding GTase, MTase, VP4, and VP5 were amplified from a complimentary deoxyribonucleic acid (cDNA) library [6] using the corresponding PCR primer pairs (Sangon, Shanghai, China) (listed in Table 1) and cloned into a pMal-C2X vector (NEB (Beijing), Beijing, China) as N-terminal MBP fusion proteins. The recombinant plasmids were transformed into *Escherichia coli* BL21 cells (Invitrogen, Carlsbad, CA, USA), and induced by isopropyl-β-d-thiogalactopyranoside for protein production. The soluble fusion proteins were purified by amylose resin column separation (NEB) according to the manufacture's instruction manual.

(ii) *Bm*ALP. The truncated *alp* gene fragment without the transmembrane domain encoding amino acids 1–519 (GenBank: NM_001044071.3) was amplified from *B. mori* midgut total RNA by reverse transcriptase PCR with the primer pairs *alp*-F: 5′-CGGAATTCATGTCTACATGGTGGTTAGTTGTG-3′

and *alp*-R: 5′-CCCAAGCTTTTAGCGGCCCGGGC-3′ and then cloned into a pET28a vector (Invitrogen) with a 6 × His-tag at the C-terminal. The cell lysate from the transformed *E. coli* BL21 cells (Invitrogen) was purified by a Ni-nitrilotriacetic acid (NTA) agarose column (Invitrogen). Purified ALP was used to produce polyclonal antibodies in an ALP-immunized rabbit.

Table 1. Primers used in this study.

Primer	Primer Sequence (5′→3′)
gtase-F	CGGGATCCATGTGGCATTATACGAGTATCAAC (*Bam*HI Site underlined)
gtase-R	CCCAAGCTTTTAGGAAATATAATTCGCGGTGAT (*Hind*III Site underlined)
mtase-F	CGGAATTCGAGCCAATGAGCATAGCT (*Eco*RI Site underlined)
mtase-R	CGGGATCCTTACGAGCGTACAAGTTTGATC (*Bam*HI Site underlined)
vp4-F	CGGAATTCATGTTCGCAATCGATCCACT (*Eco*RI Site underlined)
vp4-R	CGGAATTCATGTTCGCAATCGATCCACT (*Bam*HI Site underlined)
vp5-F	CGGAATTCATGTTACAACAACCAGCAGGA (*Eco*RI Site underlined)
vp5-R	CGGGATCCTCATAGGATGTCATCTGAGTGC (*Bam*HI Site underlined)

2.3. Purification of DpCPV Virions

DpCPV occlusion bodies (OBs) were isolated from infected *S. exigua* larvae by differential centrifugation. To obtain DpCPV virions, OB suspensions were lysed with 0.2 M Na_2CO_3-$NaHCO_3$ (pH 10.8) and then pH was adjusted to 7.5 with 1 M Tris-Cl (pH 6.8), then purified by linear 20%–60% (*w/v*) sucrose gradient centrifugation [16]. The concentration of the purified virions was determined by a UV-31000PC spectrophotometer (Mapada, Shanghai, China) at 280 nm.

2.4. Preparation of BBMVs

Midgut brush border membrane vesicles (BBMVs) were prepared from the dissected midguts of fourth-instar *S. exigua* or *B. mori* larvae by the differential magnesium precipitation method [18] and then stored in MET buffer (0.3 M Mannitol, 17 mM Tris-HCl, 5 mM EGTA, pH 7.5) at −80 °C. The protein concentration was determined by the Bradford method with bovine serum albumin (BSA) as the standard.

2.5. Binding Characteristics of S. exigua BBMVs and Competition Assays

ELISA plates (96-well) were coated with 10 µg/mL of *S. exigua* BBMVs (*Se*BBMVs) in carbonate buffer at 4 °C overnight and blocked with 3% BSA for 2 h. Increasing amounts of purified DpCPV virions (0.01–100 µg) or MBP-fusion proteins (0.02–4 µM) in a total volume of 100 µL were added to the wells and incubated for 1 h. The wells were washed with PBST (4 mM KH_2PO_4, 16 mM Na_2HPO_4, 115 mM NaCl, 0.05% Tween 20, pH 7.4) and incubated with a rabbit anti-DpCPV virion antibody or a mouse anti-MBP mAb (Sigma-Aldrich, St. Louis, MO, USA) followed by horseradish peroxidase (HRP) labeled anti-rabbit IgG (Sigma-Aldrich) or HRP labeled anti-mouse IgG (Proteintech, Wuhan, China). The reactions were visualized using TMB reagent (Beyotime, Nantong, China), stopped by addition of 2 M H_2SO_4, and the absorbance was measured at 450 nm using an ELx808 absorbance reader (BioTek, Winooski, VT, USA). The non-specific binding of BSA coated wells, instead of BBMVs, was subtracted from the total binding value. For the competition assays, the concentrations of the MBP-fusion proteins were kept stable (0.3 μM), while the DpCPV virions were serially diluted (0–10 μg). The protein-virion mixtures were added to the wells; after absorption for 1 h, the bound MBP-fusion proteins were detected as described above.

2.6. Inhibition of Viral Binding by Antibodies

For the inhibition assays, DpCPV virions (final amounts, 10 μg each) were pretreated with different concentrations (1–100 μg/mL) of anti-MTase antibody, anti-VP4 antibody, a mixture of anti-MTase and anti-VP4 antibodies or the pre-immune antibody at 37 °C for 1 h. The individual mixtures were added to SeBBMVs (10 μg/mL) coated on 96-well ELISA plates. After absorption for 1 h, the bound virions were detected as described above. The binding rates for each concentration of the three antibody treatments were compared by One-way ANOVA followed by least significant difference (LSD) t-tests after Arcsin square root transformation.

2.7. Far-Western Blotting

B. mori BBMVs (BmBBMVs) (10 μg/lane) were separated by 10% sodium dodecyl (lauryl) sulfate-polyacrylamide gel electrophoresis (SDS-PAGE), transferred to a polyvinylidene difluoride (PVDF) membrane (Millipore, Billerica, MA, USA) and blocked in 5% BSA in phosphate buffered saline (PBS) for 2 h. The membranes were incubated with 5 μg/mL of purified MBP-MTase, MBP-VP4, or MBP proteins (as the negative control) for 1 h at room temperature. After washing with PBST buffer, the membranes were incubated with a mouse anti-MBP mAb (Sigma-Aldrich) followed by HRP-labeled anti-mouse IgG (Proteintech), and visualized using ECL reagent (Beyotime).

2.8. Mass Spectrometry Analysis

A coomassie brilliant blue stained protein band of 65 kDa was excised from a 10% SDS-PAGE gel and subjected to in-gel digestion using trypsin at 37 °C overnight. Liquid chromatography- tandem mass spectrometry (LC-MS/MS) was performed, and the obtained peptide sequences were analyzed by the MASCOT program based on the database of Bombyx.

2.9. Co-Immunoprecipitation (Co-IP) Assays

Protein G Dynabeads (Thermo Scientific, Waltham, MA, USA) (20 μL) were immobilized with 5 μL of mouse anti-MBP mAb (Sigma-Aldrich) overnight at 4 °C, mixed with 20 μg of MBP or MBP-MTase with 20 μg of ALP or BSA in a total volume of 500 μL PBS and then incubated at 4 °C for 1 h. The beads were washed five times with 500 μL of PBS with a magnet, suspended in 30 μL of 2 × sodium dodecyl (lauryl) sulfate (SDS) sample buffer, boiled for 5 min, and loaded on a 10% SDS-PAGE gel for Western blotting analysis using a mixture of mouse anti-MBP (Sigma-Aldrich) and mouse anti-His (Proteintech) mAbs. Prokaryotically expressed BmALP protein was used as the control.

2.10. Inhibition of Viral Binding

(i) By anti-ALP antibody. BmBBMVs proteins (10 μg/mL) coated on 96-well ELISA plates were pretreated with different concentrations of the anti-ALP or pre-immune antibodies (0.2–20 μg/mL) at 37°C for 1 h, then incubated with DpCPV virions (final amounts, 10 μg) at 37 °C for a further 1 h. The wells were washed with PBST and the bound virions were detected using a rabbit anti-DpCPV virion antibody followed by HRP-labeled anti-rabbit IgG (Sigma-Aldrich).

(ii) By BmALP proteins. DpCPV virions (final amounts, 10 μg) were pretreated with different concentrations of BmALP protein (0.02–2 μM) or BSA at 37 °C for 1 h, and then added to the BmBBMVs (10 μg/mL) coated in 96-well ELISA plates. After absorption for 1 h, the bound virions were detected as described above.

2.11. In Vivo Neutralization Tests

Neutralization bioassays were conducted using a modified version of the droplet-feeding method [19]. Second instar *S. exigua* larvae were starved for 16 h before treatment. The larvae were divided into four groups: group 1 were fed PBS, group 2 DpCPV virions, group 3 DpCPV virions plus BSA, and group 4 DpCPV virions plus *Bm*ALP. The final amount of DpCPV virions was 10 µg in all the groups, and the concentration of *Bm*ALP or BSA proteins was 20 µg/mL. The virion-protein mixtures were incubated for 1 h, and then mixed with 40% sucrose and Erioglaucine disodium salt (food coloring) in a total volume of 1 mL. The larvae fed with the resulting mixture were judged by the uptake of the blue stain, and then transferred to the individual wells of a 24-well culture plate containing fresh artificial diet. The experiment was conducted with 48 larvae in each group and replicated in duplicate. The number of dead insects was recorded every day until all the larvae died or pupated (15 days post inoculation). The survival function of the larvae among the groups was estimated by the Kaplan-Meier method and compared using the log-rank test [20]. The data from two replicates were pooled for survival analysis as long as the two replicates did not differ significantly.

3. Results

3.1. DpCPV Virions Binding to SeBBMVs

To confirm whether DpCPV virions can bind to *Se*BBMVs, an ELISA was performed in which increasing amounts of the purified virions were incubated with *Se*BBMVs coated on 96-well ELISA plates. The results showed that the DpCPV virions bound to the *Se*BBMVs in a dose-dependent manner (Figure 1), and no binding was observed when BSA was the coating control.

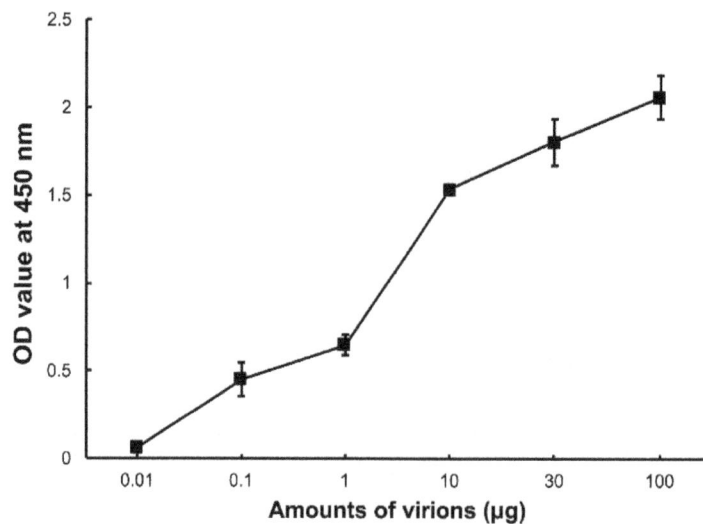

Figure 1. Binding of *Dendrolimus punctatus cypovirus* (DpCPV) virions to *Spodoptera exigua* midgut brush border membrane vesicles (*Se*BBMVs). *Se*BBMVs (10 µg/mL) were coated on 96-well plates and incubated with the amount of purified DpCPV virions indicated. Virion binding was determined using a rabbit anti-DpCPV virion antibody followed by horseradish peroxidase (HRP)-labeled anti-mouse IgG as the secondary antibody. Nonspecific binding was determined with bovine serum albumin (BSA) coated wells instead of those coated with BBMVs.

3.2. Binding Characteristics of the MBP Fusion Proteins to SeBBMVs

To determine which protein in DpCPV mediates efficient binding to susceptible host midgut BBMVs, the GTase and MTase domains of VP3, and the full-length VP4 and VP5 proteins were co-expressed with MBP as soluble proteins along with MBP. The sizes of purified MBP, MBP-GTase,

MBP-MTase, MBP-VP4, and MBP-VP5 were 42 kDa, 94 kDa, 123 kDa, 112 kDa, and 98 kDa, respectively (Figure 2A), which is consistent with the predicted sizes of these proteins plus the N-terminal MBP.

Next, the binding efficiency of the DpCPV turret proteins to SeBBMVs was tested. MBP-GTase, MBP-MTase, and MBP-VP4 binding to SeBBMVs were all dose-dependent, while MBP-VP5 bound at a relatively low level (Figure 2B). No binding was detected with the MBP-incubated wells. To further test the binding specificity, GTase, MTase, or VP4 proteins were mixed with different amounts of purified DpCPV virions and then the mixtures were added to the SeBBMVs. This treatment reduced the binding of MBP-MTase and MBP-VP4 in a dose-dependent manner (Figure 2C), suggesting that the MTase domain and VP4 competed with the virions for the same cellular receptor. In contrast, MBP-GTase binding to the SeBBMVs was unaffected by virion addition (Figure 2C).

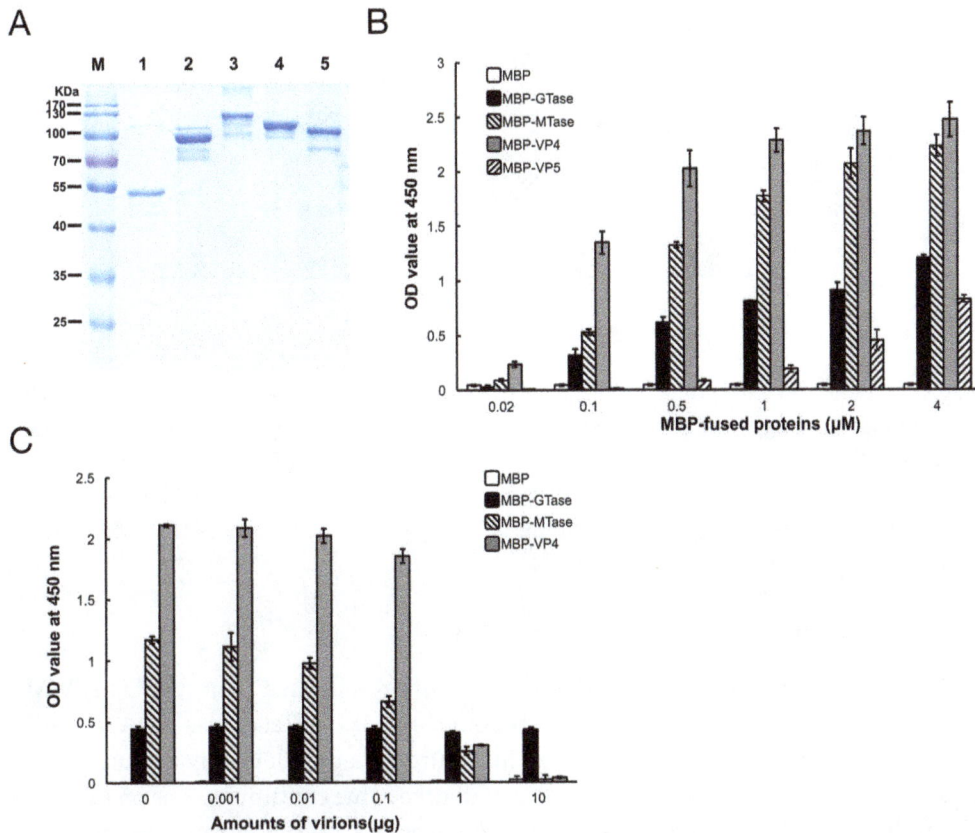

Figure 2. Maltose binding protein (MBP)-fusion protein binding to SeBBMVs. (**A**) Expression and purification of MBP-fusion proteins. MBP (lane 1), MBP-galactosyltransferase (GTase) (lane 2), MBP-methyltransferase (MTase) (lane 3), MBP-VP4 (lane 4), and MBP-VP5 (lane 5) were separated by 12% sodium dodecyl (lauryl) sulfate-polyacrylamide gel electrophoresis (SDS-PAGE); (**B**) SeBBMV proteins were coated on 96-well plates and incubated with the concentrations of MBP-fusion proteins indicated. The binding efficiency was determined with an anti-MBP mAb and HRP-labeled IgG as the secondary antibody. Nonspecific binding was determined with BSA coating instead of SeBBMVs; (**C**) MBP-fusion proteins (0.3 μM) were mixed with the amounts of purified DpCPV virions indicated, then added to the SeBBMVs-coated 96-well plates, and absorption was allowed for 1 h. Protein binding was determined as described above.

3.3. Inhibition of Viral Binding by Anti-MTase and Anti-VP4 Antibodies

Because the DpCPV virions bound specifically to the SeBBMVs, we next tested whether the viral attachment could be blocked by antibodies to MTase and/or VP4. For this, purified virions (fixed 10 μg amounts) were incubated with different concentrations of anti-MTase and/or anti-VP4 antibodies for 1 h prior to inoculation to the SeBBMVs. At all higher concentration treatments (4, 10, 20, 40, and 100 µg/mL), the mixture of anti-MTase and anti-VP4 antibodies reduced the viral attachment more effectively than the antibodies used alone ($F_{2,6}$ = 60.3, 93.7, 245.1, 78.3, and 38.7, respectively, $p < 0.001$; Figure 3). For example, at the highest concentration of antibodies (100 µg/mL), while the single anti-MTase or anti-VP4 antibodies inhibited viral attachment to 48.4% ± 3.1% and 41.4% ± 3.9%, respectively, the mixed antibodies inhibited viral attachment to 27.2% ± 1.9% in comparison with the control (pre-immune antibody); this treatment was significantly more effective than the anti-MTase antibody (LSD-t = 8.607, p = 0.00014) or the anti-VP4 antibody alone (LSD-t = 5.676, p = 0.00108).

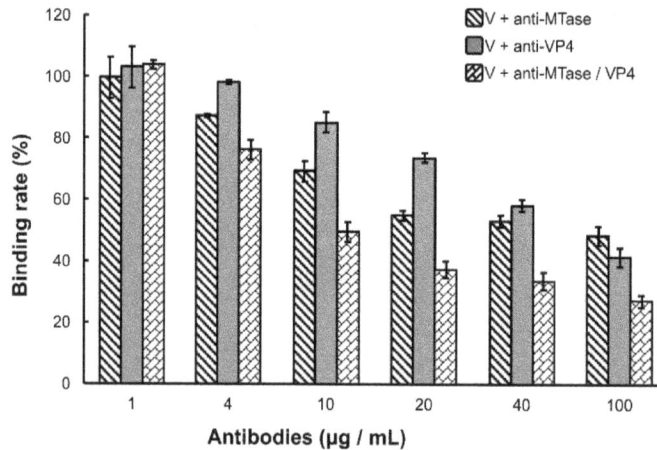

Figure 3. Inhibition of viral binding by anti-MTase and/or anti-VP4 antibodies. Purified DpCPV virions (fixed amounts of 10 µg) were pretreated with the concentrations of different antibodies indicated before inoculation into SeBBMVs (10 µg/mL) and coated on 96-well ELISA plates. Viral binding was determined by ELISA as described above. No significant differences were observed between the virion alone treatment and the virion with pre-immune antibody treatment regardless of the dilution factor. The error bars represent standard deviations.

3.4. Identification of MTase Binding Proteins in Host BBMVs

Far-Western blotting was utilized to identify the host proteins that bound to DpCPV MTase and VP4. A 65 kDa protein from BmBBMVs, which bound to MBP-MTase, was detected, while other membrane proteins also bound to the MBP protein control in a parallel assay (Figure 4A). However, MBP-VP4 showed no interaction with BmBBMVs under the same conditions. The 65 kDa protein was excised from the gel and analyzed by mass spectrometry, the identified result is shown in Table 2. Since B. mori ALP (gi | 189332880) was consistent with the protein molecular weight of 65 kDa and was a component of midgut membranes, BmALP was further investigated. Recombinant BmALP (1–519 aa; without the transmembrane domain) was expressed in E. coli (Figure 4B), a rabbit anti-ALP antibody was raised and Western blotting showed that the anti-ALP antibody reacted with BmALP.

The interaction between MBP-MTase and BmALP was validated by Co-IP. The results showed that BmALP was only co-precipitated following incubation with MBP-MTase, not MBP alone (Figure 4C, lanes 1 and 2), suggesting that MTase bound directly to BmALP.

Table 2. Identification of host proteins by LC-MS/MS and MASCOT program.

Accession	Similar Protein (Protein Name/Organism)	MASCOT Score	Nominal Mass (Da)	Sequence Coverage (%)
gi | 3721840	aminopeptidase N/Bombyx mori	41	109,624	16
gi | 227072221	troponin I variant D/Bombyx mandarina	38	23,711	7
gi | 356582739	TSSK/Bombyx mori	33	39,640	9
gi | 189332880	alkaline phosphatase/Bombyx mori	32	60,682	9

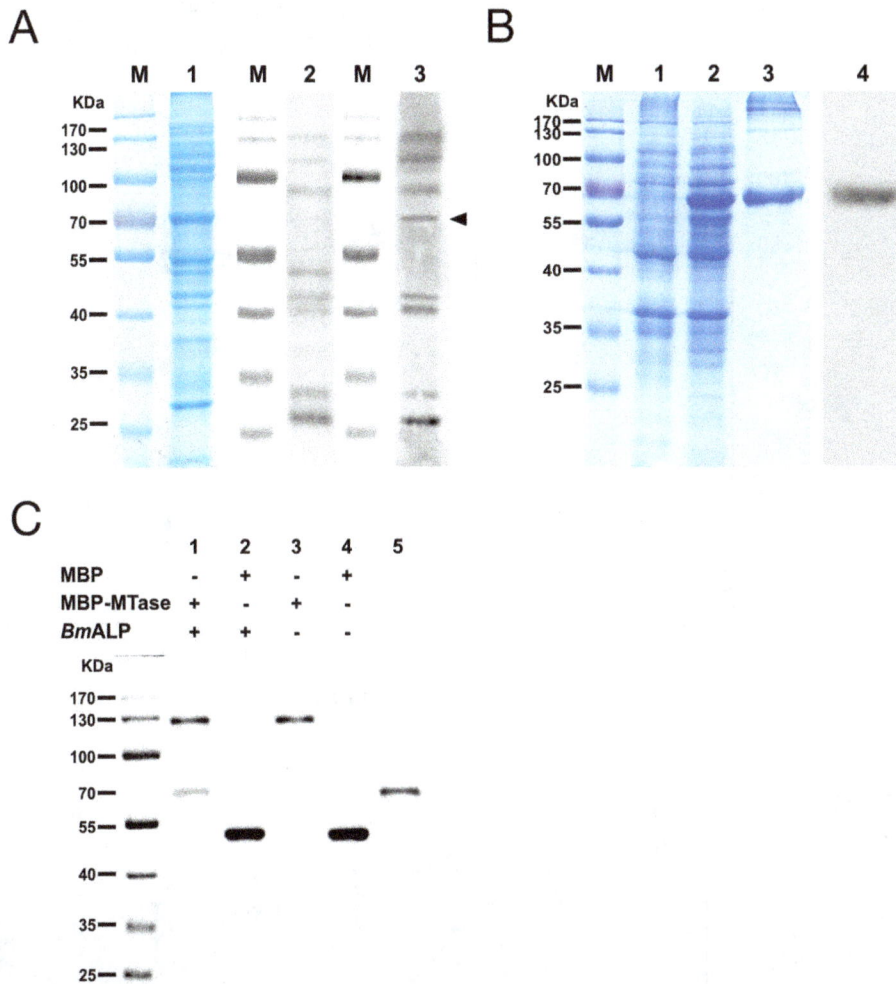

Figure 4. The MTase binding protein was identified as *B. mori* alkaline phosphatase protein (ALP). (**A**) Far-Western blotting of *Bm*BBMVs with MBP-MTase. Blot of *Bm*BBMVs were separated by 10% SDS-PAGE (lane 1), incubated with MBP (lane 2) or MBP-MTase (lane 3), and probed with an anti-MBP mAb followed by HRP-labeled anti-mouse IgG as the secondary antibody; (**B**) Expression and purification of recombinant *Bm*ALP protein separated by 12% SDS-PAGE. Lanes: M, protein molecular mass marker; 1, non-induced bacterial cell lysate; 2, induced bacterial cell lysate; 3, *Bm*ALP purified by a Ni-NTA agarose column; 4, *Bm*ALP detected with a rabbit anti-ALP antibody; (**C**) Co-immunoprecipitation (Co-IP) validated the interaction between MBP-MTase and *Bm*ALP. The immunoprecipitated products (lane 1 to lane 4) and *Bm*ALP protein (lane 5) were separated by SDS-PAGE, transferred to a PVDF membrane, and were incubated with a mixture of mouse anti-MBP mAb and mouse anti-His mAb as the primary antibody and HRP-labeled anti-mouse IgG as the secondary antibody.

3.5. Inhibition of Viral Binding to BmBBMVs

To investigate the function of ALP in DpCPV binding, *Bm*BBMVs coated in the ELISA plate were pretreated with different concentrations of the anti-ALP antibody prior to viral inoculation. Treatment with the anti-ALP antibody significantly reduced the attachment of DpCPV virions in a dose-dependent manner, and the final rate was 46.7% ± 3.0% of the control where no antibody was added (Figure 5A). In contrast, pre-immune antibody treatment did not reduce viral attachment.

To further study the role of ALP, DpCPV virions were incubated with different concentrations of *Bm*ALP before inoculation into *Bm*BBMVs. The results showed that viral binding was reduced as the concentration of *Bm*ALP increased, and the final rate was 42.1% ± 5.8% of the control where no protein was added (Figure 5B). However, BSA (as a control) had no effect on the binding.

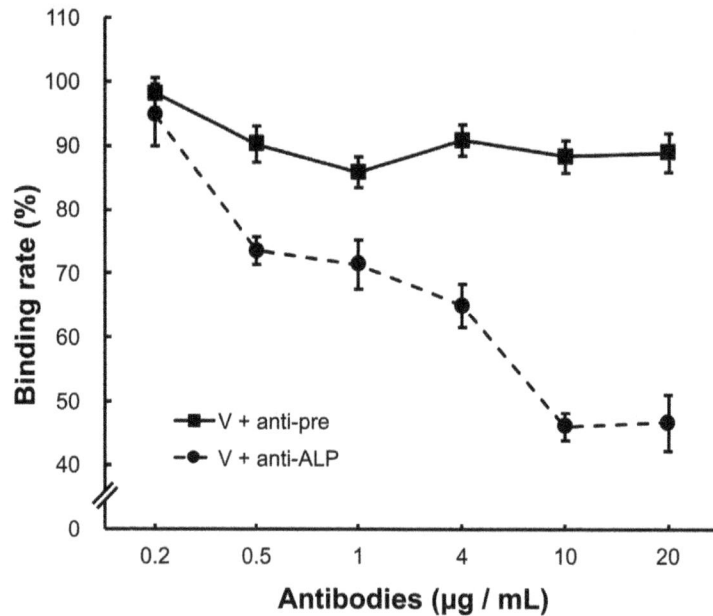

(A)

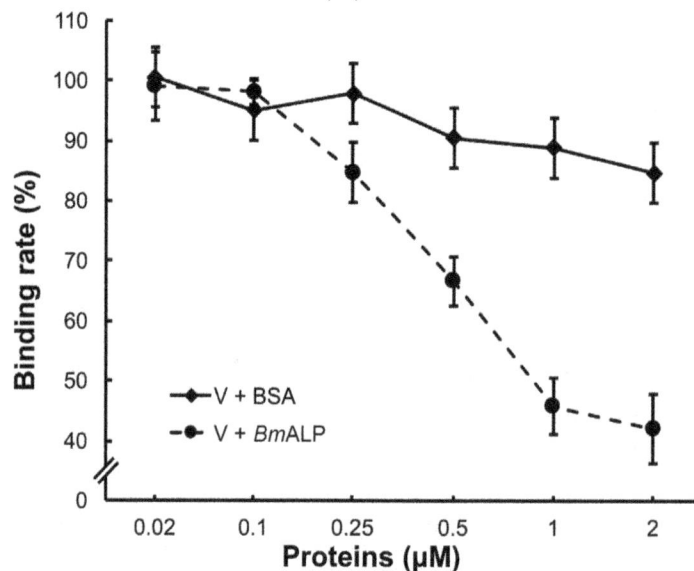

(B)

Figure 5. Inhibition of viral binding by anti-ALP antibody or *Bm*ALP protein. (**A**) *Bm*BBMVs (10 μg/mL) coated on 96-well plate were incubated with different concentrations of anti-ALP or pre-immune antibodies (0.2–20 μg/mL). After washing, DpCPV virions (10 μg) were added to each well. Viral binding was determined as described in the main text; (**B**) DpCPV virions (10 μg) were pre-incubated with different concentrations of *Bm*ALP or BSA protein (0.02–2 μM) before inoculation into the *Bm*BBMVs-coated 96-well plate. Viral binding was determined as described in the main text.

3.6. In Vivo Neutralization with ALP in S. exigua

To ascertain whether ALP mediates DpCPV infectivity, an in vivo neutralization test was performed. Purified DpCPV virions were incubated with increasing concentrations of *Bm*ALP prior to inoculation of *S. exigua*. At 4 days post inoculation, the larvae started to exhibit signs of CPV infection, such as appetite loss, slow movement, and bradygenesis. From days 7 to 15 post inoculation, the survival rates of treated larvae decreased steadily, and the final survival rates of larvae treated with virions alone and virions plus BSA were 33.3% and 34.4%, respectively (n = 96) (Figure 6). However, the survival rate of larvae treated with virions plus BmALP was 58.3% (n = 96). Overall, the survival function of the larvae treated with virions and BmALP was significantly different from that of the larvae treated with virions alone (χ^2 = 27.274, p < 0.0001) or virions plus BSA (χ^2 = 12.822, p = 0.000343). The dead larvae did not liquefy and their cuticles were intact, displaying a typical feature of CPV infection. In contrast, most larvae pupated in the PBS group. Although BmALP could not completely block DpCPV infection in $S.$ $exigua$, the results imply that ALP contributes to the infectivity of DpCPV in larvae.

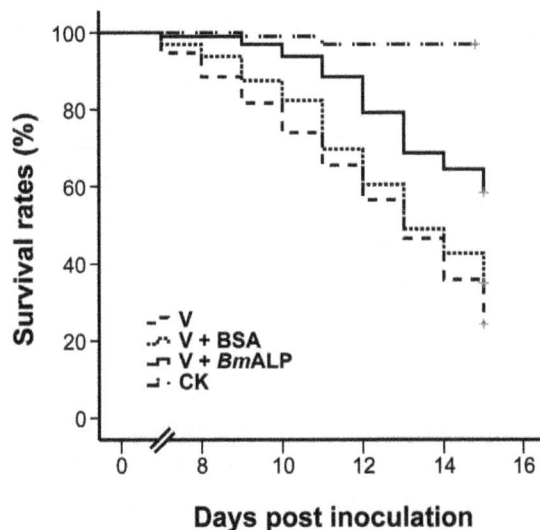

Figure 6. In vivo neutralization with BmALP in $S.$ $exigua$. Survival rate data represent the results pooled from two replications (n = 96 for each treatment).

4. Discussion

CPVs, the second major group of insect viral pathogens, have been isolated from more than 250 different insect species reared in laboratories or from the field. In recent years, studies of CPV have focused extensively on its characterization and structure [5,21,22], but little is known about the molecular mechanisms of CPV infection because a reliable cell culture system to measure viral infection efficiency is lacking. In this study, we used midgut BBMV proteins instead of susceptible cell lines. Based on the consensus that spike proteins are the typical sites for viral interaction with host cells [9,11,12], we selected the turret proteins of DpCPV for investigation.

Initially, we confirmed that DpCPV particles could bind to SeBBMVs (Figure 1). Then, we found that the MTase domain of VP3 (B-spike) and VP4 (A-spike) bound efficiently to the SeBBMVs, whereas VP5 binding was at a relatively low level (Figure 2B). The structural homolog proteins of CPV VP5 are commonly present in the inner capsids of other turreted reoviruses (e.g., orthoreovirus λ2 and aquareovirus VP6), and they also function as molecular clamps to stabilize the inner capsids [23]. Furthermore, the competition assays (Figure 2C) suggested that the C-terminal of VP3 (MTase) rather than the N-terminal (GTase) participates in viral attachment to host midguts. This finding is consistent with the fact that the GTase domain is an interior turret part situated underneath the capsid shell [1,4], making it inaccessible to host molecules during the virus entry procedure. The potential roles for MTase and VP4 proteins in DpCPV attachment were investigated by blocking tests (Figure 3). The results showed that the mixture of anti-MTase and anti-VP4 antibodies was more effective at blocking the viral attachment than the anti-MTase and anti-VP4 antibodies used alone, indicating that both MTase (B-spike) and VP4 (A-spike) were involved in the viral attachment.

In vivo, DpCPV can infect a wide range of insects of different genera and the infections are mostly limited to the midgut [24]. These characteristics suggest that DpCPV can bind to common targets on midgut cells in a variety of hosts. Thus, we conducted far-Western blotting to identify the host factors that interact with DpCPV MTase and VP4. *Bm*BBMVs was used because *B. mori* was a model organism with a clear genetic background and could be infected by DpCPV. A 65 kDa protein in *Bm*BBMVs showed a specific interaction with MTase (Figure 4A) and was later identified as a glycosylphosphatidylinositol (GPI)-anchored ALP of *B. mori*. However, we failed to identify the MTase binding protein in *Se*BBMVs using the same conditions. Because ALP is a housekeeping gene in insects [25] and *B. mori* ALP shares 65% amino-acid sequence identity with *S. exigua* ALP (KM048197.1), we investigated *B. mori* ALP further.

ALP is a glycoprotein anchored to the cell membrane by a GPI anchor. It is abundant in midgut BBMVs and is estimated to account for 15 to 20% of the total midgut protein [25]. Immunofluorescence studies [26] have shown that ALPs are located primarily in the posterior midgut epithelial cells, the region in which DpCPV infection occurs. ALP is found at higher levels in the first and second larval instars [27], and it appears that in early instars, ALP plays an important role in the viral infection. It has been reported that it acts as a receptor for *Bacillus thuringiensis* Cry1Ac in *Manduca sexta* and *Heliothis virescens*, and as a Cry11Aa receptor in *Aedes aegypti* [25].

In the current study, Co-IP assays verified that *Bm*ALP specifically bound to the MTase (Figure 4C). Importantly, pretreatment of *Bm*BBMVs with an anti-ALP antibody or incubation of DpCPV virions with soluble *Bm*ALP resulted in significantly reduced attachment of virions to *Bm*BBMVs (Figure 5), indicating that ALP affected the attachment of DpCPV to *Bm*BBMVs. The results of in vivo neutralization assays (Figure 6) suggested that ALP was essential in the early steps of DpCPV infection. However, ALP on its own is not able to completely block viral attachment; thus, we presume that additional host factors act in concert with ALP to promote viral infection. For EV71, human P-selectin glycoprotein ligand-1 and human scavenger receptor class B member 2 (hSCARB2) have been identified as receptors. SCARB2 is expressed ubiquitously on most susceptible cell types. However, antibody blocking of SCARB2 can only partially reduce the entry of EV71 into some cell lines [28,29]. A more recent study found that Annexin II, a cellular adherent factor, also interacts with EV71 and enhances viral infectivity [30].

Based on our data, we hypothesize that DpCPV may enter host cells under the combined effects of A-spikes and B-spikes. Initially, DpCPV may attach to an unknown cellular receptor through A-spikes and then, following polyhedra disruption in the midgut, the A-spikes fall away from the B-spike of the virus particle. When this occurs, a more stable B-spike is exposed, and this attaches to the secondary receptor (ALP) on the midgut membrane. This entry pattern is similar to reoviruses, which rely on the coordinated engagement of multiple receptors to mediate the initial viral entry [10,13].

In summary, this is the first study to show that both the MTase domain (B-spike) and VP4 (A-spike) are responsible for CPV attachment. We also found that MTase interacted with *Bm*ALP, and this interaction affected the viral infection, although other host factors may also contribute to infection with DpCPV. In this regard, it would be pertinent to investigate the functional receptor for VP4 and how ALP, in conjunction with the additional receptor, behaves during CPV entry into cells. Our data facilitate better understanding of CPV pathogenesis and open the way for novel methods to study the interactions between CPV and hosts.

Acknowledgments: This work was supported by the WIV "One-Three-Five" Strategic Programs: Y602111SA1. We thank Core Facility Center and Technical Support, Wuhan Institute of Virology for technical support.

Author Contributions: Xiulian Sun conceived and designed the experiments. Lan Su, Congrui Xu, Chuangang Cheng, and Chengfeng Lei performed the experiments. Lan Su prepared the manuscript. Xiulian Sun and Lan Su analyzed the data and revised the manuscript.

References

1. Yu, X.; Jin, L.; Zhou, Z.H. 3.88 Å structure of cytoplasmic polyhedrosis virus by cryo-electron microscopy. *Nature* **2008**, *453*, 415–419. [CrossRef] [PubMed]

2. Mori, H.; Metcalf, P. *Cypoviruses. Insect Virology*; Asgari, S., Johnson, K., Eds.; Caister Academic Press Publisher: Norfolk, UK, 2010; pp. 307–323.

3. Zhang, Y.; Cao, G.; Zhu, L.; Chen, F.; Zar, M.S.; Wang, S.; Hu, X.; Wei, Y.; Xue, R.; Gong, C. Integrin beta and receptor for activated protein kinase C are involved in the cell entry of *Bombyx mori cypovirus*. *Appl. Microbiol. Biotechnol.* **2017**. [CrossRef] [PubMed]

4. Cheng, L.; Sun, J.; Zhang, K.; Mou, Z.; Huang, X.; Ji, G.; Sun, F.; Zhang, J.; Zhu, P. Atomic model of a cypovirus built from cryo-EM structure provides insight into the mechanism of mRNA capping. *Proc. Natl. Acad. Sci. USA* **2011**, *108*, 1373–1378. [CrossRef] [PubMed]

5. Yu, X.K.; Ge, P.; Jiang, J.S.; Atanasov, I.; Zhou, Z.H. Atomic model of CPV reveals the mechanism used by this single-shelled virus to economically carry out functions conserved in multishelled reoviruses. *Structure* **2011**, *19*, 652–661. [CrossRef] [PubMed]

6. Cheng, C.G.; Shao, Y.P.; Su, L.; Zhou, Y.; Sun, X.L. Interactions among *dendrolimus punctatus* cypovirus proteins and identification of the genomic segment encoding its A-spike. *J. Gen. Virol.* **2014**, *95*, 1532–1538. [CrossRef] [PubMed]

7. Chen, J.; Sun, J.; Atanasov, I.; Ryazantsev, S.; Zhou, Z.H. Electron tomography reveals polyhedrin binding and existence of both empty and full cytoplasmic polyhedrosis virus particles inside infectious polyhedra. *J. Virol.* **2011**, *85*, 6077–6081. [CrossRef] [PubMed]

8. Nibert, M.L.; Baker, T.S. CPV, a stable and symmetrical machine for mRNA synthesis. *Structure* **2003**, *11*, 605–607. [CrossRef]

9. Xu, G.; Wilson, W.; Mecham, J.; Murphy, K.; Zhou, E.M.; Tabachnick, W. VP7: An attachment protein of bluetongue virus for cellular receptors in *Culicoides variipennis*. *J. Gen. Virol.* **1997**, *78*, 1617–1623. [CrossRef] [PubMed]

10. Konopka-Anstadt, J.L.; Mainou, B.A.; Sutherland, D.M.; Sekine, Y.; Strittmatter, S.M.; Dermody, T.S. The Nogo receptor NgR1 mediates infection by mammalian reovirus. *Cell Host Microbe* **2014**, *15*, 681–691. [CrossRef] [PubMed]

11. Ludert, J.E.; Feng, N.; Yu, J.H.; Broome, R.L.; Hoshino, Y.; Greenberg, H.B. Genetic mapping indicates that VP4 is the rotavirus cell attachment protein in vitro and in vivo. *J. Virol.* **1996**, *70*, 487–493. [PubMed]

12. Furlong, D.B.; Nibert, M.L.; Fields, B.N. Sigma 1 protein of mammalian reoviruses extends from the surfaces of viral particles. *J. Virol.* **1988**, *62*, 246–256. [PubMed]

13. Kirchner, E.; Guglielmi, K.M.; Strauss, H.M.; Dermody, T.S.; Stehle, T. Structure of reovirus Sigma 1 in complex with its receptor junctional adhesion molecule-A. *PLoS Pathog.* **2008**, *4*, e1000235. [CrossRef] [PubMed]

14. Schulz, W.L.; Haj, A.K.; Schiff, L.A. Reovirus uses multiple endocytic pathways for cell entry. *J. Virol.* **2012**, *86*, 12665–12675. [CrossRef] [PubMed]

15. Arias, C.F.; Isa, P.; Guerrero, C.A.; Méndez, E.; Zárate, S.; López, T.; Espinosa, R.; Romero, P.; López, S. Molecular biology of rotavirus cell entry. *Arch. Med. Res.* **2002**, *33*, 356–361. [CrossRef]

16. Jin, L.; Wang, H.X.; Liu, J.J.; Wang, J.C.; Zhang, L.L.; Yang, G.; Zhang, X.H.; Zheng, G.H. Polyclonal antibody preparation and immune activity of structural proteins VP1, VP3 and VP5 of *dendrolimus puntatus* cytoplasmic polyhedrosis virus. *Chin. J. Appl. Environ. Biol.* **2014**, *20*, 345–350. (In Chinese)

17. Jin, L.; Dai, C.W.; Qin, T.C.; Sun, X.L. Molecular characterization of protein p50 of *dendrolimus punctatus* cytoplasmic polyhedrosis virus. *J. Basic Microbiol.* **2013**, *53*, 37–44. [CrossRef] [PubMed]

18. Wolfersberger, M.; Luthy, P.; Maurer, A.; Parenti, P.; Sacchi, V.F.; Giordana, B.; Hanozet, G.M. Preparation and partial characterization of amino acid transporting brush border membrane vesicles from the larval midgut of the cabbage butterfly (*pieris brassicae*). *Comp. Biochem. Physiol.* **1987**, *86*, 301–308. [CrossRef]

19. Zhou, Y.; Qin, T.C.; Xiao, Y.Z.; Qin, F.J.; Lei, C.F.; Sun, X.L. Genomic and biological characterization of a new cypovirus isolated from *dendrolimus punctatus*. *PLoS ONE* **2014**, *9*, e113201. [CrossRef] [PubMed]

20. Kalbfleisch, J.D.; Prentice, R.D. *The Statistical Analysis of Failure Time Data*; Wiley: New York, NY, USA, 1980.

21. Zhu, B.; Yang, C.W.; Liu, H.R.; Cheng, L.P.; Song, F.; Zeng, S.J.; Huang, X.J.; Ji, G.; Zhu, P. Identification of the active sites in the methyltransferases of a transcribing dsRNA virus. *J. Mol. Biol.* **2014**, *426*, 2167–2174. [CrossRef] [PubMed]

22. Coulibaly, F.; Chiu, E.; Ikeda, K.; Gutmann, S.; Haebel, P.W.; Schulze-Briese, C.; Mori, H.; Metcalf, P. The molecular organization of cypovirus polyhedra. *Nature* **2007**, *446*, 97–101. [CrossRef] [PubMed]

23. Yang, J.; Cheng, Z.; Zhang, S.; Xiong, W.; Xia, H.; Qiu, Y.; Wang, Z.; Wu, F.; Qin, C.F.; Yin, L.; et al. A cypovirus VP5 displays the RNA chaperone-like activity that destabilizes RNA helices and accelerates strand annealing. *Nucleic Acids Res.* **2014**, *42*, 2538–2554. [CrossRef] [PubMed]

24. Kolliopoulou, A.; Nieuwerburgh, V.F.; Stravopodis, D.J.; Deforce, D.; Swevers, L.; Smagghe, G. Transcriptome analysis of *bombyx mori* larval midgut during persistent and pathogenic cytoplasmic polyhedrosis virus infection. *PLoS ONE* **2015**, *10*, e0121447. [CrossRef] [PubMed]

25. Pigott, C.R.; Ellar, D.J. Role of receptors in *bacillus thuringiensis* crystal toxin activity. *Microbiol. Mol. Biol. Rev.* **2007**, *71*, 255–281. [CrossRef] [PubMed]

26. Fernandez, L.E.; Aimanova, K.G.; Gill, S.S.; Bravo, A.; Soberón, M. A GPI-anchored alkaline phosphatase is a functional midgut receptor of cry11Aa toxin in *Aedes aegypti* larvae. *Biochem. J.* **2006**, *394*, 77–84. [CrossRef] [PubMed]

27. Arenas, I.; Bravo, A.; Soberón, M.; Gómez, I. Role of alkaline phosphatase from *manduca sexta* in the mechanism of action of *bacillus thuringiensis* Cry1Ab toxin. *J. Biol. Chem.* **2010**, *285*, 12497–12503. [CrossRef] [PubMed]

28. Yamayoshi, S.; Yamashita, Y.; Li, J.; Hanagata, N.; Minowa, T.; Takemura, T.; Koike, S. Scavenger receptor B2 is a cellular receptor for enterovirus 71. *Nat. Med.* **2009**, *15*, 798–801. [CrossRef] [PubMed]

29. Nishimura, Y.; Shimojima, M.; Tano, Y.; Miyamura, T.; Wakita, T.; Shimizu, H. Human P-selectin glycoprotein ligand-1 is a functional receptor for enterovirus 71. *Nat. Med.* **2009**, *15*, 794–797. [CrossRef] [PubMed]

30. Yang, S.L.; Chou, Y.T.; Wu, C.N.; Ho, M.S. Annexin II binds to capsid protein VP1 of enterovirus 71 and enhances viral infectivity. *J. Virol.* **2011**, *85*, 11809–11820. [CrossRef] [PubMed]

12

The Envelope Gene of Hepatitis B Virus is Implicated in Both Differential Virion Secretion and Genome Replication Capacities between Genotype B and Genotype C Isolates

Haodi Jia [1], Yanli Qin [2], Chaoyang Chen [1], Fei Zhang [1], Cheng Li [1], Li Zong [1], Yongxiang Wang [1], Jiming Zhang [2], Jisu Li [3], Yumei Wen [1] and Shuping Tong [1,3,*]

[1] Key Lab of Medical Molecular Virology, School of Basic Medical Sciences, Fudan University, Shanghai 200032, China; hdjia14@fudan.edu.cn (H.J.); 13111010065@fudan.edu.cn (C.C.); 13817459252@126.com (F.Z.); 15211010047@fudan.edu.cn (C.L.); zongli1226@163.com (L.Z.); boldawing@aliyun.com (Y.W.); ymwen@shmu.edu.cn (Y.W.)

[2] Department of Infectious Diseases, Huashan Hospital, Fudan University, Shanghai 200032, China; qyl2122@hotmail.com (Y.Q.); jmzhang@fudan.edu.cn (J.Z.)

[3] Liver Research Center, Rhode Island Hospital, Warren Alpert School of Medicine, Brown University, Providence, RI 02903, USA; ji_su_li_md@brown.edu

* Correspondence: shuping_tong_md@brown.edu

Academic Editor: Joanna Parish

Abstract: Chronic infection by hepatitis B virus (HBV) genotype C is associated with a prolonged replicative phase and an increased risk of liver cancer, compared with genotype B infection. We previously found lower replication capacity but more efficient virion secretion by genotype C than genotype B isolates. Virion secretion requires interaction between core particles and ENVELOPE proteins. In the present study, chimeric constructs between genotype B and genotype C clones were generated to identify the structural basis for differential virion secretion. In addition to dimeric constructs, we also employed 1.1mer constructs, where the cytomegalovirus (CMV) promoter drove pregenomic RNA transcription. Through transient transfection experiments in Huh7 cells, we found that exchanging the entire *envelope* gene or just its *S* region could enhance virion secretion by genotype B clones while diminishing virion secretion by genotype C. Site-directed mutagenesis established the contribution of genotype-specific divergence at codons 108 and 115 in the *preS1* region, as well as codon 126 in the *S* region, to differential virion secretion. Surprisingly, exchanging the *envelope* gene or just its *S* region, but not the *core* gene or 3′ *S* region, could markedly increase intracellular replicative DNA for genotype C clones but diminish that for genotype B, although the underlying mechanism remains to be clarified.

Keywords: *core* gene; *envelope* gene; genome replication; genotype; hepatitis B virus; virion secretion

1. Introduction

The hepatitis B virus (HBV) causes acute and chronic infection of the liver, with the latter being a global public health problem because of its widespread distribution and severe sequelae, including liver cirrhosis and hepatocellular carcinoma (HCC) [1,2]. HBV can be classified into ten genotypes according to the nucleotide sequence divergence of its genomic DNA [3–6]. Genotypes B and C are responsible for the majority of chronic HBV infections in East Asian countries. Chronic HBV infection is a dynamic process involving the interaction between the virus, hepatocytes, and immune cells. Numerous epidemiological studies suggest that genotype C is more pathogenic than genotype B due to a prolonged hepatitis B e antigen (HBeAg) positive phase characterized by a higher viral load [7–12]. Moreover, adulthood infection with genotype C has a greater risk of becoming chronic [13]. On the other hand, genotype B infection is associated with a higher risk of fulminant hepatitis and acute exacerbation of chronic infection [14–16]. Comparative studies of the biological properties between these two major HBV genotypes could help gain a better understanding of the molecular basis for their different clinical outcomes. In a previous study, we cloned full-length genotype B and genotype C genomes from Chinese and U.S. patients. Transient transfection of these HBV genomes into Huh7 cells, a human hepatoma cell line, revealed that most genotype C clones (or isolates) replicated less efficiently than genotype B clones (or isolates) but possessed higher virion secretion efficiency [17].

In this regard, HBV is an enveloped virus with a relaxed circular DNA (rcDNA) genome of 3.2 kb. The rcDNA is converted to covalently closed circular (ccc) DNA in the nucleus of infected hepatocytes, which serves as a transcriptional template [18]. Four genes are arranged in a circular and overlapping manner; *precore/core, polymerase (P), preS1/preS2/S (envelope)*, and *X*, which generate seven viral proteins through alternative translation initiation from the *precore/core* and *preS1/preS2/S* genes. Thus, HBeAg and CORE protein are products of the *precore/core* gene and *core* gene alone, respectively. Translation initiation from the *preS1, preS2*, and *S* region AUG codons in the *envelope* gene generate LARGE (L: *preS1/preS2/S*), MEDIUM (M: *preS2/S*), and SMALL (S: *S* domain alone) ENVELOPE proteins, respectively. Expression of the seven viral proteins is ensured by the transcription of four size forms of co-terminal RNAs of 3.5 kb (HBeAg, CORE, and P), 2.4 kb (L), 2.1 kb (M and S), and 0.7 kb (HBx) [18]. The 3.5-kb RNAs are over genome length and hence terminally redundant. Only the 3.5-kb pregenomic RNA (pgRNA) is required for genome replication. First, it serves as mRNA for both CORE and P proteins. Second, it is packaged together with the P protein inside the core particle assembled from the CORE protein [19,20]. Genome replication as catalyzed by the P protein involves minus strand DNA synthesis using pgRNA as the template, followed by pgRNA degradation and plus strand DNA synthesis using the minus stranded DNA as a template. L and S proteins are essential for virion secretion but play distinct functions [21,22]. The L protein interacts with core particles and retains the S protein towards the formation of the 42-nm virions. Otherwise the default function of the S protein is release as the 22-nm noninfectious subviral particles. Subviral particles constitute the bulk of hepatitis B surface antigen (HBsAg) and can reach 10,000- to 1,000,000- fold higher concentration than virions in the blood of infected individuals [1,21].

In the present study, we attempted to identify the structural basis for differential virion secretion efficiencies between clones of genotype B and genotype C. As virion formation requires the interaction between core particles and ENVELOPE proteins, we focused our attention on the *core* and *envelope* genes. Cloning of the full-length HBV genome to a vector disrupts continuity in the HBV genome and prevents transcription of the terminally redundant pgRNA. In our original study, we used a recircularized HBV genome or a tandem dimer of the HBV genome cloned to a pUC18 vector via the unique *SphI* site (*SphI* dimer) for transfection experiments [17]. Considering that differential replication capacities between clones of the two HBV genotypes complicate data interpretation, in the present study we also cloned 1.1 copies of the HBV genome (the DNA equivalent of pgRNA) to pcDNA3.1zeo (−) vector to overproduce the pgRNA through the exogenous cytomegalovirus (CMV) promoter [23]. We hoped that such a 1.1mer construct would show much higher replication capacity than SphI dimer, thus increasing virion secretion. Moreover, similar levels of the pgRNA would diminish the difference in the replication capacities between the two genotypes, thus simplifying data interpretation.

2. Materials and Methods

2.1. The 1.1mer over-Length HBV Construct and SphI Dimer for Genome Replication

HBV clones 22.5 (GenBank accession number: KU964112) and 24.6 (KU964143) of genotype B, as well as 17.3 (KU964036) and 27.2 (KU964186) of genotype C have been previously described [17]. Clone 56 (AF100309) of genotype B was a kind gift from Dr. Youhua Xie, Fudan University. A 1.1mer construct for these clones was generated by inserting nucleotide sequence 1818–3215/1–1932 into the *SacI* and *HindIII* sites of the pcDNA3.1zeo (−) vector in two sequential steps. First, HBV DNA fragment 240–1932 was amplified by polymerase chain reaction (PCR) using *SphI* dimer as the template [17], with a *HindIII* site attached to the antisense primer. The PCR product was doubly digested with *XbaI* and *HindIII*, and the resulting HBV DNA fragment was inserted into the *XbaI-HindIII* sites of the pcDNA3.1zeo (−) vector. Next, HBV DNA fragment 1818–3215/1–247 was amplified by PCR and cloned to the *SacI-XbaI* sites of the above construct in a similar way (with the *SacI* site attached to the sense primer). Fragment exchange and site-directed mutagenesis were performed on such 1.1mer constructs as detailed below. To study the biological properties of HBV under endogenous promoters, 1.1mer constructs with DNA fragment exchange or site-directed mutagenesis were remade into *SphI* dimer using a newly developed method [24].

2.2. DNA Fragment Exchange and Site-Directed Mutagenesis

To exchange the *core* gene (positions 1901–2452) between 1.1mer constructs of the two genotypes, a 1.6-kb chimeric DNA fragment covering positions 1818–3215/1–247 was generated by overlap extension PCR, with positions 1901–2452 replaced. The PCR product was doubly digested with *SacI* and *XbaI* for replacement of the cognate fragment in the 1.1mer construct. To exchange the entire *envelope* gene (position 2848–3215/1–835) or just the *S* region (position 155–835) between the 1.1mer construct, a 2.8-kb chimeric fragment covering positions 1818–3215/1–1406 was generated by overlap extension PCR, doubly digested with *SacI* and *BamHI* (position 1406), and used to replace the cognate *SacI-BamHI* fragment in the 1.1mer construct. In a similar way, a chimeric fragment covering positions 247–1406 was generated and doubly digested with *XbaI-BamHI*, so as to exchange just the 3′ half of the *S* region (position 430–835) between these two genotypes. Overlap extension PCR was also used to introduce the L108I/T115S mutations (at the amino acid level) into the *preS1* region for genotype B clones, as well as the I108L/S115T changes for genotype C clones, followed by replacement of the 1.6-kb *SacI-XbaI* restriction fragment. Similar fragment exchange was used to introduce the A152T mutation (at the nucleotide level) to reduce S protein translation [22]. The T126I mutation in the *S* domain for two genotype B clones, as well as the I126T mutation for two genotype C clones, were introduced by exchange of the *XbaI-BamHI* restriction fragment with PCR products. The 4B genome with a nonfunctional encapsidation signal (4B ε^-) served as a negative control in the transfection experiments. It contains the G1879T/T1880A double mutation in the loop of the ε signal, thus abolishing pgRNA packaging and consequently genome replication [25].

2.3. Subgenomic Expression Constructs for ENVELOPE Proteins

The 0.7mer HBV DNA construct has a 2.3-kb HBV DNA fragment encompassing nt 2721–3215/1–1770 inserted upstream of the SV40 polyadenylation signal and cloned to the pBluescript vector [22]. The 0.7mer construct used in this study is a chimeric construct with the backbone of clone 4B and the *envelope* gene of clone 6.2. It could express all the three ENVELOPE proteins under endogenous promoters and enhancers but not any other HBV proteins. A stop codon in the *preS2* region (C117A) was used to abolish L and M protein expression [22]. Alternatively, the *preS2* and *S* gene ATGs were mutated to ATA and GCG, respectively, to abolish both M and S protein expression.

2.4. Transient Transfection and Measurement of Secreted HBsAg

DNA constructs were purified by the high-speed plasmid midi kit (Macherey-Nagel, Duren, Germany), followed by phenol and chloroform extraction. The human hepatoma cell lines Huh7 and HepG2 were cultured in Dulbecco's Modified Eagle's Medium (GIBCO, Grand Island, NY, USA), supplemented with 10% fetal bovine serum (Sigma, St. Louis, MO, USA). Transient transfection was performed on cells seeded in 6-well plates using Lipofectamine 3000 reagent (Invitrogen, Carlsbad, CA, USA), with pUC18 DNA, to bring the total amount of DNA to 2 µg/well. HBsAg secreted to culture supernatant was measured by an enzyme-linked immunosorbent assay (ELISA) kit (KHB, Shanghai, China) with proper dilution (1:100–1:500) to prevent signal saturation.

2.5. Detection of HBV DNA Replication and Virion Secretion

The details of DNA analysis have been described previously [17,22,26]. Briefly, cells seeded in 6-well plates were harvested at day 4 post-transfection. Core particles were precipitated from half of the cell lysate in lysis buffer containing 0.5% NP40. Virions were immunoprecipitated from 1.4 mL of precleared culture supernatant by a mixture of custom-made polyclonal rabbit anti-preS1 antibody (Genscript, Nanjing, China) (3 µL) and rabbit anti-HBs antibody (Novus, Littleton, CO, USA) (1 µL) precojugated to 10 µL protein G-agarose beads (BioVision, Milpitas, CA, USA). The preS1 antibody targets residues 12–46 (MGTNLSVPNPLGFFPDHQLDPAFGANSNNPDWDFN) and could neutralize HBV infectivity [27,28]. Following nuclease digestion to remove transfected DNA and proteinase K digestion to disrupt core particles or virions, DNA was extracted with phenol and precipitated with ethanol using 10 µg of salmon sperm DNA as carrier. Purified DNA was separated in 1.3% agarose gel with 0.5 µg/mL ethidium bromide added to show the equal loading of carrier DNA for both replicative DNA and virion DNA (data not shown). DNA was transferred to a positively charged nylon membrane (Roche, Indianapolis, IN, USA) for hybridization. The nearly full-length (3.1-kb) HBV DNA for probe making was obtained from a cloned genome of genotype B or C by nested PCR amplification and labeled with [^{32}P] dGTP or dCTP by a random primed DNA labeling kit (Roche). Some of the blots for replicative DNA (Figure 1A, right panel; Figure 2A, right panel; Figure 4A; Figure 5A) were hybridized with a digoxigenin-labeled HBV RNA probe, as previously described [28]. Briefly, a 0.7-kb HBV DNA fragment covering positions 1266–1950 was cloned to the *KpnI-XhoI* sites of the pcDNA3 vector, and a positive-stranded HBV RNA probe was generated by in vitro transcription of the plasmid linearized at the *XhoI* site using T7 RNA polymerase (DIG-Northern Starter kit from Roche). For quantification, the grey values of signals on the blots were measured by ImageJ software [29].

3. Results

3.1. Rationale for Using 1.1mer Construct to Compare Virion Secretion between Genotypes B and C

The objective of this study was to identify the structural basis for more efficient virion secretion by genotype C than genotype B. Based on our previous findings [17], we chose two genotype B clones and two genotype C clones for further characterization. Full-length sequencing revealed that both clones 22.5 and 24.6 belong to the B2 subgenotype, while clones 17.3 and 27.2 belong to the C2 and C1 subgenotypes, respectively. Following transient transfection with the *SphI* dimer constructs into Huh7 cells, the level of intracellular replicative DNA was lower for the genotype C clones, especially 17.3 (Figure 1A, right panel). Virion secretion efficiency, as judged by the ratio of extracellular virion DNA/intracellular replicative DNA, was highest for clone 17.3, followed by 27.2 (Figure 1C, right panel). In the present study, we continued to employ the dimeric construct for making some chimeras and site-directed mutants. However, mutagenesis by restriction fragment exchange will revert a dimer into a monomer, necessitating the remaking of a dimer for each new construct. More importantly, the amount of virion DNA produced following transient transfection with *SphI* dimer constructs is rather low, as a consequence of the low replication capacity of genotype C clones. Towards this end, the 1.1mer construct was generated, with sequence 1818–3215/1–1932 inserted to the pcDNA3.1zeo (−) vector. This should lead to robust transcription of pgRNA driven by the CMV promoter [23], thus greatly enhancing HBV DNA replication and, consequently, virion secretion. It was hoped that under the common CMV promoter, a similar level of pgRNA would be transcribed to achieve a comparable level of genome replication and thus simplify the comparison of viron secretion efficiency.

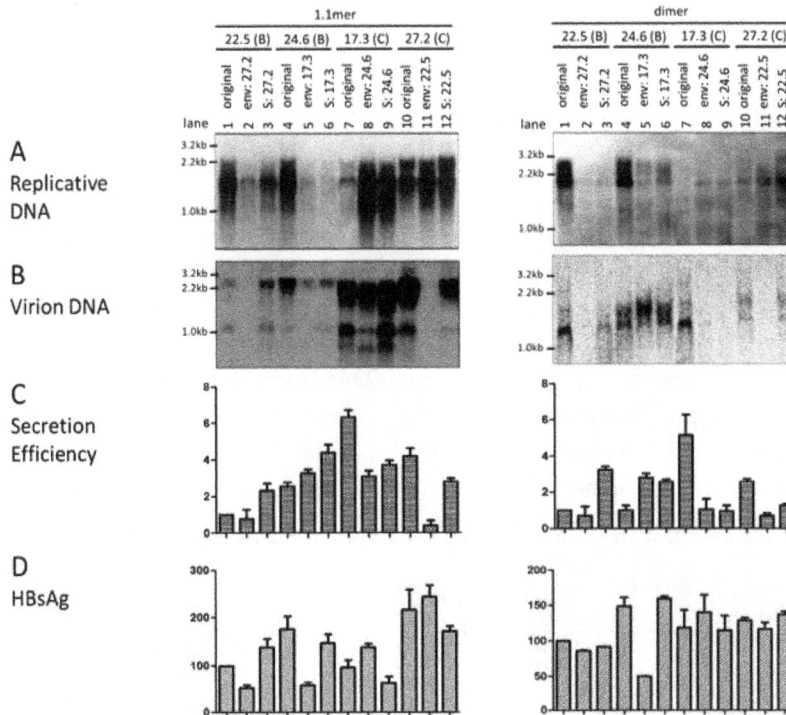

Figure 1. Impact of exchanging the entire *envelope* gene or just the *S* region between clones of genotype B and genotype C on hepatitis B virus (HBV) DNA replication and virion secretion. Both the 1.1mer construct (left panels) and *SphI* dimer (right panels) were used for the exchange of the entire *envelope* (env) gene or just the *S* region. The original and chimeric constructs (2 µg) were transfected to Huh7 cells, which were harvested four days later. (**A**) Southern blot analysis of intracellular replicative DNA (**B**) and extracellular virion-associated DNA from a representative transfection experiment. Mixed probes of genotype B and genotype C were used for hybridization. Positions of the 3.2-kb, 2.2-kb, and 1.0-kb DNA are indicated. (**C**) Ratio of extracellular virion DNA/intracellular replicative DNA. Densitometric values were obtained from Southern blots from three independent transfection experiments, using ImageJ software. The ratio for clone 22.5 was set arbitrarily at 1. (**D**) Secreted hepatitis B surface antigen (HBsAg) averaged from three transfection experiments (1:300 dilution for 1.1mer, 1:500 for dimer), with values for clone 22.5 set arbitrarily at 100.

3.2. Swapping the Envelope Gene Could Reverse the Replication Phenotype between Clones of the Two Genotypes

As L and S proteins are required for virion formation and release, the entire *envelope* gene or just the *S* region was swapped between clones 22.5 (genotype B) and 27.2 (genotype C) and also between clones 24.6 (genotype B) and 17.3 (genotype C) to evaluate the *envelope* gene as a contributor of differential virion secretion. Surprisingly, the 1.1mer construct of clone 17.3 continued to display a much lower level of replicative DNA than the two genotype B clones. A similar result was obtained when the 1.1mer construct was transiently transfected to HepG2 cells, another human hepatoma cell line supporting HBV genome replication and virion secretion (Supplementary Figure S1). Moreover, replacing its *envelope* gene or just the *S* region with that of clone 24.6 markedly increased replicative DNA (Figure 1A, left panel, compare lanes 7–9). Similar, although less dramatic effects, were observed with *SphI* dimer (Figure 1A, right panel). Conversely, inserting the *envelope* gene or *S* region of clone 17.3 into clone 24.6 drastically reduced DNA replication for both 1.1mer and dimer constructs (Figure 1A, lanes 4–6). Replacing the *envelope* gene of clone 22.5 with that of clone 27.2 also markedly reduced intracellular replicative DNA, whether for the 1.1mer construct or *SphI* dimer (Figure 1A, lanes 1 and 2). Overall, these results revealed the association of the *envelope* gene of genotype C with reduced level of replicative DNA. The impact of the *S* gene on the replication of the 1.1mer construct was further validated in HepG2 cells (Supplementary Figure S2).

3.3. Level of Replicative DNA Was Unaffected by Exchanging the Core Gene or 3′ S Region between the Two HBV Genotypes

That exchanging the entire *envelope* gene or just the *S* region could markedly alter intracellular level of replicative HBV DNA was quite unexpected. To substantiate this finding, we also exchanged the *core* gene between these two pairs of genotype B—genotype C clones. Remarkably, *core* gene replacement did not reduce replication for the two genotype B clones as either the 1.1mer construct (Figure 2A, left panel) or *SphI* dimer (Figure 2A, right panel). It also failed to increase replication for clone 17.3 of genotype C as a 1.1mer construct. This is in sharp contrast to the constructs with the entire *envelope* gene exchanged, which were transfected in parallel to serve as internal controls (Figure 2A, left panel, lanes 3, 6 and 9; right panel, lanes 3 and 6). In an initial attempt to identify the region responsible for differential replication capacity, we exchanged the 3′ *S* region only. As shown in Figure 3A, such an exchange failed to significantly alter the replication activity for either genotype B or genotype C clones.

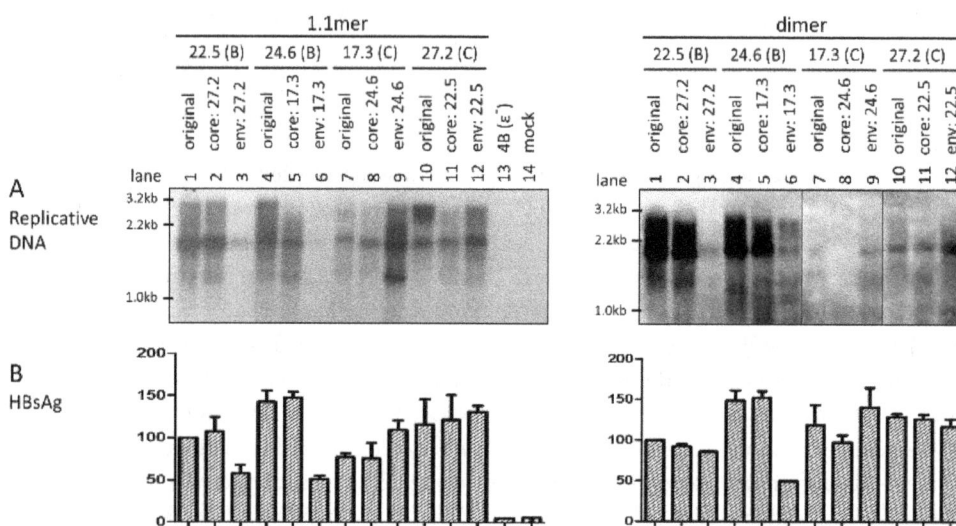

Figure 2. Exchanging the *core* gene between the two HBV genotypes did not alter the intracellular levels of replicative DNA. The *core* gene was exchanged between clones of the two HBV genotypes, as both 1.1mer construct (left panels) and dimer construct (right panels). Chimeras with the *envelope* gene exchanged were analyzed in parallel for comparison. Huh7 cells were transiently transfected with these HBV constructs and harvested four days later. (**A**) Southern blot analysis of replicative DNA from cell lysate using mixed probes of the two genotypes. A genotype A (clone 4B) mutant deficient in pregenomic RNA (pgRNA) encapsidation (ε^-) served as a negative control. (**B**) Secreted HBsAg averaged from three transfection experiments (1:300 dilution for 1.1mer, 1:500 for dimer), with values for clone 22.5 set arbitrarily at 100.

3.4. Exchanging the Entire Envelope Gene Also Affected HBsAg Level in Culture Supernatant, Especially for 1.1mer Construct

HBsAg was measured from culture supernatant of transfected Huh7 cells by ELISA, and Figure 1D, Figure 2B, and Figure 3B show the relative HBsAg titers averaged from three transfection experiments. Interestingly, replacement of the *envelope*, but not the *core* gene, for the 1.1mer construct resulted in a marked reduction of HBsAg titer for the two genotype B clones but increased the HBsAg titer for clone 17.3 of genotype C (Figure 2B, left panel, lanes 1, 3, 4, 6, 7, and 9). A similar trend was observed for *SphI* dimer, although the effect was striking only for replacement of the *envelope* gene of clone 24.6 by that of clone 17.3 (Figure 2B, right panel, lanes 4 and 6). Exchanging just the *S* region or its 3′ end did not markedly reduce HBsAg titers for the two genotype B clones or increase HBsAg for clone 17.3 of genotype C (Figures 1D and 3B), suggesting that the *preS* region of genotype B confers higher a HBsAg titer.

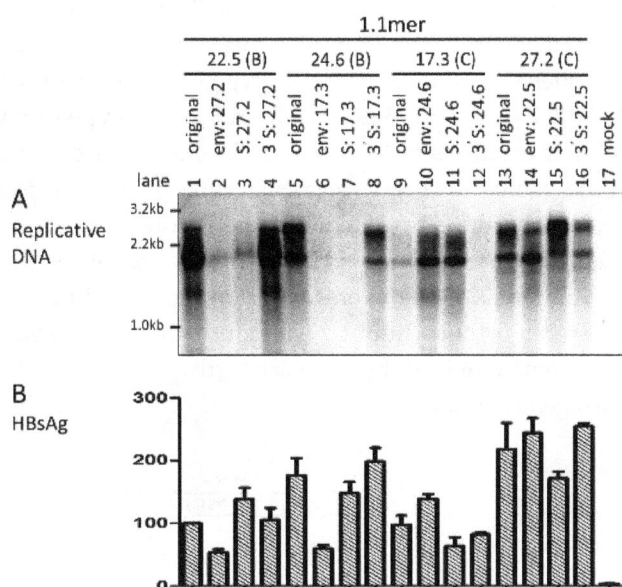

Figure 3. Impact of exchanging the 3′ *S* region on levels of intracellular replicative DNA and extracellular HBsAg. (**A**) The 1.1mer construct was used for exchange of the 3′ *S* region. The original construct and chimeric constructs with the entire *envelope* gene or the entire *S* region replaced were analyzed in parallel. Core particles were extracted from Huh7 cells at day 4 post-transfection for Southern blots of intracellular replicative DNA. Mixed probes of genotype B and genotype C were used for hybridization. (**B**) ELISA for HBsAg measured from culture supernatant from three independent experiments at the dilution of 1:300.

3.5. The Envelope Gene of Genotype C Was Associated with Higher Efficiency of Virion Secretion

Virions were immunoprecipitated from culture supernatant of transfected Huh7 cells by a combination of anti-preS1 and anti-S antibodies, followed by DNA extraction and Southern blot analysis. Due to the drastic difference in intracellular HBV DNA, we quantified the amount of both intracellular replicative DNA and extracellular virion DNA for each construct in the Southern blots, and calculated the ratio of virion DNA/replicative DNA as an indicator of virion secretion efficiency. A typical Southern blot of virion DNA from chimeric constructs of the entire *envelope* gene or just the *S* region is shown in Figure 1B, with the ratios averaged from three separate transfection experiments summarized in Figure 1C. It is evident that clone 17.3 of the C2 subgenotype had the highest rate of virion secretion, followed by clone 27.2 of subgenotype C1. Very similar results were obtained in HepG2 cells (Supplementary Figure S1). Replacing the *envelope* gene or the *S* region of clone 17.3 with that of clone 24.6 markedly reduced virion secretion (Figure 1C, compare lanes 7–9). Conversely, the *envelope* gene or the *S* region from clone 17.3 conferred more efficient virion secretion for clone 24.6 of genotype B (Figure 1C, lanes 4–6). For the pair of 22.5–27.2, the *S* region from clone 27.2 conferred more efficient virion secretion for clone 22.5; conversely the *S* region and especially the entire *envelope* gene from clone 22.5 impaired virion secretion. The impact of the *S* region on virion secretion was also tested and confirmed in HepG2 cells (Supplementary Figure S2C). Replacing the entire *envelope* gene of clone 22.5 with that of clone 27.2 rather reduced the ratio of virion DNA/replicative DNA for both the 1.1mer and dimer constructs (Figure 1C, compare lane 2 with lane 1). In the reverse experiment, replacing the *envelope* gene of clone 27.2 with that of clone 22.5 also impaired virion secretion, even for 1.1mer construct (Figure 1C, lanes 10 and 11).

3.6. Two Residues in the preS1 Domain Contributed to Genotypic Difference in Virion Secretion

The sequence from Arg103 to Ser124 in the *preS* domain of the L protein is highly conserved among different HBV isolates. These 22 amino acid residues are involved in interaction with core particles leading to virion formation [30,31]. Alignment of 2224 sequences of genotype B and 2227 sequences of genotype C available from the HBV database revealed genotype specific variations at two positions within this sequence [32], with most genotype B isolates having L108 (92.6%) and T115 (98.0%), in contrast to I108 (99.7%) and S115 (98.0%) found in genotype C isolates. To establish the impact of these two divergent positions on virion secretion, they were exchanged between clones of the two genotypes in the context of the 1.1mer construct. While the point mutations did not affect genome replication in either Huh7 or HepG2 cells (Figure 4A; Supplementary Figure S2A), the L108I/T115S substitutions increased virion DNA for the two genotype B clones in both cell lines (Figure 4C; Supplementary Figure S2C). Conversely, virion secretion in the two genotype C clones was diminished by the I108L/S115T substitutions. Therefore, the amino acid sequence difference at these two positions in *preS1* domain of L protein at least partly accounts for different virion secretion efficiencies between the two genotypes.

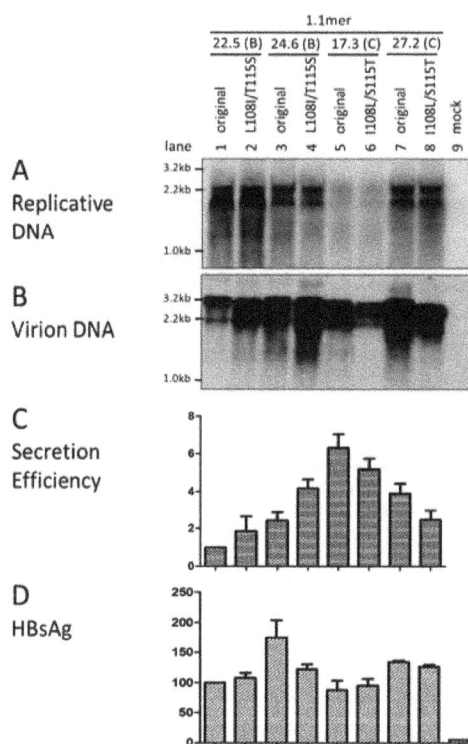

Figure 4. Effect of exchanging two codons in the *preS1* region between the two HBV genotypes on the efficiency of virion secretion. Codons 108 and 115 in the *preS1* region were exchanged between the 1.1mer constructs of genotype B and genotype C clones. Huh7 cells were transiently transfected with the parental constructs and site-directed mutants, and harvested four days later. Southern blot analysis of (**A**) intracellular replicative DNA and (**B**) virion DNA using mixed genotype B/C probes. (**C**) Calculated ratio of extracellular virion DNA/intracellular replicative DNA following densitometric analysis of the Southern blots from three independent transfection experiments. The value for clone 22.5 of genotype B was set arbitrarily at 1. (**D**) Secreted HBsAg values from three independent transfection experiments, with the value for clone 22.5 set arbitrarily at 100. Culture supernatant was diluted 1:300 for ELISA.

3.7. I126 in the S Domain Was Partly Responsible for Efficient Virion Secretion by Genotype C

While most HBV genotypes, including genotype B, have T126 in the *S* domain, genotype C is unusual in having I126 as its wild-type sequence [3]. Considering the large difference in chemical properties between threonine and isoleucine, we explored the possible contribution of this genotype-specific position to different virion secretion capacities. The T126I change failed to significantly alter virion secretion for the 1.1mer construct of the two genotype B clones (56 and 22.5) analyzed, although it moderately increased virion secretion for the *SphI* dimer of clone 56 (Figure 5B). Introducing the I126T substitution reduced virion secretion from both the 1.1mer and dimer constructs of the two genotype C clones, especially clone 17.3 (Figure 5B, C). With only one exception (1.1mer construct of clone 22.5), I126 was associated with a higher HBsAg titer than T126 (Figure 5D).

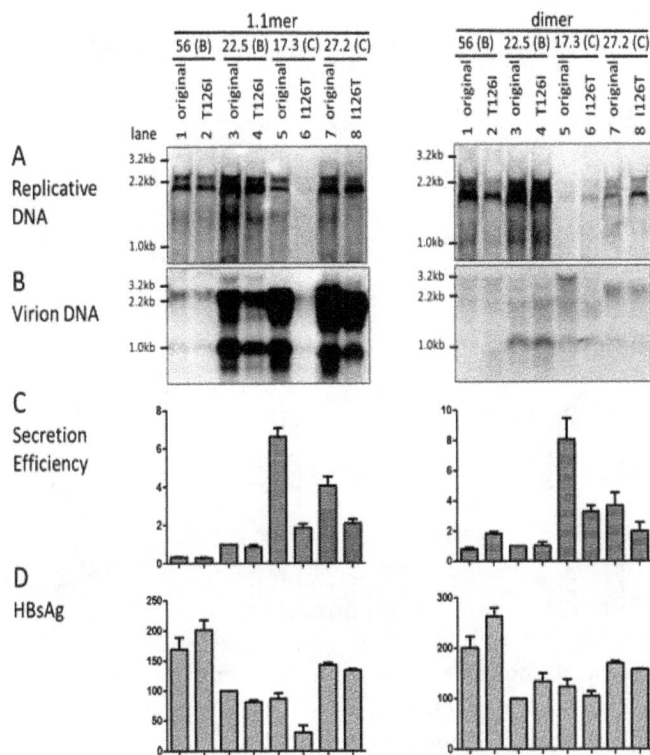

Figure 5. Impact of exchanging codon 126 in the *S* region between genotype B and genotype C clones on HBV virion secretion. The T126I mutation was introduced to two genotype B clones, while the I126T mutation was introduced to two genotype C clones. The parental constructs and the site-directed mutants as 1.1mer construct (left) or *SphI* dimer (right) were transiently transfected to Huh7 cells, which were harvested 4 days later. Shown are Southern blot analysis of (**A**) intracellular replicative DNA and (**B**) virion-associated DNA, (**C**) the ratio of extracellular virion DNA/intracellular replicative DNA as an indicator of secretion efficiency, (**D**) and secreted HBsAg. Culture supernatant was diluted 1:300 (for 1.1mer) or 1:500 (for dimer) for ELISA.

3.8. The L/S Protein Ratio Produced by the 1.1mer Construct Appeared Optimal for Virion Secretion

L protein could inhibit HBsAg secretion, and a too high or too low L/S protein ratio diminishes virion secretion [33,34]. Efficient genome replication achieved by the 1.1mer construct makes it the preferred HBV DNA construct for stable transfection into human hepatoma cell lines to generate virus particles for infection experiments [35,36]. Considering transcriptional interference among co-terminal HBV RNAs [28], we wondered whether overproduction of the 3.5-kb pgRNA from the 1.1mer construct suppresses transcription of the 2.4-kb subgenomic RNA to reduce L protein expression, thus diminishing virion secretion. In this regard, we previously found that an A152T mutation affecting the −3 position of the *S* gene translation initiation site could reduce HBsAg secretion from a 1.5mer construct of genotype A by about 70% without impairing virion secretion [22].

Introducing the A152T mutation into the 1.1mer constructs did reduce the amount of HBsAg secreted for both genotype B and genotype C clones (Figure 6A, bottom panel). However, virion secretion was diminished as well (Figure 6A, 2nd and 3rd panels). In another approach, Huh7 cells were co-transfected with fixed amount of 1.1mer construct of clone 22.5 or clone 27.2 and an increasing amount of the 0.7mer expression construct for L protein (Figure 6B) or S protein (Figure 6C). As anticipated, HBsAg secretion was suppressed by the extra L protein construct but increased by the extra S protein construct (Figure 6B,C, bottom panels). The L protein construct reduced virion secretion in a dose-dependent manner (Figure 6B, 2nd and 3rd panels), while the S construct abolished virion secretion even at the lowest dose (Figure 6C, 2nd and 3rd panels). These results suggest that the L/S protein ratio achieved in the 1.1mer construct is optimal or nearly optimal for virion secretion; further increase or reduction in L or S protein expression will rather reduce virion secretion.

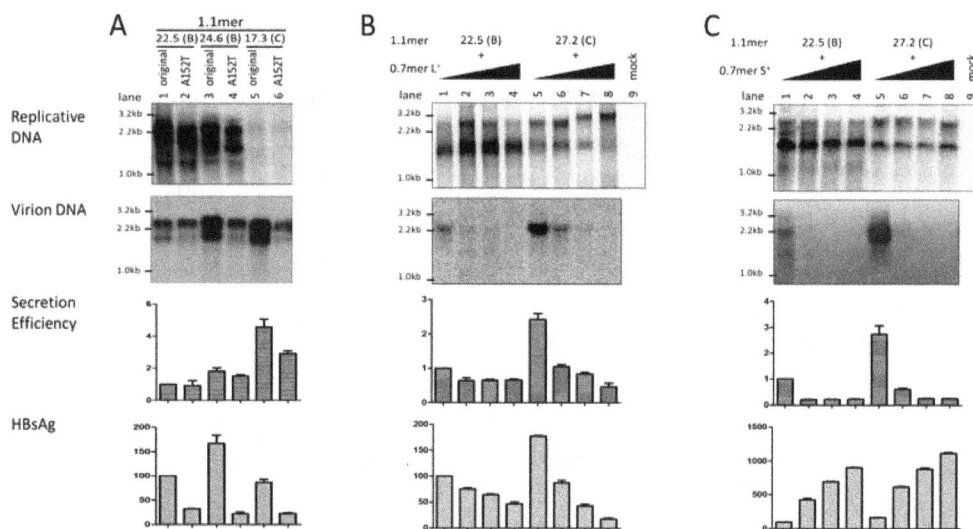

Figure 6. Impact of reducing endogenous S protein expression or providing extra L or S protein to HBV virion secretion from the 1.1mer construct. (**A**) The A152T mutation was introduced to clones of genotype B or genotype C to reduce S protein expression. Both the parental constructs and site-directed mutants were transfected to Huh7 cells. (**B**) Huh7 cells in 6-well plates were transfected with 1 µg of the 1.1mer genotype B or genotype C clone, together with 0, 0.125, 0.25, or 0.5 µg of 0.7mer expression construct for L protein. Variable amounts of pUC18 DNA were added to make the total amount of DNA be 2 µg. (**C**) Huh7 cells in 6-well plates were transfected with 1 µg of the 1.1mer genotype B or genotype C clone, together with 0, 0.25, 0.5, or 0.75 µg of 0.7mer expression construct for S protein, as well as variable amounts of pUC18 DNA for a total of 2 µg. For all three panels, shown from top are intracellular replicative DNA, virion DNA, calculated virion secretion efficiency (from three transfection experiments), and secreted HBsAg (averaged from three transfection experiments; 1:300 dilution for panel A; 1:100 dilution for panel B; 1:500 dilution for panel C).

4. Discussion

HBV genotypes B and C co-circulate in East Asia and target hosts of similar ethnic backgrounds. Moreover, chronic infection by both genotypes is primarily attributed to perinatal transmission from HBeAg-positive mothers. However, genotype C isolates are more likely implicated in breakthrough infections of newborns despite combined active/passive immunization. Moreover, adulthood infection with genotype C has greater risk to become chronic [13]. On the other hand, genotype B infection is associated with a higher risk of fulminant hepatitis and acute exacerbation of chronic infection [14–16], and genotype B isolates respond to interferon therapy more favorably than genotype C isolates.

Independent studies demonstrated that genotype C patients seroconvert from HBeAg to anti-HBe about 10 years later than genotype B patients, and, consequently, the prolonged active viral DNA replication and protein expression increase the lifelong risk for liver cirrhosis and HCC [7–12]. Based on our studies, wild-type genotype C isolates show a more reduced replication capacity than genotype B isolates but more efficient virion secretion [17]. We propose that the low replication capacity of wild-type genotype C isolates, meaning low expression levels of CORE protein, a strong immune target, may mitigate immune attack and delay HBeAg seroconversion. The low replication capacity may also serve as a driving force for the emergence of core promoter mutations during the immune clearance phase, which increase genome replication [26]. Indeed, genotype C isolates are more likely to develop core promoter mutations than genotype B isolates [8], which serve as independent risk factors for HCC development. The higher virion secretion efficiency of genotype C most likely compensates for its low replication capacity to promote virus transmission, both between different hosts and among hepatocytes inside the same liver. This will promote the establishment of persistent infection without triggering too strong an immune attack.

Our previous comparative study of the biological properties of genotype B and genotype C isolates was based on serum samples of Chinese and U.S. patients [17]. The full-length HBV genome was amplified by PCR and cloned to the *HindIII-SacI* sites of the pUC18 vector. All the genotype B isolates belonged to the B2 subgenotype, while the genotype C isolates were of the C2 subgenotype, except for a few C1 isolates. Two alternative approaches were taken afterwards. For the Chinese samples, the whole transformation product without plating was grown in liquid culture, followed by plasmid DNA preparation to generate a population pool. The 3.2-kb HBV genome was released from plasmid DNA by *BspQI* digestion, gel purified, and recircularized to mimic cccDNA. For the U.S. samples, individual PCR clones were converted to *SphI* dimer. Transient transfection experiments using both types of HBV DNA revealed the higher replication capacity of genotype B isolates/clones without core promoter mutations compared to the corresponding genotype C isolates/clones. Virion secretion was analyzed for the population pools (Chinese samples), with genotype C isolates showing much more efficient secretion (extracellular virion DNA/intracellular replicative DNA) than genotype B isolates.

The present study, as a direct extension of our reported work, was aimed at identifying the structural basis for differential virion secretion by the two HBV genotypes. Since the clone pools from the Chinese samples would complicate mutational analysis, we resorted to the individual clones from the U.S. samples with well-defined replication phenotype. Two clones of the B2 subgenotype (22.5 and 24.6) and one clone each of the C1 and C2 subgenotypes (27.2 and 17.3) were selected for the generation of chimeric constructs and site-directed mutants. Among the four clones, 22.5 and 24.6 have high (and comparable) replication capacities, while 17.3 has the lowest replication capacity [17]. Besides the *SphI* dimer, we also employed 1.1mer construct to markedly increase genome replication and, consequently, virion secretion. While that approach largely elevated the amount of intracellular replicative DNA for clone 27.2, clone 17.3 continued to display much less replicative DNA than the two genotype B clones (Figure 1A; Supplementary Figure S1A). Furthermore, some chimeric constructs showed greatly altered levels of replicative DNA than did the parental construct; thus, we had to resort to the ratio of virion DNA/replicative DNA as an indicator of virion secretion efficiency. We primarily used the Huh7 cell line to characterize the biological properties of these two HBV genotypes due to its higher transfection efficiency compared to HepG2 cells. Nevertheless, some key findings were further tested in HepG2 cells, and concordant results were obtained.

For both the 1.1mer and dimer constructs, clone 17.3 (C2) was most effective at virion secretion, followed by clone 27.2 (C1). In contrast, clone 22.5 had the lowest secretion efficiency (Figures 1, 4, and 6). Similar results were obtained in the HepG2 cell line. Functional characterization of additional clones of the C1 and C2 subgenotypes is needed to establish whether the C2 subgenotype has more efficient virion secretion than the C1 subgenotype. Reciprocal fragment exchange was then carried out between clones 17.3 and 24.6 and also between 27.2 and 22.5. We found that replacing the *core* gene between clones of the two HBV genotypes did not alter the efficiency of virion secretion (data not shown). In contrast, exchanging the entire *envelope* gene or just the *S* region between 17.3 and 24.6 could largely reverse the virion secretion phenotype (Figure 1C). Therefore, the *S* region, and probably also the *preS* region, harbor determinants for differential virion secretion. Similarly, exchanging the *S* region could increase virion secretion for clone 22.5, while diminishing virion secretion for clone 27.2. However, for unknown reasons, exchanging the entire *envelope* gene diminished virion secretion for 22.5 (Figure 1C, lane 2). One possibility is the incompatibility between the *preS* region and the *core* gene of these two clones. Considering that, with the 1.1mer construct, virion secretion was highest for clone 17.3 but lowest for clone 22.5, it will be interesting to generate chimeric constructs between these two clones to more effectively identify the determinants of differential virion secretion.

The L protein is engaged in core particle interaction, as well as the retention of the S protein for participation in virion morphogenesis, while the S protein is primarily responsible for particle (subviral particle and virion) formation and release. The M protein is not essential for virion secretion, but loss of M protein expression reduces the efficiency of virion secretion while increasing the maturity of the genome inside virions [22]. The *preS1*, *preS2*, and *S* domains of most HBV genotypes contain 119, 55, and 226 residues, respectively, of which 12, 12, and 20 positions are divergent between genotype B and genotype C isolates (Table 1). Previous studies identified a linear sequence at the boundary of the *preS1* and *preS2* domains (residues 103–124) as responsible for interaction with core particles [30,31]. Within that linear sequence, two residues are different between the two HBV genotypes; L108 and T115 for genotype B versus I108 and S115 for genotype C. Site-directed mutagenesis of both genotype B clones and both genotype C clones clearly demonstrated higher efficiency of I108/S115 than L108/T115 in mediating virion formation or release (Figure 4; Supplementary Figure S2). Interestingly, genotype A is similar to genotype C in having I108/S115, whereas genotype D has the same sequences as genotype B. Whether such a difference affects virion secretion for genotypes A and D remains to be determined.

Table 1. Divergent amino acid (aa) positions between genotype B and genotype C isolates in the *preS1*, *preS2*, and *S* domains.

	preS1 aa Position											
	10	35	39	45	48	51	54	57	60	84	108	115
Genotype B (2246 isolates)	K	K	E	L	H	N	D	K	V	L	L	T
Genotype C (2227 isolates)	Q	G	N	F	N	H	E	Q	A	I	I	S
	preS2 aa Position											
	130	132	138	152	154	155	156	158	160	165	167	172
Genotype B (2419 isolates)	T	Q	A	S	A	Q	N	V	A	L	K	V
Genotype C (2377 isolates)	A	L	G	N	V	P	T	A	P	F	R	A
	S aa Position											
	4	5	8	24	45	47	49	56	57	59		
Genotype B (4224 isolates)	I	A	L	K	T	V	L	Q	I	S		
Genotype C (7183 isolates)	T	T	F	R	A	T	P	P	T	N		
	64	85	110	113	126	143	160	161	200	213		
Genotype B (4224 isolates)	C	C	I	S	T	T	K	Y	F	M		
Genotype C (7183 isolates)	S	F	L	T	I	S	R	F	Y	L		

Twenty residues in the S domain are divergent between the two genotypes (Table 1). The current study focused on residue 126 for several reasons. First, genotype C is unique in having I126 rather than T126, which is found in other HBV genotypes. Second, residue 126 is part of the so-called 'a' determinant (residues 124–147), which is exposed on the virion surface and constitutes a major target of neutralizing antibodies. Indeed, immune escape mutations such as G145R frequently arise during the late stages of HBV infection or in association with vaccine escape or occult HBV infection [37–40]. In this regard, residue 126 is frequently mutated during the late stage of HBV infection, such as

I126T/S/N for genotype C and T126A for other HBV genotypes [41,42]. There was a report that the I126S mutation can cause decreased HBsAg detection and occult HBV infection [43]. Third, many immune escape mutants are impaired in virion secretion [37,43,44]. Our transfection experiments confirmed that I126 supported virion secretion from genotype C clones better than T126, although whether the T126I mutation improves virion secretion for genotype B remains inconclusive (Figure 5C). We also observed the association of I126 with a higher HBsAg titer than T126 in culture supernatant (Figure 5D), which is consistent with the result in a recent study [41]. Certainly, whether this is genuine or an artifact of the preferred recognition of I126 over T126 by our ELISA kit remains to be clarified.

In our initial study, the higher replication capacity of genotype B isolates/clones compared to genotype C isolates/clones correlated with higher level of the 3.5-kb RNA, suggesting control at the transcriptional level [17]. In the follow-up investigation, genotype B clones were found to possess a stronger enhancer II and/or core promoter region (these two elements overlap), and replacement of that region (1627–1866) was sufficient to reverse the replication phenotype [45]. In this regard, it was quite unexpected that replacement of the *envelope* gene or the *S* region of the 1.1mer construct of clone 17.3 with that of clone 24.6 could markedly increase the level of intracellular replicative DNA, whereas the reverse exchange could greatly reduce replication for 24.6 (Figure 1A, left panel). Moreover, similar findings can be made with *SphI* dimer, which represents a more physiological form of HBV DNA (Figure 1A, right panel). Similarly, replacement of the *envelope* gene of clone 22.5 with that of clone 27.2 suppressed replicative DNA for both 1.1mer and dimer constructs. However, enhancer I, enhancer II, or core promoter, which can all regulate levels of the 3.5-kb pgRNA, lie outside the *envelope* gene. Northern blot analysis or primer extension assay will be needed to establish whether the altered level of replicative DNA correlates with a changed level of the 3.5-kb RNA or pgRNA.

It is worth mentioning that clone 17.3, which showed the lowest level of intracellular replicative DNA in the 1.1mer form, had the highest efficiency of virion secretion. A similar correlation can be made for the chimeric constructs; those with increased intracellular replicative DNA had reduced virion secretion and vice versa. Thus, one intriguing possibility is that increased interaction between core particles and HBV ENVELOPE proteins will commit such particles towards the secretory pathway, although a large fraction will be degraded rather than secreted. The outcome is both a high level of virion DNA and a low level of intracellular replicative DNA. It will be very interesting to determine whether preventing the expression of ENVELOPE proteins from the 1.1mer genome of clone 17.3 or chimeric constructs with altered replication capacity will revert the phenotype of replicative DNA and whether suppressing the proteasome or lysosome degradation pathway can increase level of intracellular HBV DNA for clones with 'low replication' phenotypes such as 17.3.

Another interesting observation is that the replacement of the entire *envelope* gene of the two genotype B clones with genotype C sequences also reduced HBsAg in culture supernatant (Figure 3B). The opposite effect was seen when the *envelope* gene of the two genotype C clones was replaced with that of genotype B, although to lesser extent. No such effect was seen when just the *S* region was exchanged. It will be very helpful to measure the HBsAg titer from cell lysate to calculate the extracellular/intracellular HBsAg ratio to determine whether HBsAg production or its secretion was diminished. If HBsAg production was reduced, then genotype C isolates may have a weaker SPII promoter, which directs the transcription of the 2.1-kb RNA for M/S proteins and is located in the *preS* region. Indeed 28 positions within the SPII promoter are divergent between genotypes B and C (data not shown). If HBsAg secretion was reduced, then the L protein of genotype C might be more efficient at retaining the S protein (and possibly promoting virion formation through the same mechanism).

Acknowledgments: This study was supported by a grant 81371822 from the National Science Foundation of China and also by NIH grants AI103648, AI107618, AI113394, and AI116639.

Author Contributions: H.J., Y. Wang, J.L., Y. Wen, and S.T. designed the experiments; H.J., Y.Q., C.C., F.Z., C.L., and L.Z. performed the experiments and analysis; J.Z. contributed reagents/materials/analysis tools; and H.J. and S.T. wrote the manuscript. All authors reviewed the manuscript.

References

1.   Ganem, D.; Prince, A.M. Hepatitis B virus infection–natural history and clinical consequences. *N. Engl. J. Med.* **2004**, *350*, 1118–1129. [CrossRef] [PubMed]

2.   Trépo, C.; Chan, H.L.Y.; Lok, A. Hepatitis B virus infection. *Lancet* **2014**, *384*, 2053–2063. [CrossRef]

3.   Norder, H.; Courouce, A.M.; Coursaget, P.; Echevarria, J.M.; Lee, S.D.; Mushahwar, I.K.; Robertson, B.H.; Locarnini, S.; Magnius, L.O. Genetic diversity of hepatitis B virus strains derived worldwide: Genotypes, subgenotypes, and HBsAg subtypes. *Intervirology* **2004**, *47*, 289–309. [CrossRef] [PubMed]

4.   Schaefer, S. Hepatitis B virus: Significance of genotypes. *J. Viral Hepat.* **2005**, *12*, 111–124. [CrossRef] [PubMed]

5.   Kramvis, A. Genotypes and genetic variability of hepatitis B virus. *Intervirology* **2013**, *57*, 141–150. [CrossRef] [PubMed]

6.   Tong, S.; Revill, P. Overview of hepatitis B viral replication and genetic variability. *J. Hepatol.* **2016**, *64*, S4–S16. [CrossRef] [PubMed]

7.   Yuen, M.F.; Tanaka, Y.; Shinkai, N.; Poon, R.T.; But, Y.K.; Fong, Y.T.; Fung, J.; Wong, K.H.; Yuen, C.H.; Mizokami, M. Risk for hepatocellular carcinoma with respect to hepatitis B virus genotypes B/C, specific mutations of enhancer II/core promoter/precore regions and HBV DNA levels. *Gut* **2008**, *57*, 98–102. [CrossRef] [PubMed]

8.   Yuen, M.F.; Sablon, E.; Yuan, H.J.; Wong, K.H.; Hui, C.K.; Wong, C.Y.; Chan, O.O.; Lai, C.L. Significance of hepatitis B genotype in acute exacerbation, HBeAg seroconversion, cirrhosis-related complications, and hepatocellular carcinoma. *Hepatology* **2003**, *37*, 562–567. [CrossRef] [PubMed]

9.   Orito, E.; Ichida, T.; Sakugawa, H.; Sata, M.; Horiike, N.; Hino, K.; Okita, K.; Okanoue, T.; Iino, S.; Tanaka, E. Geographic distribution of hepatitis B virus (HBV) genotype in patients with chronic HBV infection in Japan. *Hepatology* **2001**, *34*, 590–594. [CrossRef] [PubMed]

10. Chu, C.J.; Hussain, M.; Lok, A.S. Hepatitis B virus genotype B is associated with earlier HBeAg seroconversion compared with hepatitis B virus genotype c. *Gastroenterology* **2002**, *122*, 1756–1762. [CrossRef] [PubMed]

11. Chan, H.L.; Hui, A.Y.; Wong, M.L.; Tse, A.M.; Hung, L.C.; Wong, V.W.; Sung, J.J. Genotype c hepatitis B virus infection is associated with an increased risk of hepatocellular carcinoma. *Gut* **2005**, *53*, 1494–1498. [CrossRef] [PubMed]

12. Chu, C.M.; Liaw, Y.F. Genotype C hepatitis B virus infection is associated with a higher risk of reactivation of hepatitis B and progression to cirrhosis than genotype b: A longitudinal study of hepatitis B e antigen-positive patients with normal aminotransferase levels at base. *J. Hepatol.* **2005**, *43*, 411–417. [CrossRef] [PubMed]

13. Zhang, H.W.; Yin, J.H.; Li, Y.T.; Li, C.Z.; Ren, H.; Gu, C.Y.; Wu, H.Y.; Liang, X.S.; Zhang, P.; Zhao, J.F. Risk factors for acute hepatitis B and its progression to chronic hepatitis in Shanghai, China. *Gut* **2008**, *57*, 1713–1720. [CrossRef] [PubMed]

14. Imamura, T. Distribution of hepatitis B viral genotypes and mutations in the core promoter and precore regions in acute forms of liver disease in patients from Chiba, Japan. *Gut* **2003**, *52*, 1630–1637. [CrossRef] [PubMed]

15. Ozasa, A.; Tanaka, Y.; Orito, E.; Sugiyama, M.; Kang, J.H.; Hige, S.; Kuramitsu, T.; Suzuki, K.; Tanaka, E.; Okada, S. Influence of genotypes and precore mutations on fulminant or chronic outcome of acute hepatitis B virus infection. *Hepatology* **2006**, *44*, 326–334. [CrossRef] [PubMed]

16. Ren, X.; Xu, Z.; Liu, Y.; Li, X.; Bai, S.; Ding, N.; Zhong, Y.; Wang, L.; Mao, P.; Zoulim, F. Hepatitis B virus genotype and basal core promoter/precore mutations are associated with hepatitis B-related acute-on-chronic liver failure without pre-existing liver cirrhosis. *J. Viral Hepat.* **2010**, *17*, 887–895. [CrossRef] [PubMed]

17. Qin, Y.; Tang, X.; Garcia, T.; Hussain, M.; Zhang, J.; Lok, A.; Wands, J.; Li, J.; Tong, S. Hepatitis B virus genotype c isolates with wild-type core promoter sequence replicate less efficiently than genotype B isolates but possess higher virion secretion capacity. *J. Virol.* **2011**, *85*, 10167–10177. [CrossRef] [PubMed]

18. Seeger, C.; Mason, W.S. Molecular biology of hepatitis b virus infection. *Virology* **2015**, *479–480*, 672–686. [CrossRef] [PubMed]

19. Hu, J.; Boyer, M. Hepatitis B virus reverse transcriptase and epsilon RNA sequences required for specific interaction in vitro. *J. Virol.* **2006**, *80*, 2141–2150. [CrossRef] [PubMed]

20. Tavis, J.E.; Ganem, D. Evidence for activation of the hepatitis B virus polymerase by binding of its RNA template. *J. Virol.* **1996**, *70*, 5741–5750. [PubMed]

21. Bruss, V.; Ganem, D. The role of envelope proteins in hepatitis B virus assembly. *Proc. Natl. Acad. Sci. USA* **1991**, *88*, 1059–1063. [CrossRef] [PubMed]

22. Garcia, T.; Li, J.; Sureau, C.; Ito, K.; Qin, Y.; Wands, J.; Tong, S. Drastic reduction in the production of subviral particles does not impair hepatitis B virus virion secretion. *J. Virol.* **2009**, *83*, 11152–11165. [CrossRef] [PubMed]

23. Junker, M.; Galle, P.; Schaller, H. Expression and replication of the hepatitis B virus genome under foreign promoter control. *Nucleic Acids Res.* **1988**, *15*, 10117–10132. [CrossRef]

24. Zong, L.; Qin, Y.; Jia, H.; Zhou, H.; Chen, C.; Qiao, K.; Zhang, J.; Wang, Y.; Li, J.; Tong, S. Two-way molecular ligation for efficient conversion of monomeric hepatitis B virus DNA constructs into tandem dimers. *J. Virol. Methods* **2016**, *233*, 46–50. [CrossRef] [PubMed]

25. Bang, G.; Kim, K.H.; Guarnieri, M.; Zoulim, F.; Kawai, S.; Li, J.; Wands, J.; Tong, S. Effect of mutating the two cysteines required for HBe antigenicity on hepatitis B virus DNA replication and virion secretion. *Virology* **2005**, *332*, 216–224. [CrossRef] [PubMed]

26. Parekh, S.; Zoulim, F.; Ahn, S.H.; Tsai, A.; Li, J.; Kawai, S.; Khan, N.; Trepo, C.; Wands, J.; Tong, S. Genome replication, virion secretion, and e antigen expression of naturally occurring hepatitis B virus core promoter mutants. *J. Virol.* **2003**, *77*, 6601–6612. [CrossRef] [PubMed]

27. Li, J.; Zong, L.; Sureau, C.; Barker, L.; Wands, J.R.; Tong, S. Unusual features of sodium taurocholate cotransporting polypeptide as a hepatitis B virus receptor. *J. Virol.* **2016**, *90*, 8302–8313. [CrossRef] [PubMed]

28. Chen, C.; Jia, H.; Zhang, F.; Qin, Y.; Zong, L.; Yuan, Q.; Wang, Y.; Xia, N.; Li, J.; Wen, Y. Functional characterization of hepatitis B virus core promoter mutants revealed transcriptional interference among co-terminal viral mRNAs. *J. Gen. Virol.* **2016**, *97*, 1–9.

29. Schindelin, J.; Rueden, C.T.; Hiner, M.C.; Eliceiri, K.W. The imagej ecosystem: An open platform for biomedical image analysis. *Mol. Reprod. Dev.* **2015**, *82*, 518. [CrossRef] [PubMed]

30. Bruss, V. A short linear sequence in the pre-S domain of the large hepatitis B virus envelope protein required for virion formation. *J. Virol.* **1997**, *71*, 9350–9357. [PubMed]

31. Bruss, V. Envelopment of the hepatitis B virus nucleocapsid. *Virus Res.* **2004**, *106*, 199–209. [CrossRef] [PubMed]

32. The Hepatitis B Virus Database (HBVdb). Available online: https://hbvdb.ibcp.fr/HBVdb/HBVdbIndex (accessed on 24 March 2017).

33. Ou, J.H.; Rutter, W.J. Regulation of secretion of the hepatitis B virus major surface antigen by the preS-1 protein. *J. Virol.* **1987**, *61*, 782–786. [PubMed]

34. Persing, D.H.; Varmus, H.E.; Ganem, D. Inhibition of secretion of hepatitis B surface antigen by a related presurface polypeptide. *Science* **1986**, *234*, 1388–1391. [CrossRef] [PubMed]

35. Ladner, S.K.; Otto, M.J.; Barker, C.S.; Zaifert, K.; Wang, G.H.; Guo, J.T.; Seeger, C.; King, R.W. Inducible expression of human hepatitis B virus (HBV) in stably transfected hepatoblastoma cells: A novel system for screening potential inhibitors of HBV replication. *Antimicrob. Agents Chemother.* **1997**, *41*, 1715–1720. [PubMed]

36. Guo, H.; Jiang, D.; Zhou, T.; Cuconati, A.; Block, T.M.; Guo, J.T. Characterization of the intracellular deproteinized relaxed circular DNA of hepatitis B virus: An intermediate of covalently closed circular DNA formation. *J. Virol.* **2007**, *81*, 12472–12484. [CrossRef] [PubMed]

37. Kalinina, T.; Iwanski, A.; Will, H.; Sterneck, M. Deficiency in virion secretion and decreased stability of the hepatitis B virus immune escape mutant G145R. *Hepatology* **2003**, *38*, 1274–1281. [CrossRef] [PubMed]

38. Waters, J.A.; Kennedy, M.; Voet, P.; Hauser, P.; Petre, J.; Carman, W.; Thomas, H.C. Loss of the common "A" determinant of hepatitis B surface antigen by a vaccine-induced escape mutant. *J. Clin. Investig.* **1992**, *90*, 2543–2547. [CrossRef] [PubMed]

39. Raimondo, G.; Caccamo, G.; Filomia, R.; Pollicino, T. Occult HBV infection. *Semin. Immunopathol.* **2013**, *35*, 39–52. [CrossRef] [PubMed]

40. Carman, W.F.; Karayiannis, P.; Waters, J.; Thomas, H.; Zanetti, A.; Manzillo, G.; Zuckerman, A.J. Vaccine-induced escape mutant of hepatitis B virus. *Lancet* **1990**, *336*, 325–329. [CrossRef]

41. Xiang, K.H.; Michailidis, E.; Ding, H.; Peng, Y.Q.; Su, M.Z.; Li, Y.; Liu, X.E.; Dao Thi, V.L.; Wu, X.F.; Schneider, W.M.; et al. Effects of amino acid substitutions in hepatitis B virus surface protein on virion secretion, antigenicity, HBsAg and viral DNA. *J. Hepatol.* **2016**, *66*, 288–296. [CrossRef] [PubMed]

42. Ren, F.; Tsubota, A.; Hirokawa, T.; Kumada, H.; Yang, Z.; Tanaka, H. A unique amino acid substitution, T126I, in human genotype C of hepatitis B virus S gene and its possible influence on antigenic structural change. *Gene* **2006**, *383*, 43–51. [CrossRef] [PubMed]

43. Huang, C.H.; Yuan, Q.; Chen, P.J.; Zhang, Y.L.; Chen, C.R.; Zheng, Q.B.; Yeh, S.H.; Yu, H.; Xue, Y.; Chen, Y.X. Influence of mutations in hepatitis B virus surface protein on viral antigenicity and phenotype in occult HBV strains from blood donors. *J. Hepatol.* **2012**, *57*, 720–729. [CrossRef] [PubMed]

44. Kwei, K.; Tang, X.; Lok, A.S.; Sureau, C.; Garcia, T.; Li, J.; Wands, J.; Tong, S. Impaired virion secretion by hepatitis B virus immune escape mutants and its rescue by wild-type envelope proteins or a second-site mutation. *J. Virol.* **2013**, *87*, 2352–2357. [CrossRef] [PubMed]

45. Qin, Y.; Zhou, X.; Jia, H.; Chen, C.; Zhao, W.; Zhang, J.; Tong, S. Stronger enhancer ii/core promoter activities of hepatitis B virus isolates of B2 subgenotype than those of C2 subgenotype. *Sci. Rep.* **2016**, *6*, 30374. [CrossRef] [PubMed]

13

Porcine Epidemic Diarrhea Virus Induces Autophagy to Benefit its Replication

Xiaozhen Guo [1,2], **Mengjia Zhang** [1,2], **Xiaoqian Zhang** [1,2], **Xin Tan** [1,2], **Hengke Guo** [1,2], **Wei Zeng** [1,2], **Guokai Yan** [2,3], **Atta Muhammad Memon** [1,2], **Zhonghua Li** [1,2], **Yinxing Zhu** [1,2], **Bingzhou Zhang** [1,2], **Xugang Ku** [1,2], **Meizhou Wu** [1,2], **Shengxian Fan** [1,2,]*** and **Qigai He** [1,2,]***

[1]   State Key Laboratory of Agricultural Microbiology, College of Veterinary Medicine, Huazhong Agricultural University, Wuhan 430070, China; guoxiaozhen@webmail.hzau.edu.cn (X.G.); zhangmengjia@webmail.hzau.edu.cn (M.Z.); 632071039@webmail.hzau.edu.cn (X.Z.); wstx1992@163.com (X.T.); 15071194245@163.com (H.G.); aiyouwei0620@163.com (W.Z.); Memonatta80@webmail.hzau.edu.cn (A.M.M.); lzh1990@webmail.hzau.edu.cn (Z.L.); yingzizhu10@163.com (Y.Z.); abing0313@webmail.hzau.edu.cn (B.Z.); kuxugang84@163.com (X.K.); wumeizhou@mail.hzau.edu.cn (M.W.)

[2]   The Cooperative Innovation Center for Sustainable Pig Production, Wuhan 430070, China; gkyangk@163.com

[3]   College of Animal Sciences and Technology, Huazhong Agricultural University, Wuhan 430070, China

*   Correspondence: fanshengxian@mail.hzau.edu.cn (S.F.); he628@mail.hzau.edu.cn (Q.H.)

Academic Editors: Linda Dixon and Simon Graham

Abstract: The new porcine epidemic diarrhea (PED) has caused devastating economic losses to the swine industry worldwide. Despite extensive research on the relationship between autophagy and virus infection, the concrete role of autophagy in porcine epidemic diarrhea virus (PEDV) infection has not been reported. In this study, autophagy was demonstrated to be triggered by the effective replication of PEDV through transmission electron microscopy, confocal microscopy, and Western blot analysis. Moreover, autophagy was confirmed to benefit PEDV replication by using autophagy regulators and RNA interference. Furthermore, autophagy might be associated with the expression of inflammatory cytokines and have a positive feedback loop with the NF-κB signaling pathway during PEDV infection. This work is the first attempt to explore the complex interplay between autophagy and PEDV infection. Our findings might accelerate our understanding of the pathogenesis of PEDV infection and provide new insights into the development of effective therapeutic strategies.

Keywords: autophagy machinery; PEDV replication; inflammatory responses; apoptosis

1. Introduction

The new porcine epidemic diarrhea (PED) outbreaks caused by porcine epidemic diarrhea virus (PEDV) variant has been documented in China since late 2010 and is now distributed all over the world. PED is characterized by acute enteric infection and high mortality in sucking piglets, causing enormous economic losses to the swine industry [1–4]. PEDV is an enveloped, single-stranded positive-sense RNA virus of the Coronaviridae family. The viral genome is approximately 28 kb, arranged with at least seven open reading frames (ORFs), ORF1a, ORF1b, S, ORF3, E, M, and N. ORF1a and ORF1b are further processed into 16 nonstructural proteins, nsp1 to nsp16. The S, E, M, and N genes encode four structural proteins, whereas ORF3 encodes an accessory protein [5–7]. Despite the elucidation of PEDV pathogenesis in some aspects, the underlying mechanism of PEDV replication is still largely unknown.

Autophagy is an evolutionarily highly conserved intracellular degradation process in which double-membrane vesicles (termed autophagosomes) are generated, and the long-lived proteins and damaged organelles are delivered to lysosomes for degradation and recycling [8,9]. Autophagy can be induced by diverse intracellular and extracellular stimuli, such as nutrient starvation, endoplasmic reticulum (ER) stress, pathogen-associated molecular patterns (PAMPs), and virus infection [10]. Increasing evidence indicates that autophagy plays both anti-viral and pro-viral roles in the life cycles and pathogenesis of a broad range of viruses [11]. Specifically, autophagy is an intrinsic host defense mechanism that inhibits viral replication or eliminates viruses by delivering them to the lysosomal compartment for degradation. Meanwhile, viruses develop many mechanisms to block autophagy or even hijack it for their own benefit, such as human cytomegalovirus (HCMV) and herpes simplex virus type 1 (HSV-1) [9,12–14]. However, autophagy is also believed to serve as a platform for viral replication, especially for RNA viruses, such as classical swine fever virus (CSFV), porcine reproductive and respiratory syndrome virus (PRRSV), and rotavirus (RV), utilizing the membranes of the autophagosome-like vesicles for their replication [15–17]. These polar characteristics reveal the complicated relationship between autophagy and viral infection.

A previous proteomic study indicated that more differentially expressed proteins were mapped to the autophagy pathway, and the microtubule-associated protein 1B, a useful biomarker protein for autophagy was up-regulated in PEDV-infected Vero cells [18]. In addition, our previous study demonstrated that mTOR (the mammalian target of rapamycin) pathway, which was closely associated with cellular autophagy, was down-regulated, and that the autophagy associated protein ATG5 was up-regulated during PEDV infection [19]. These studies indicated that autophagy might participate in PEDV infection, but the specific function of autophagy in the process of PEDV infection has not been elucidated. In the present study, we demonstrated for the first time that autophagy was triggered in Vero cells during PEDV infection to promote its replication. Moreover, autophagy might mediate the inflammatory responses induced by PEDV infection and have a positive correlation with the NF-κB signaling pathway.

2. Materials and Methods

2.1. Cells and Viruses

African green monkey kidney cell lines, Vero-E6 cells, were cultured in Dulbecco's modified Eagle's medium (DMEM), supplemented with 10% fetal bovine serum (Invitrogen, Carlsbad, CA, USA) at 37 °C with 5% CO_2. The PEDV variant strain CH/YNKM-8/2013 (Accession no. KF761675) was isolated from a sucking piglet with acute diarrhea. To obtain replication-incompetent PEDV, virus suspension was irradiated with UV light for 1 h. The absence of virus infectivity was confirmed by $TCID_{50}$ and real-time PCR [16].

2.2. Virus Infection

For autophagy induction and inhibition experiments, Vero cells were pretreated with rapamycin (1 μg/mL, Santa Cruz, CA, USA), 3-methyladenine (3-MA, 5 μm, Sigma, St. Louis, USA), Chloroquine (CQ, 50 μm, Sigma), and BAY 11-7082 (10 μm, Sigma) for the indicated time, and were then infected with PEDV at a MOI (multiplicity of infection) of 0.1. After 1 h incubation at 37 °C, unbound viruses were removed by washing three times with PBS, followed by incubation with serum-free DMEM with 8 μg/mL trypsin (Invitrogen) containing varying concentrations of rapamycin, 3-MA, CQ, BAY 11-7082, or DMSO.

2.3. Quantitative Real-Time PCR

Total RNA was extracted from Vero cells using the TRIzol reagent (Invitrogen) according to the manufacturer's protocol, and was then reverse-transcribed into cDNA using oligo (dT) as the primer (Invitrogen). Relative and absolute quantitative real-time PCR were performed in an Applied Biosystems ViiA 7 real-time PCR system as previously described [19]. The primers and probe used are listed in Table 1.

Table 1. Primers used for real-time real time PCR.

Primer	Sequence (5'–3')
PEDV-F	CGTACAGGTAAGTCAATTAC
PEDV-R	GATGAAGCATTGACTGAA
PEDV-probe-M	TTCGTCACAGTCGCCAAGG
ATG5-F	TTCACGCTATATCAGGAT
ATG5-R	ATCTCACTAATGTCTTCTTG
Beclin1-F	TGGCACAATCAATAACTTC
Beclin1-R	CAAGCAGCATTAATCTCAT
IL-6-F	TGTGAAAGCAGCAAAGAG
IL-6-R	AGTGTCCTCATTGAATCCA
IL-1β-F	GCGGCAACGAGGATGACTT
IL-1β-R	TGGCTACAACAACTGACACGG
IL-8-F	GGAACCATCTCGCTCTGTGTAA
IL-8-R	GGTCCACTCTCAATCACTCTCAG
CCL5-F	ACGCCTCGCTGTCATCCT
CCL5-R	GCACTTGCCACTGGTGTAGAA
TNF-α-F	CACCACGCTCTTCTGTCT
TNF-α-R	AGATGATCTGACTGCCTGAG
MCP-1-F	CTTCTGTGCCTGCTGCTCATA
MCP-1-R	ACTTGCTGCTGGTGATTCTTCT
GAPDH-F	ACATCATCCCTGCCTCTACTG
GAPDH-R	CCTGCTTCACCACCTTCTTG
β-actin-F	TTAGTTGCGTTACACCCTTTC
β-actin-R	ACCTTCACCGTTCCAGTT

2.4. Transmission Electron Microscopy

Vero cells were mock infected or infected with PEDV at 0.1 MOI and collected at 24 h post-infection (hpi) for ultrastructural analysis. Ultra-thin sections were viewed on a Hitachi H-7650 transmission electron microscope (Hitachi Ltd., Tokyo, Japan). Autophagosome-like vesicles were defined as double- or single-membrane vesicles measuring 0.3 to 2.0 μm in diameter with clearly recognizable cytoplasmic contents.

2.5. Confocal Fluorescence Microscopy

Vero cells were seeded on coverslips and transfected with GFP-LC3 or mRFP-GFP-LC3. After transfection for 24 h, the cells were infected with PEDV and fixed with cold 4% paraformaldehyde. After permeabilization and blocking, the cells were then incubated with mouse monoclonal antibody directed against the PEDV S protein (made in our laboratory), and were then inoculated with Alexa Fluor 594 Donkey Anti Mouse IgG (H+L) antibody (Ant Gene). Cell nuclei were counterstained with 0.01% 4′,6-diamidino-2-phenylindole (DAPI, Invitrogen). The fluorescent images were examined under a confocal laser scanning microscope (LSM 510 Meta, Carl Zeiss, Munich, Germany).

2.6. Western Blot Analysis

Vero cells were lysed in lysis buffer containing 50 mM Tris-HCl (pH 6.8), 10% glycerol, and 2% SDS [20]. The protein concentration was quantified by the BCA protein assay kit and equal amounts of protein samples were mixed with 5× sample loading buffer and boiled for 10 min, and then separated by 12% sodium dodecyl sulfate polyacrylamide gel electrophoresis (SDS-PAGE). The proteins were electro-transferred to 0.45 μm PVDF membranes (Millipore, Mississauga, ON, Canada). Membranes were blocked with 5% (w/v) skim milk-TBST at room temperature for 2 h and then incubated overnight at 4 °C with primary antibodies. The blots were then incubated with corresponding horseradish peroxidase (HRP) conjugated secondary antibodies (ABclonal, Wuhan, China). The protein bands were visualized using the Clarity™ Western ECL Blotting Substrate (Bio-Rad, Hercules, CA, USA). The protein blots were quantified by Image J software (National Institutes of Health, Bethesda, MD, USA).

2.7. RNA Interference

Vero cells grown to 60% confluence were transfected separately with Beclin1 or ATG5 and the corresponding scrambled siRNA with Lipofectamine 2000 (Invitrogen) according to the manufacturer's guidelines. The silencing efficiency was determined by Western blot and real-time PCR. Twenty-four hours after transfection, the cells were infected with PEDV as described above.

The siRNA was designed by and obtained from GenePharma (Shanghai, China). *Beclin1* siRNA sequence: 5'-CCCAGUGUUCCCGUAGAAUUAUUCUACGGGAACACUGGGTT-3'; *ATG5* siRNA sequence: 5'-GCAACUCUGGAUGGGAUUAUUUAAUCCCAUCCAGAGUUGCTT-3'.

2.8. Cell Viability Assay

The cytotoxic effects of reagents on Vero cells were determined using the MTT (3-[4, 5-dimethylthiazol-2-yl]-2, 5-diphenyl-2H-tetra-zolium bromide) assay as previously described [21]. Briefly, Vero cells were inoculated in a 96-well plate and treated with different concentrations of pharmacological drugs. Then, the cells were inoculated with MTT and the resulting formazan crystals were dissolved in dimethyl sulfoxide (DMSO). The absorbance was measured by a microplate spectrophotometer at a wavelength of 490 nm.

2.9. Statistical Analysis

All experiments were performed independently three times, and variables are expressed as the means with SEM. Statistical analyses were performed using student's *t*-test. A *p*-value < 0.05 was considered as statistically significant.

3. Results

3.1. PEDV Infection Increases the Levels of Autophagy in Vero Cells

Whether PEDV infection can activate the autophagy machinery was investigated by examining the formation of autophagosome-like vesicles in Vero cells at 24 h post PEDV infection through transmission electron microscopy (TEM). A large number of double- or single-membrane vesicles were observed in PEDV-infected cells, which contained cytosolic components or sequestered organelles. However, autophagosome-like vesicles were rarely observed in the mock-infected cells (Figure 1A). It is well known that coronavirus infection can induce a large number of double-membrane vesicles (DMVs), and the functional link between autophagic DMVs and coronavirus-induced replication-associated DMVs remains controversial [22–24]. In this study, these two different DMVs might co-exist in PEDV-infected cells [25].

In addition, the GFP-LC3 tandem plasmid was transfected into Vero cells to verify the response of cellular autophagy to PEDV infection. As is well-known, the LC3, a protein, is selectively recruited to autophagic vesicles, which can be considered as its redistribution from a diffuse cytoplasmic localization to a distinctive punctate cytoplasmic pattern during autophagy [16]. GFP-LC3 positive cells treated with rapamycin showed high punctate LC3 accumulation. Additionally, large amounts of punctate GFP-LC3 proteins were observed in PEDV-infected cells at 18 hpi, while GFP-LC3 was detected as a diffuse distribution in mock-infected cells (Figure 1B), indicating that the accumulation of GFP-LC3 dots was induced by PEDV infection. The quantitative analysis of the percentage of punctate GFP-LC3 cells in the total GFP-positive cells was performed and 100 GFP-positive cells were detected in each sample. The percentage of punctate GFP-LC3 cells in PEDV infected cells was nearly 70%, which was obviously higher than that in the mock-treated cells (Figure 1C).

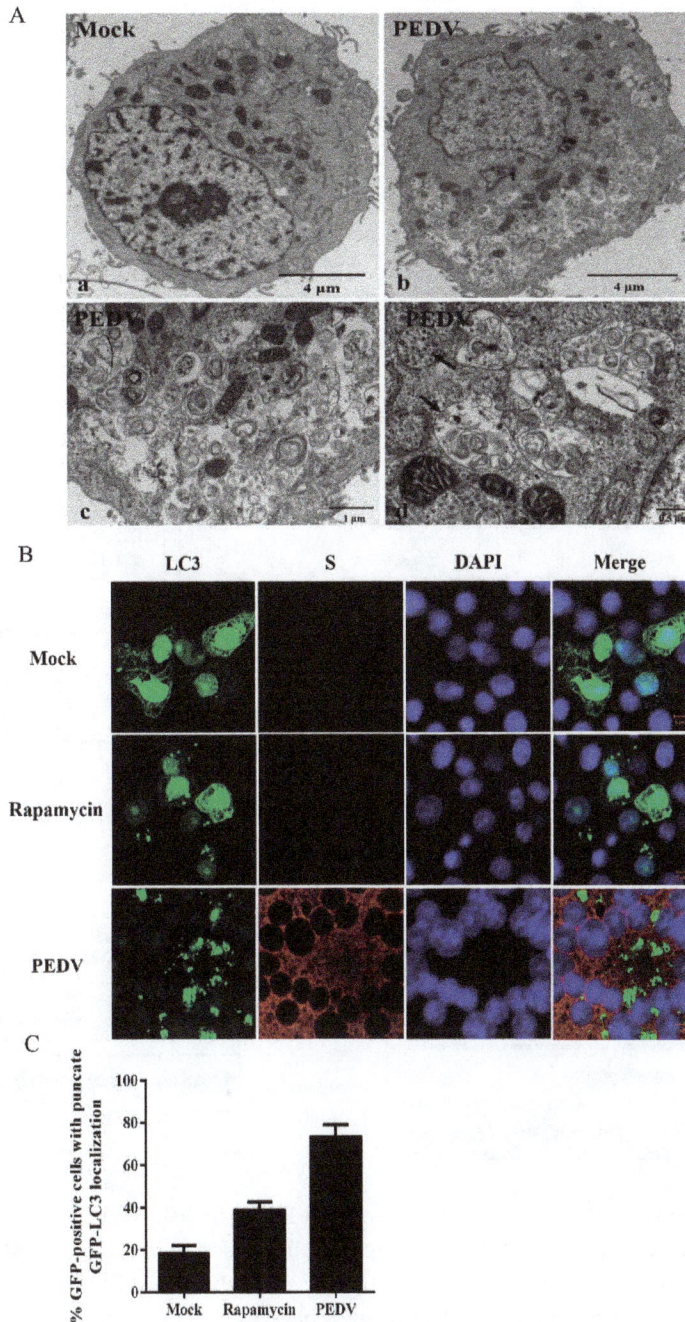

Figure 1. Porcine epidemic diarrhea virus (PEDV) infection increases the formation of autophagosome-like vesicles. (**A**) TEM observation. Vero cells were mock-treated (**a**) or infected with PEDV at 0.1 MOI for 24 h (**b**). Scale bar, 4 μm (**a,b**). (**c**) higher-magnification views of (**b**). Scale bar, 1 μm. (**d**) enlargement of the autophagosome-like structure. Scale bar, 0.5 μm. (**B**) Confocal microscope. The redistribution of GFP-LC3 was induced by PEDV infection. Vero cells were transfected with the plasmid GFP-LC3. Twenty-four hours later, the transfected cells were infected or mock-infected with PEDV at 0.1 MOI for 18 h. Meanwhile, cells pretreated with rapamycin for 4 h served as a positive control. PEDV infection was detected with the monoclonal antibody against PEDV S and cell nuclei were counterstained with 4',6-diamidino-2-phenylindole (DAPI). Scale bar, 5 μm. (**C**) The relative number of cells with punctate GFP-LC3 locations relative to all green fluorescent protein-positive cells. The data were presented as mean ± SEM of three independent experiments.

To further analyze whether autophagy was induced by PEDV infection, we examined the level of autophagy marker proteins in PEDV-infected cells by using immunoblotting. The conversion from LC3-I to LC3-II was monitored at 6, 12, 18, 24, and 30 h post PEDV infection. As shown in Figure 2A,B, a significant conversion of LC3-I to LC3-II was observed during the progression of PEDV infection, which was tracked by the PEDV N protein. Meanwhile, PEDV infection increased the expression of ATG5 and beclin1 in Vero cells relative to the mock-infected cells (Figure 2C). The results further supported that autophagy was induced by the PEDV infection.

Whether viral replication was required in PEDV-induced autophagy was confirmed by an experiment with ultraviolet (UV)-inactivated PEDV. The results demonstrated that no obvious detectable conversion from LC3-I to LC3-II was observed at 24 hpi, while autophagy was triggered normally by native PEDV (Figure 2D). The results indicated that viral replication was required for PEDV-induced autophagy.

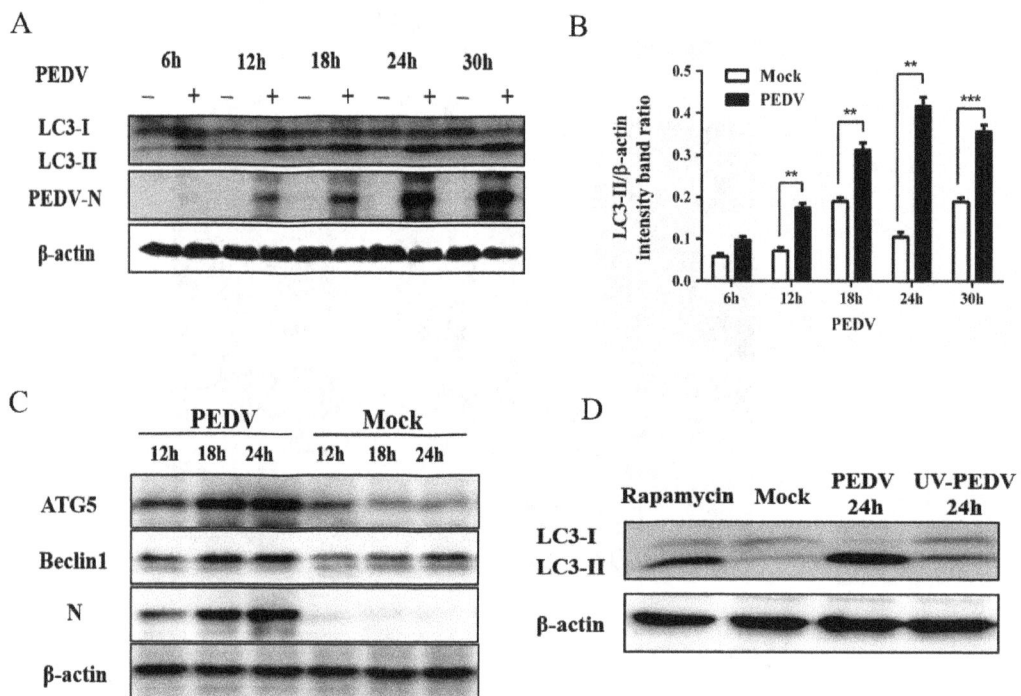

Figure 2. Expression of autophagy marker proteins in PEDV infected Vero cells. (**A**) Western blot analysis of the turnover of LC3-I to LC3-II in Vero cells at the indicated time points post PEDV infection using a polyclonal antibody against LC3 or a monoclonal antibody against PEDV N. β-actin expression was used as a protein loading control. (**B**) The intensity band ratio of LC3-II to β-actin was analyzed by using ImageJ software. The data were presented as mean ± SEM of three independent experiments (*t*-test, * $p < 0.05$, ** $p < 0.01$, *** $p < 0.001$). (**C**) Western blot analysis of the level of ATG5 and Beclin1 in Vero cells at 12, 18, and 24 hpi. β-actin expression was used as a protein loading control. (**D**) The turnovers of LC3-I to LC3-II were detected for mock-treated, rapamycin-treated, native PEDV, and UV-inactivated PEDV (MOI = 0.1) infection.

3.2. PEDV Infection Can Enhance Autophagy Flux

The degradation of SQSTM1 (p62) was recognized as an indicator for assessing autophagy flux. Whether a complete autophagic process was triggered by the PEDV infection was first determined by the degradation of p62 through immunoblotting analysis. As shown in Figure 3A,B, p62 was slightly accumulated during the early life cycle of PEDV infection, but degraded at the later stages. Meanwhile, p62 showed no obvious change from 6 to 30 h in the mock-infected cells.

The autophagy flux upon PEDV infection was further verified by measuring the levels of LC3-II and p62 through the treatment with chloroquine (CQ), which can inhibit the fusion of the autophagosome with lysosome. As shown in Figure 3C, CQ elevated the levels of LC3-II and p62 markedly at 24 h post PEDV infection, compared to the mock-treated cells.

Furthermore, the PEDV-induced autophagy flux was also confirmed using a tandem-reporter construct, GFP-mRFP-LC3. GFP is sensitive to lysosomal proteolysis and may diminish quickly in acidic pH, whereas RFP (red fluorescent protein) retains its fluorescence even at acidic pH. Our results showed that PEDV infection resulted in a partially red fluorescence at 24 hpi (Figure 3D), indicating the elevated level of autophagic flux. Taken together, this substantial evidence suggests that PEDV infection can enhance autophagy flux in Vero cells.

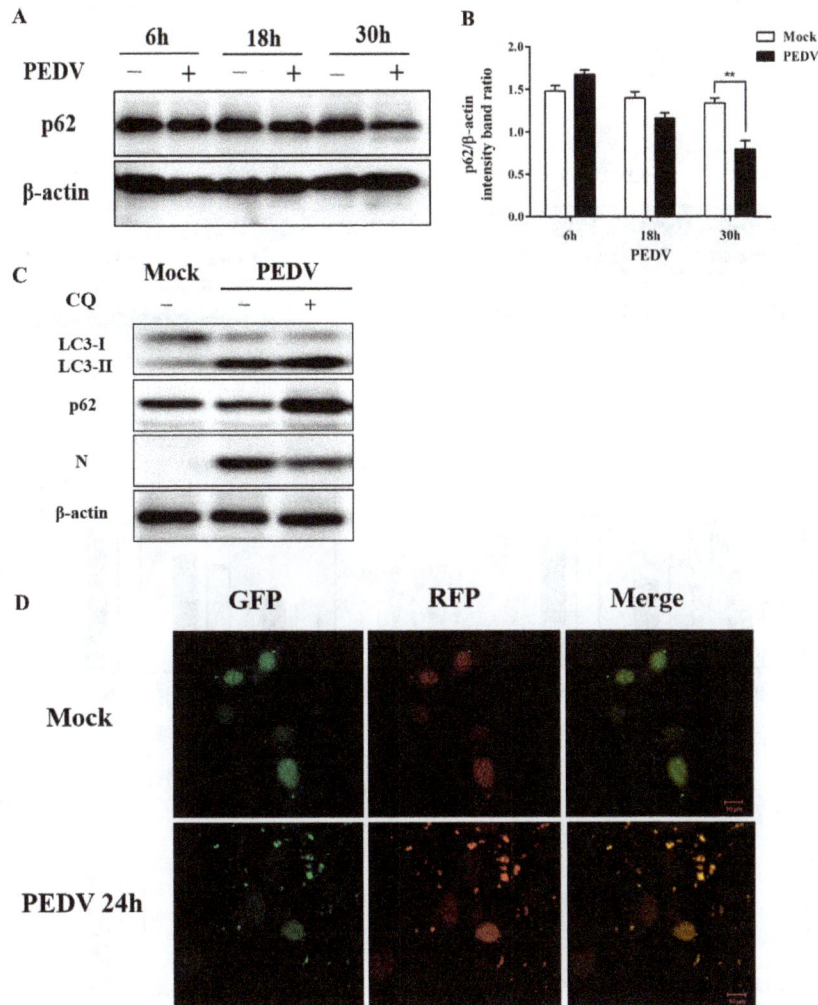

Figure 3. PEDV infection enhances autophagy flux. (**A**) Vero cells were mock-infected or infected with PEDV (0.1 MOI) for 6, 18, and 30 h. The cells were then analyzed by Western blot with antibodies against p62 and β-actin, separately. (**B**) The intensity band ratio of p62 to β-actin was analyzed by using ImageJ software. The data were presented as mean ± SEM of three independent experiments (*t*-test, * $p < 0.05$, ** $p < 0.01$, *** $p < 0.001$). (**C**) Vero cells were pretreated with CQ (50 μm) for 4 h, prior to PEDV (0.1 MOI) infection. After PEDV adsorption for 1 h, the cells were further cultured in fresh medium in the absence or presence of CQ. At 24 hpi, cell samples were detected by Western blot with antibodies against LC3, p62, N, and β-actin. (**D**) Vero cells were transfected with mRFP-GFP-LC3. Twenty-four hours later, the cells were mock-infected or infected with PEDV (0.1 MOI), then collected and visualized at 24 hpi, respectively. Scale bar, 10 μm.

3.3. Pharmacological Inhibition of Autophagy Decreases Viral Yield

The specific role of autophagy machinery on PEDV replication was explored by exposing Vero cells to 3-MA, which can inhibit autophagy at the early stage by suppressing the formation of autophagosomes [26]. As shown in Figure 4A,C,E, when compared to the mock-treated cells, 3-MA treatment not only reduced the level of LC3-II, but also significantly decreased the virus titer at different time points. In addition, CQ treatment (described above) reduced the expression of N protein, although it elevated the level of LC3-II (Figure 3C). These data indicated that autophagy inhibition could block PEDV infection. Similar results were also found in PEDV-infected ST cells.

Meanwhile, the role of autophagy machinery on PEDV replication was confirmed by rapamycin treatment, an inducer of autophagy through inhibition of the mTOR signaling pathway [8,17]. As shown in Figure 4B,D,F, induction of autophagy with rapamycin increased the LC3-II level and elevated the virus titer, compared to the mock-treated cells. The results suggested that autophagy induction could facilitate PEDV replication.

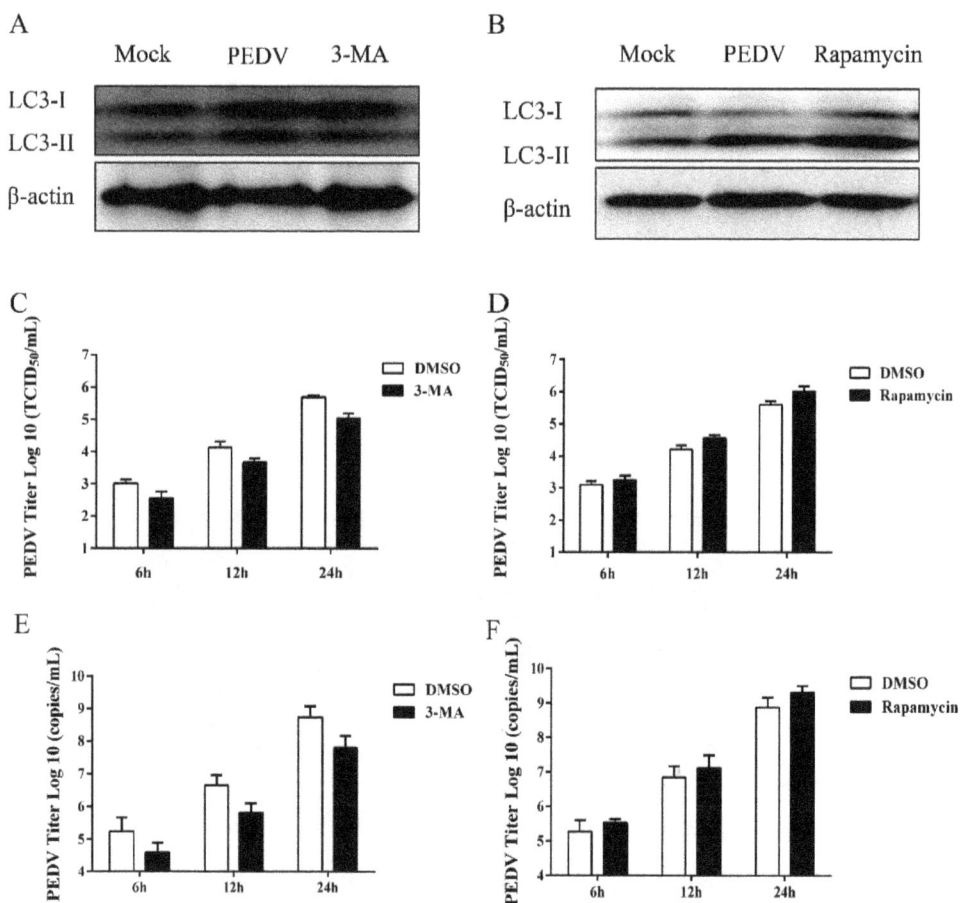

Figure 4. Pharmacological inhibition of autophagy decreases viral yield. (**A,B**) Vero cells were pretreated separately with 3-MA (5 mM) (**A**) or rapamycin (1 μg/mL) (**C**) for 4 h prior to PEDV (0.1 MOI) infection. After PEDV adsorption for 1 h, the cells were further cultured in fresh medium in the absence or presence of 3-MA or rapamycin. DMSO was used as a control. At 24 hpi, cell samples were detected by Western blot with antibodies against LC3 and β-actin. (**C,D**) The cells were collected separately at 6, 12, and 24 hpi to determine the viral titer. The data were presented as mean ± SEM of three independent experiments. (**E,F**) The cells were collected separately at 6, 12, and 24 hpi. The virus copy number was determined by real time PCR. The data were presented as mean ± SEM of three independent experiments.

3.4. Silencing Endogenous Beclin1 or ATG5 Gene Reduces the PEDV Titer

The relationship between autophagy and PEDV replication was further confirmed through gene-silencing experiments, with the endogenous *Beclin1* or *ATG5* gene specifically silenced, which was verified at both the transcriptional and translational levels (Figure 5A,B). Data from Figure 5C,D demonstrated that suppression of Beclin1 or ATG5 expression obviously decreased the viral titer and virus copy number compared to the control group. All the aforementioned data indicated that the autophagy mechanism was triggered by PEDV infection to facilitate its replication.

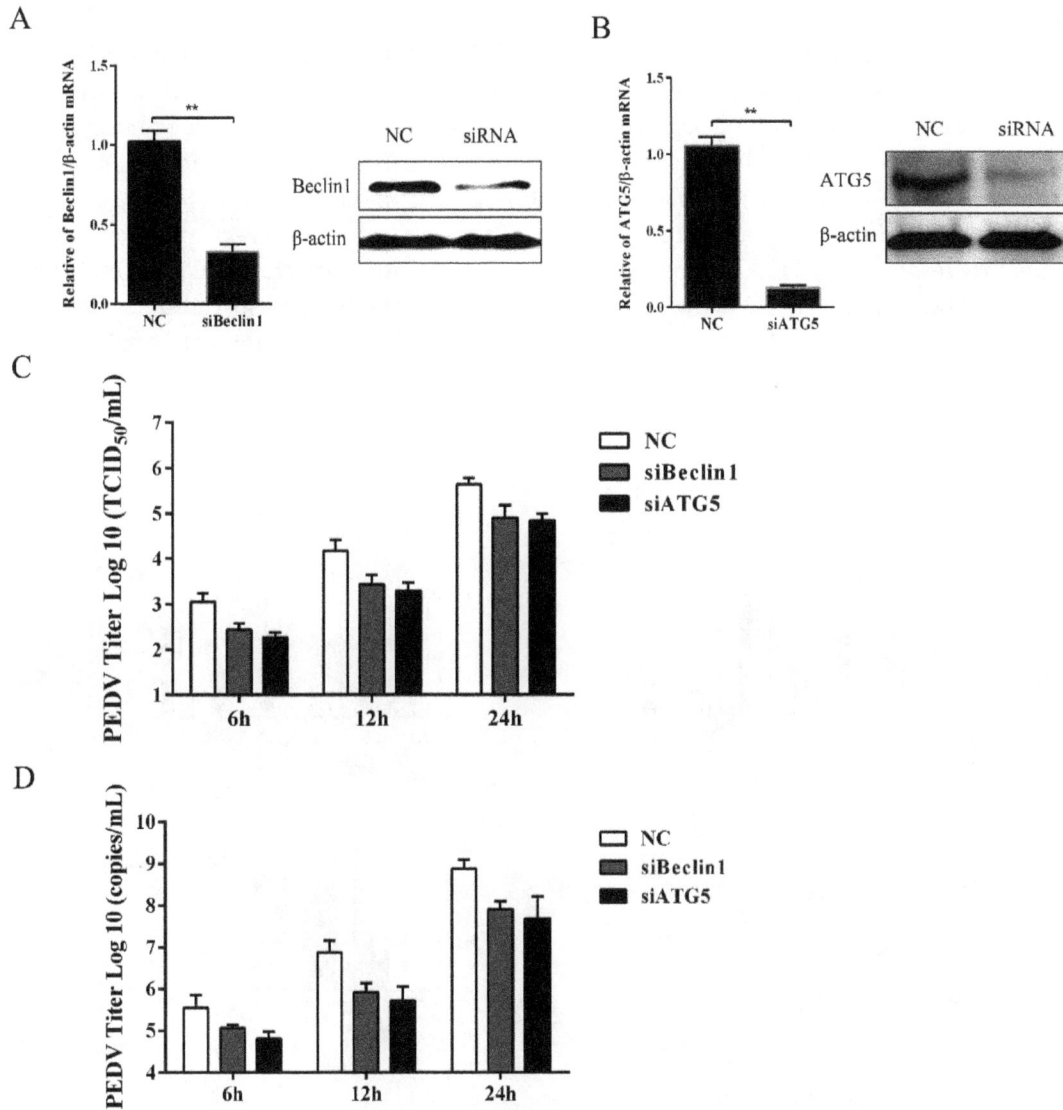

Figure 5. Inhibition of autophagy with specific siRNA targeting *Beclin1* or *ATG5* reduces PEDV replication. (**A**,**B**) Vero cells were transfected with siRNA targeting *Beclin1*, *ATG5*, or negative control (NC) for 48 h. The silencing efficiency was determined separately by quantitative real-time PCR and Western blot. (**C**) At 24 h post-transfection, cells were mock-infected or infected with PEDV for another 6, 12, and 24 h. The virus titer was determined by TCID$_{50}$. The data were presented as mean $\pm$ SEM of three independent experiments. (**D**) The cells were treated as described in (**C**) and collected separately at 6, 12, and 24 hpi, respectively. The virus copy number was determined by qRT-PCR. The data were presented as mean $\pm$ SEM of three independent experiments.

3.5. Autophagy Has a Positive Correlation with NF-κB Signaling Pathway

Accumulating data revealed a strong association between autophagy and the host immune response in the progression of virus infection [27,28]. Thus, the role of autophagy in PEDV induced inflammatory responses was investigated first by measuring the level of inflammatory cytokines under the circumstances when autophagy was suppressed by silencing the expression of Beclin1 and ATG5 proteins. Data showed that the inflammatory cytokines (such as IL-1β, IL-6, IL-8, CCL5, TNF-α, and MCP-1) were significantly down-regulated at 24 hpi when autophagy was inhibited in PEDV infected cells (Figure 6A,B), suggesting the potential involvement of autophagy in PEDV induced inflammation. It is well-known that the NF-κB signaling pathway plays a pivotal role in PEDV induced inflammatory response and Vero cells are interferon deficient [19,29]. Therefore, the influence of autophagy on the NF-κB pathway was determined under the deficient expression of Beclin1 and ATG5 at 24 h upon PEDV infection by immunoblotting analysis. The level of LC3-II decreased, and the phosphorylation of p65 was also down-regulated at the same time (Figure 6C). The results indicated that autophagy might participate in PEDV induced inflammation through the NF-κB pathway.

To further explore the relationship between autophagy and the NF-κB pathway mediated inflammatory response, the NF-κB pathway was attenuated by administrating BAY 11-7082, an NF-κB inhibitor. As shown in Figure 6D,E, the administration of BAY 11-7082 abolished the elevated level of LC3-II at 16, 20, and 24 h post PEDV infection, which was especially significantly at 24 hpi. These data indicated that a potential positive feedback loop between autophagy and the NF-κB signaling pathway might exist during PEDV infection.

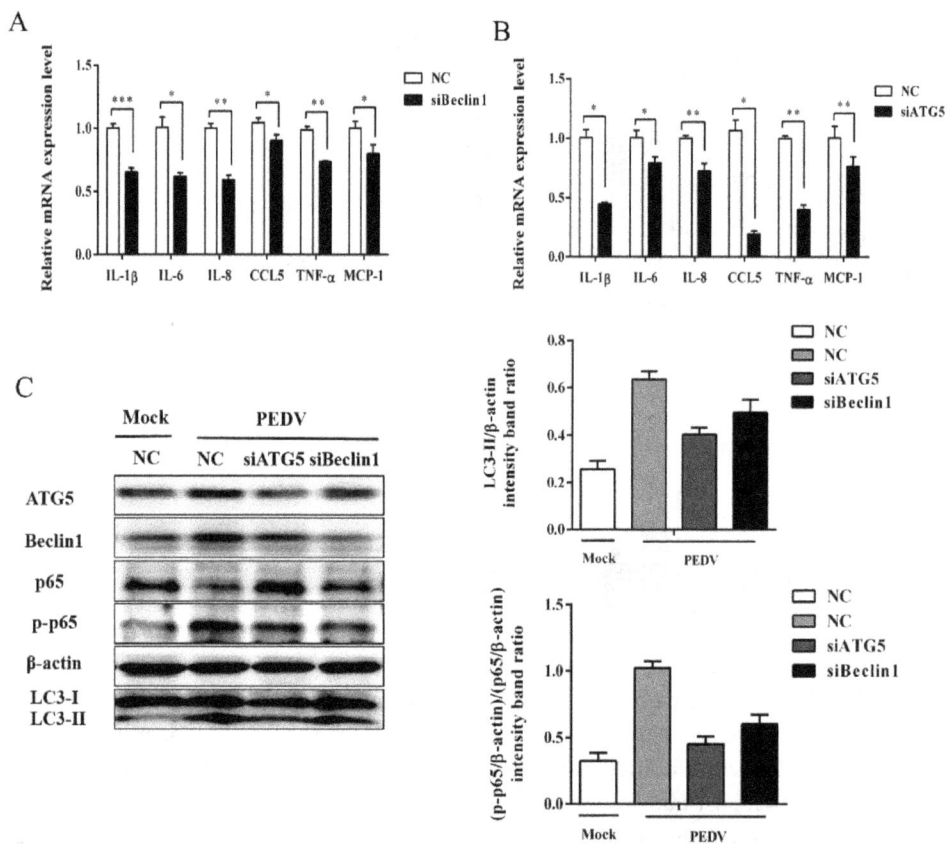

Figure 6. *Cont.*

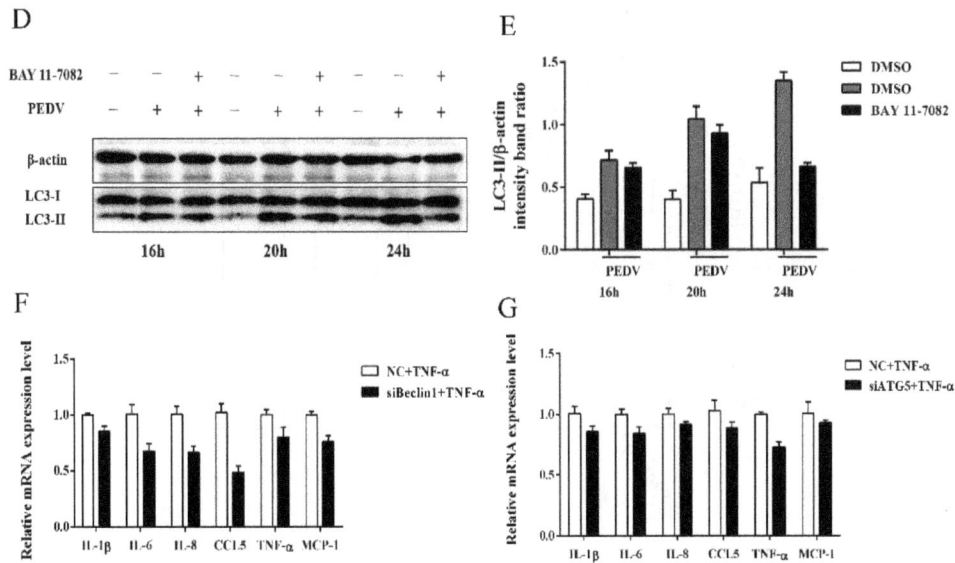

Figure 6. Autophagy mediates the production of inflammatory cytokines and correlates with the NF-κB signaling pathway in PEDV infected Vero cells. (**A,B**) The expression of inflammatory cytokines. Vero cells were transfected with siRNA targeting *Beclin1*, *ATG5*, or negative control (NC) for 24 h, followed by PEDV infection (0.1 MOI). The mRNA levels of cytokines were determined at 24 hpi by quantitative real-time PCR. (**C**) The level of LC3-II, p65, or phospho-p65 was also examined separately at 24 hpi with the corresponding antibodies by Western blot. The intensity band ratios of LC3-II to β-actin and p-p65 to p65 were analyzed by using ImageJ software. (**D**) Vero cells were pretreated with 10 μm BAY11-7082 for 12 h, then followed by PEDV infection for 16, 20, and 24 h, separately. The cell samples were collected for LC3-II detection by Western blot. (**E**) The intensity band ratio of LC3-II to β-actin was analyzed by using ImageJ software. All data were presented as mean ± SEM of three independent experiments. (**F,G**) Vero cells were transfected with siATG5, siBeclin1, or negative control. Twenty-four hours post-transfection, the cells were treated with TNF-α for 4 h, and then the expression of inflammatory cytokines were determined by qRT-PCR. The data were presented as mean ± SEM of three independent experiments.

Considering that autophagy inhibition can block PEDV replication, a TNF-α induction of inflammatory cytokines experiment in autophagy-deficiency cells was carried out to exclude the effect of PEDV replication on cytokine production. From Figure 6F,G, it can be seen that the expression of inflammatory cytokines induced by TNF-α in ATG5-or Beclin1-deficient cells was significantly down-regulated, when compared to the transfected cells of the negative control. These results also supported the potential positive relationship between autophagy and the NF-κB signaling pathway induced by PEDV infection.

3.6. Pharmacological Regulation of Autophagy Does Not Affect Cell Viability

The effect of the autophagy regulators on the capability of PEDV replication by changing the cell viability was tested by the MTT assay. No significant effects on cell viability were observed from the treatment with CQ, 3-MA, rapamycin, or BAY11-7082, respectively ($p > 0.05$) (Figure 7), indicating that pharmacological regulation of autophagy does not affect the cell viability.

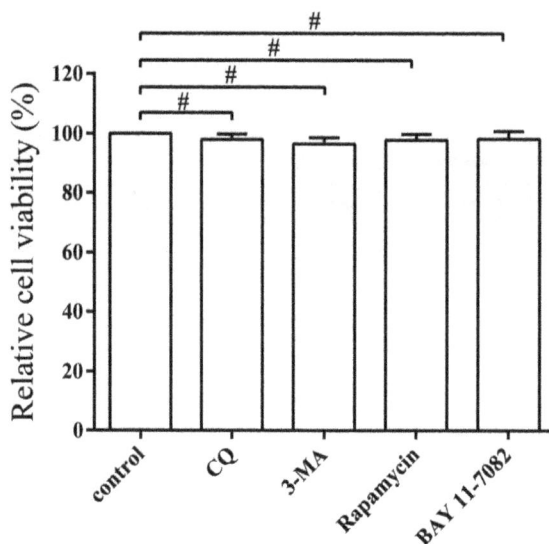

Figure 7. Pharmacological regulation of autophagy does not affect cell viability. The cell viability was determined by MTT assay after treatment with CQ (50 μm), 3-MA (5 μm), rapamycin (1 μg/mL), and BAY 11-7082 (10 μm) for 48 h, respectively. The data represent the mean ± SEM of three independent experiments. (*t*-test, [#] $p > 0.05$).

4. Discussion

In recent years, autophagy has been widely investigated due to its important role in the pathogenesis of many diseases [9,30]. The relationship between autophagy and viral infection, including coronavirus, has attracted the attention of an increasing number of researchers [11,31,32]. For instance, the formation of double membrane-bound MHV replication complexes was found essential for the autophagy induced by mouse hepatitis virus (MHV) infection [33]. A further study revealed that the autophagy-like process induced by infectious bronchitis virus (IBV) infection might not be important for virus replication [34]. In addition, autophagy was confirmed to negatively regulate transmissible gastroenteritis virus (TGEV) replication [35]. However, the specific role of autophagy on PEDV infection has not been reported until now. In this study, we provide the first strong evidence that PEDV infection can trigger autophagy to facilitate its replication.

In the present study, we firstly found that PEDV infection promoted the formation of DMVs by TEM, which is a hallmark of coronavirus replication and might potentially provide a platform for viral RNA synthesis [23,36]. In coronavirus infection, the formation of DMVs is usually believed to be associated with the autophagy pathway. This observation was further supported by the accumulation of GFP-LC3 dots in virus induced syncytia and the increased expression of autophagic marker proteins from the immunoblotting analysis. These data indicated that autophagy activity might be triggered by PEDV infection [37]. In addition, autophagy was documented to play a role in syncytia formation, which could further increase autophagy in multinucleated cells as well. Therefore, we inferred that autophagy might closely associate with the syncytia formation in Vero cells during PEDV infection [38,39]. Moreover, the UV inactivation test implied that the effective replication of PEDV was required for autophagy induction. Previous reports indicated that autophagosome accumulation might be attributed to their de novo formation or a block in trafficking to lysosomes for maturation. The cellular state of autophagy can be measured by detecting the degradation of SQSTM1 (p62), using a lysosome inhibitor, or a tandem reporter construct mRFP-GFP-LC3 [40]. Our results from these tests indicated that a complete autophagy process might be triggered to modulate PEDV infection in Vero cells.

Recent studies have demonstrated that the replication mechanisms may vary among different members of the *Coronaviridae* family. For example, TGEV infection induces autophagy to negatively regulate its replication [35]. IBV can induce autophagy but does not require autophagy for its replication [34]. The specific effect of autophagy induction on MHV infection is controversial. Prentice et al. reported that MHV utilized the autophagy process to form DMVs to enhance its replication [33]. However, another study showed that autophagy was not important for MHV replication [41]. In the present study, the concrete effect of autophagy upon PEDV infection was examined by administrating pharmacological regulators. The yield of PEDV was found to be suppressed whenever the autophagy process was inhibited by 3-MA at the early stage, or by CQ at the late stage. Additionally, the induction of autophagy by rapamycin also enhanced the viral titer. The effect of autophagy modulation on PEDV replication was further evaluated by silencing the two essential endogenous components *Beclin1* and *ATG5* [26,42]. Beclin1 (ATG6), a critical component in the class III PI3 kinase complex (PI3KC3), is involved in the initial step of autophagosome formation, while the ATG12-ATG5 conjugate, a key regulator, participates in autophagosome maturation [43]. The abolishment of their expression reduced the PEDV titer. These results suggested that autophagy induction might benefit PEDV replication. Autophagy has also been documented to have a complex interaction with apoptosis during viral infection [44–46]. For instance, influenza A virus was documented to induce apoptosis in autophagy protein deficient cells [47]. Autophagy was also reported to postpone apoptotic cell death during porcine reproductive and respiratory syndrome virus (PRRSV) infection through Bad-Beclin1 interaction [48,49]. The interplay of autophagy and apoptosis during PEDV infection needs to be further elucidated for a better understanding of the autophagy mechanism by which PEDV benefits its infection.

Recently, autophagy has been identified as an important component of the immune system, functioning from the elimination of infectious agents to the modulation of inflammatory responses [50,51]. The crosstalk between autophagy and inflammation has drawn much attention from researchers in recent years, while the relationship between autophagy and PEDV induced inflammatory responses is largely unclear. In the present study, we showed the expression of inflammatory cytokines and the activation of the NF-κB signaling pathway were both restrained in PEDV infected autophagy deficient cells, implying that autophagy might participate in PEDV induced inflammatory responses, especially for the NF-κB signaling pathway. Meanwhile, Vero cells with a NF-κB pathway deficiency exhibited a decreased level of LC3-II after PEDV infection, suggesting that the NF-κB signaling pathway might also have a positive role in autophagy induction. It was worth mentioning that the PEDV yield was reduced in both autophagy deficient or NF-κB pathway deficient cells. The decreased expression of inflammatory cytokines induced by TNF-α in autophagy deficient cells might provide new clues for deciphering the complex relationship between autophagy and inflammation. The cause-and-effect relationship between the restriction of PEDV yield reduction on the expression of inflammatory cytokines and the restraint of inflammation inhibition on PEDV replication in autophagy deficient cells needs further elucidation. A previous study reported that the positive feedback loop between autophagy and the NF-κB signaling cascade might exacerbate the inflammation induced by H5N1 pseudotyped viral particles [28]. The inhibition of NF-κB was also documented to induce autophagy suppression, leading to apoptosis in pancreatic ductal adenocarcinoma cells [52]. We speculate that autophagy and inflammatory cascade reactions might be positively regulated by PEDV infection to exacerbate inflammation. Furthermore, apoptosis might play a pivotal role in the complex relationship between autophagy and inflammation.

In summary, this is the first report concerning the potential induction of autophagy in PEDV infected Vero cells to facilitate its replication. Autophagy induction might be manipulated by PEDV to mediate the inflammatory cascade responses, and have a positive feedback loop with the NF-κB signaling pathway. Our knowledge regarding the interplay between PEDV infection and autophagy is still insufficient. The integrated data facilitate our understanding of the pathogenesis of PEDV infection and provide novel insights into the development of effective therapeutic strategies. Further research can focus on the complex relationship between autophagy modulation and immune responses, as well as apoptosis.

Acknowledgments: This work was supported by grants from the National Key Research and Development Program of China (2016YFD0500702) and the China Agricultural Research System (CARS-36). We thank prof. Xiang Mao from the Shanghai Veterinary Research Institute, for providing the GFP-mRFP-LC3 tandem reporter construct.

Author Contributions: Xiaozhen Guo had full access to all of the data in the study and takes responsibility for the integrity of the data and the accuracy of the data analysis; Study conception and design: Xiaozhen Guo, Mengjia Zhang, Xiaoqian Zhang, Zhonghua Li, Meizhou Wu, Atta Muhan mMad Memon, and Qigai He; Acquisition of data: Xiaozhen Guo, Xin Tan, Hengke Guo, Wei Zeng, Guokai Yan; Analysis and interpretation of data: Xiaozhen Guo, Yinxing Zhu, Bingzhou Zhang, Xugang Ku, Shengxian Fan.

References

1. Li, W.; Li, H.; Liu, Y.; Pan, Y.; Deng, F.; Song, Y.; Tang, X.; He, Q. New variants of porcine epidemic diarrhea virus, China, 2011. *Emerg. Infect. Dis.* **2012**, *18*, 1350–1353. [CrossRef] [PubMed]

2. Crawford, K.; Lager, K.M.; Kulshreshtha, V.; Miller, L.C.; Faaberg, K.S. Status of vaccines for porcine epidemic diarrhea virus in the United States and Canada. *Virus Res.* **2016**, *226*, 108–116. [CrossRef] [PubMed]

3. Islam, M.T.; Kubota, T.; Ujike, M.; Yahara, Y.; Taguchi, F. Phylogenetic and antigenic characterization of newly isolated porcine epidemic diarrhea viruses in Japan. *Virus Res.* **2016**, *222*, 113–119. [CrossRef] [PubMed]

4. Dastjerdi, A.; Carr, J.; Ellis, R.J.; Steinbach, F.; Williamson, S. Porcine epidemic diarrhea virus among farmed pigs, ukraine. *Emerg. Infect. Dis.* **2015**, *21*, 2235–2237. [CrossRef] [PubMed]

5. Duarte, M.; Gelfi, J.; Lambert, P.; Rasschaert, D.; Laude, H. Genome organization of porcine epidemic diarrhoea virus. *Adv. Exp. Med. Biol.* **1993**, *342*, 55–60. [PubMed]

6. Song, D.; Park, B. Porcine epidemic diarrhoea virus: A comprehensive review of molecular epidemiology, diagnosis, and vaccines. *Virus Genes* **2012**, *44*, 167–175. [CrossRef] [PubMed]

7. Zhang, Q.; Shi, K.; Yoo, D. Suppression of type I interferon production by porcine epidemic diarrhea virus and degradation of creb-binding protein by NSP1. *Virology* **2016**, *489*, 252–268. [CrossRef] [PubMed]

8. Kim, H.J.; Lee, S.; Jung, J.U. When autophagy meets viruses: A double-edged sword with functions in defense and offense. *Semin. Munopathol.* **2010**, *32*, 323–341. [CrossRef] [PubMed]

9. Sun, Y.; Yu, S.; Ding, N.; Meng, C.; Meng, S.; Zhang, S.; Zhan, Y.; Qiu, X.; Tan, L.; Chen, H.; et al. Autophagy benefits the replication of newcastle disease virus in chicken cells and tissues. *J. Virol.* **2014**, *88*, 525–537. [CrossRef] [PubMed]

10. Kroemer, G.; Marino, G.; Levine, B. Autophagy and the integrated stress response. *Mol. Cell* **2010**, *40*, 280–293. [CrossRef] [PubMed]

11. Kudchodkar, S.B.; Levine, B. Viruses and autophagy. *Rev. Med. Virol.* **2009**, *19*, 359–378. [CrossRef] [PubMed]

12. Dreux, M.; Chisari, F.V. Viruses and the autophagy machinery. *Cell Cycle* **2010**, *9*, 1295–1307. [CrossRef] [PubMed]

13. Chaumorcel, M.; Souquere, S.; Pierron, G.; Codogno, P.; Esclatine, A. Human cytomegalovirus controls a new autophagy-dependent cellular antiviral defense mechanism. *Autophagy* **2008**, *4*, 46–53. [CrossRef] [PubMed]

14. Talloczy, Z.; Jiang, W.; Virgin, H.W.T.; Leib, D.A.; Scheuner, D.; Kaufman, R.J.; Eskelinen, E.L.; Levine, B. Regulation of starvation- and virus-induced autophagy by the eIF2alpha kinase signaling pathway. *Proc. Natl. Acad. Sci. USA* **2002**, *99*, 190–195. [CrossRef] [PubMed]

15. Pei, J.; Zhao, M.; Ye, Z.; Gou, H.; Wang, J.; Yi, L.; Dong, X.; Liu, W.; Luo, Y.; Liao, M.; et al. Autophagy enhances the replication of classical swine fever virus in vitro. *Autophagy* **2014**, *10*, 93–110. [CrossRef] [PubMed]

16. Sun, M.X.; Huang, L.; Wang, R.; Yu, Y.L.; Li, C.; Li, P.P.; Hu, X.C.; Hao, H.P.; Ishag, H.A.; Mao, X. Porcine reproductive and respiratory syndrome virus induces autophagy to promote virus replication. *Autophagy* **2012**, *8*, 1434–1447. [CrossRef] [PubMed]

17. Berkova, Z.; Crawford, S.E.; Trugnan, G.; Yoshimori, T.; Morris, A.P.; Estes, M.K. Rotavirus nsp4 induces a novel vesicular compartment regulated by calcium and associated with viroplasms. *J. Virol.* **2006**, *80*, 6061–6071. [CrossRef] [PubMed]

18. Sun, D.; Shi, H.; Guo, D.; Chen, J.; Shi, D.; Zhu, Q.; Zhang, X.; Feng, L. Analysis of protein expression changes of the vero E6 cells infected with classic pedv strain CV777 by using quantitative proteomic technique. *J. Virol. Methods* **2015**, *218*, 27–39. [CrossRef] [PubMed]

19. Guo, X.; Hu, H.; Chen, F.; Li, Z.; Ye, S.; Cheng, S.; Zhang, M.; He, Q. Itraq-based comparative proteomic analysis of vero cells infected with virulent and CV777 vaccine strain-like strains of porcine epidemic diarrhea virus. *J. Proteom.* **2016**, *130*, 65–75. [CrossRef] [PubMed]

20. Cheng, S.; Yan, W.; Gu, W.; He, Q. The ubiquitin-proteasome system is required for the early stages of porcine circovirus type 2 replication. *Virology* **2014**, *456–457*, 198–204. [CrossRef] [PubMed]

21. Kim, Y.; Lee, C. Porcine epidemic diarrhea virus induces caspase-independent apoptosis through activation of mitochondrial apoptosis-inducing factor. *Virology* **2014**, *460–461*, 180–193. [CrossRef] [PubMed]

22. Reggiori, F.; Monastyrska, I.; Verheije, M.H.; Cali, T.; Ulasli, M.; Bianchi, S.; Bernasconi, R.; de Haan, C.A.; Molinari, M. Coronaviruses hijack the LC3-I-positive EDEmosomes, ER-derived vesicles exporting short-lived ERAD regulators, for replication. *Cell Host Microbe* **2010**, *7*, 500–508. [CrossRef] [PubMed]

23. Knoops, K.; Kikkert, M.; Worm, S.H.; Zevenhoven-Dobbe, J.C.; van der Meer, Y.; Koster, A.J.; Mo mMaas, A.M.; Snijder, E.J. Sars-coronavirus replication is supported by a reticulovesicular network of modified endoplasmic reticulum. *PLoS Biol.* **2008**, *6*, e226. [CrossRef] [PubMed]

24. Cottam, E.M.; Maier, H.J.; Manifava, M.; Vaux, L.C.; Chandra-Schoenfelder, P.; Gerner, W.; Britton, P.; Ktistakis, N.T.; Wileman, T. Coronavirus NSP6 proteins generate autophagosomes from the endoplasmic reticulum via an omegasome intermediate. *Autophagy* **2011**, *7*, 1335–1347. [CrossRef] [PubMed]

25. Chen, Q.; Fang, L.; Wang, D.; Wang, S.; Li, P.; Li, M.; Luo, R.; Chen, H.; Xiao, S. Induction of autophagy enhances porcine reproductive and respiratory syndrome virus replication. *Virus Res.* **2012**, *163*, 650–655. [CrossRef] [PubMed]

26. Seglen, P.O.; Gordon, P.B. 3-methyladenine: Specific inhibitor of autophagic/lysosomal protein degradation in isolated rat hepatocytes. *Proc. Natl. Acad. Sci. USA* **1982**, *79*, 1889–1892. [CrossRef] [PubMed]

27. Shrivastava, S.; Raychoudhuri, A.; Steele, R.; Ray, R.; Ray, R.B. Knockdown of autophagy enhances the innate i mMune response in hepatitis C virus-infected hepatocytes. *Hepatology* **2011**, *53*, 406–414. [CrossRef] [PubMed]

28. Pan, H.; Zhang, Y.; Luo, Z.; Li, P.; Liu, L.; Wang, C.; Wang, H.; Li, H.; Ma, Y. Autophagy mediates avian influenza H5N1 pseudotyped particle-induced lung infla mMation through NF-kappab and p38 MAPK signaling pathways. *Am. J. Physiol. Lung Cell. Mol. Physiol.* **2014**, *306*, L183–L195. [CrossRef] [PubMed]

29. Cao, L.; Ge, X.; Gao, Y.; Ren, Y.; Ren, X.; Li, G. Porcine epidemic diarrhea virus infection induces NF-kappab activation through the TLR2, TLR3 and TLR9 pathways in porcine intestinal epithelial cells. *J. Gen. Virol.* **2015**, *96*, 1757–1767. [CrossRef] [PubMed]

30. Ke, P.Y.; Chen, S.S. Autophagy in hepatitis C virus-host interactions: Potential roles and therapeutic targets for liver-associated diseases. *World J. Gastroenterol.* **2014**, *20*, 5773–5793. [CrossRef] [PubMed]

31. Dreux, M.; Gastaminza, P.; Wieland, S.F.; Chisari, F.V. The autophagy machinery is required to initiate hepatitis c virus replication. *Proc. Natl. Acad. Sci. USA* **2009**, *106*, 14046–14051. [CrossRef] [PubMed]

32. Orvedahl, A.; Alexander, D.; Talloczy, Z.; Sun, Q.; Wei, Y.; Zhang, W.; Burns, D.; Leib, D.A.; Levine, B. HSV-1 ICP34.5 confers neurovirulence by targeting the beclin 1 autophagy protein. *Cell Host Microbe* **2007**, *1*, 23–35. [CrossRef] [PubMed]

33. Prentice, E.; Jerome, W.G.; Yoshimori, T.; Mizushima, N.; Denison, M.R. Coronavirus replication complex formation utilizes components of cellular autophagy. *J. Biol. Chem.* **2004**, *279*, 10136–10141. [CrossRef] [PubMed]

34. Maier, H.J.; Cottam, E.M.; Stevenson-Leggett, P.; Wilkinson, J.A.; Harte, C.J.; Wileman, T.; Britton, P. Visualizing the autophagy pathway in avian cells and its application to studying infectious bronchitis virus. *Autophagy* **2013**, *9*, 496–509. [CrossRef] [PubMed]

35. Klionsky, D.J.; Abdelmohsen, K.; Abe, A.; Abedin, M.J.; Abeliovich, H.; Acevedo Arozena, A.; Adachi, H.; Adams, C.M.; Adams, P.D.; Adeli, K.; et al. Guidelines for the use and interpretation of assays for monitoring autophagy (3rd edition). *Autophagy* **2016**, *12*, 1–222. [CrossRef] [PubMed]

36. Ulasli, M.; Verheije, M.H.; de Haan, C.A.; Reggiori, F. Qualitative and quantitative ultrastructural analysis of the membrane rearrangements induced by coronavirus. *Cell Microbiol.* **2010**, *12*, 844–861. [CrossRef] [PubMed]

37. Mizushima, N.; Yoshimori, T.; Levine, B. Methods in ma mMalian autophagy research. *Cell* **2010**, *140*, 313–326. [CrossRef] [PubMed]

38. Richetta, C.; Gregoire, I.P.; Verlhac, P.; Azocar, O.; Baguet, J.; Flacher, M.; Tangy, F.; Rabourdin-Combe, C.; Faure, M. Sustained autophagy contributes to measles virus infectivity. *PLoS Pathogens* **2013**, *9*, e1003599. [CrossRef] [PubMed]

39. Buckingham, E.M.; Carpenter, J.E.; Jackson, W.; Grose, C. Autophagy and the effects of its inhibition on varicella-zoster virus glycoprotein biosynthesis and infectivity. *J. Virol.* **2014**, *88*, 890–902. [CrossRef] [PubMed]

40. Klionsky, D.J.; Abdalla, F.C.; Abeliovich, H.; Abraham, R.T.; Acevedo-Arozena, A.; Adeli, K.; Agholme, L.; Agnello, M.; Agostinis, P.; Aguirre-Ghiso, J.A.; et al. Guidelines for the use and interpretation of assays for monitoring autophagy. *Autophagy* **2012**, *8*, 445–544. [CrossRef] [PubMed]

41. Zhao, Z.; Thackray, L.B.; Miller, B.C.; Lynn, T.M.; Becker, M.M.; Ward, E.; Mizushima, N.N.; Denison, M.R.; Virgin, H.W.T. Coronavirus replication does not require the autophagy gene ATG5. *Autophagy* **2007**, *3*, 581–585. [CrossRef] [PubMed]

42. Petiot, A.; Ogier-Denis, E.; Blo mMaart, E.F.; Meijer, A.J.; Codogno, P. Distinct classes of phosphatidylinositol 3'-kinases are involved in signaling pathways that control macroautophagy in HT-29 cells. *J. Biol. Chem.* **2000**, *275*, 992–998. [CrossRef] [PubMed]

43. Wang, J. Beclin 1 bridges autophagy, apoptosis and differentiation. *Autophagy* **2008**, *4*, 947–948. [CrossRef] [PubMed]

44. Wang, K. Autophagy and apoptosis in liver injury. *Cell Cycle* **2015**, *14*, 1631–1642. [CrossRef] [PubMed]

45. Gougeon, M.L.; Piacentini, M. New insights on the role of apoptosis and autophagy in hiv pathogenesis. *Apoptosis* **2009**, *14*, 501–508. [CrossRef] [PubMed]

46. Xin, L.; Xiao, Z.; Ma, X.; He, F.; Yao, H.; Liu, Z. Coxsackievirus b3 induces crosstalk between autophagy and apoptosis to benefit its release after replicating in autophagosomes through a mechanism involving caspase cleavage of autophagy-related proteins. *Infect. Genet. Evol.* **2014**, *26*, 95–102. [CrossRef] [PubMed]

47. Gannage, M.; Dormann, D.; Albrecht, R.; Dengjel, J.; Torossi, T.; Ramer, P.C.; Lee, M.; Strowig, T.; Arrey, F.; Conenello, G.; et al. Matrix protein 2 of influenza a virus blocks autophagosome fusion with lysosomes. *Cell Host Microbe* **2009**, *6*, 367–380. [CrossRef] [PubMed]

48. Zhou, A.; Li, S.; Khan, F.A.; Zhang, S. Autophagy postpones apoptotic cell death in prrsv infection through bad-beclin1 interaction. *Virulence* **2016**, *7*, 98–109. [CrossRef] [PubMed]

49. Li, S.; Zhou, A.; Wang, J.; Zhang, S. Interplay of autophagy and apoptosis during prrsv infection of marc145 cell. *Infect. Genet. Evol.* **2016**, *39*, 51–54. [CrossRef] [PubMed]

50. Valdor, R.; Macian, F. Autophagy and the regulation of the I mMune response. *Pharmacol. Res.* **2012**, *66*, 475–483. [CrossRef] [PubMed]

51. Levine, B.; Mizushima, N.; Virgin, H.W. Autophagy in I mMunity and infla mMation. *Nature* **2011**, *469*, 323–335. [CrossRef] [PubMed]

52. Papademetrio, D.L.; Lompardia, S.L.; Simunovich, T.; Costantino, S.; Mihalez, C.Y.; Cavaliere, V.; Alvarez, E. Inhibition of survival pathways MAPK and NF-kb triggers apoptosis in pancreatic ductal adenocarcinoma cells via suppression of autophagy. *Target Oncol.* **2016**, *11*, 183–195. [CrossRef] [PubMed]

PERMISSIONS

LIST OF CONTRIBUTORS

Hanne Merethe Haatveit, Øystein Wessel, Turhan Markussen, Ingvild Berg Nyman, Stine Braaen and Espen Rimstad
Department of Food Safety and Infectious Biology, Faculty of Veterinary Medicine, Norwegian University of Life Sciences, 0454 Oslo, Norway

Morten Lund and Maria Krudtaa Dahle
Department of Immunology, Norwegian Veterinary Institute, 0454 Oslo, Norway

Bernd Thiede
Department of Biosciences, University of Oslo, 0316 Oslo, Norway

Guo-rong Sun, Yan-ping Zhang, Hong-chao Lv, Lin-yi Zhou, Hong-yu Cui, Yu-long Gao, Xiao-le Qi, Yong-qiang Wang, Kai Li, Li Gao, Qing Pan, Xiao-mei Wang and Chang-jun Liu
Division of Avian Immunosuppressive Diseases, State Key Laboratory of Veterinary Biotechnology, Harbin Veterinary Research Institute, Chinese Academy of Agricultural Sciences, Harbin 150069, China

Qinhai Ma, Dedong Liang, Shuai Song, Qintian Yu, Chunyu Shi, Xuefeng Xing and Jia-Bo Luo
School of Traditional Chinese Medical Science, Southern Medical University, Guangzhou 510515, China
Guangdong Provincial Key Laboratory of Chinese Medicine Pharmaceutics, Southern Medical University, Guangzhou 510515, China

Torill Vik Johannessen
Vaxxinova Norway AS, Kong Christian Frederiks plass 3, 5006 Bergen, Norway

Aud Larsen
Uni Research Environment, N-5008 Bergen, Norway

Gunnar Bratbak, António Pagarete and Ruth-Anne Sandaa
Department of Biology, University of Bergen, N-5020 Bergen, Norway

Bente Edvardsen and Elianne D. Egge
Department of Biosciences, University of Oslo, 0316 Oslo, Norway

Robert A. Kozak, Larissa Hattin, Betty-Anne McBey, Jason Morgenstern and Evan Lusty
David Leishman, Byram Bridle and Éva Nagy, Juan C. Corredor and Scott Walsh

Department of Pathobiology, Ontario Veterinary College, University of Guelph, Guelph, ON N1G 2W1, Canada

Mia J. Biondi, Vera Cherepanov and Jordan J. Feld
Sandra Rotman Centre for Global Health, University of Toronto, Toronto, ON M5G 1L7, Canada

Max Xue-Zhong, Justin Manuel and Ian D. McGilvray
Multi-Organ Transplant Program, Department of Surgery, University of Toronto, Toronto General Hospital, Toronto, ON M5G 2C4, Canada

Hui-Wen Zheng, Ming Sun, Lei Guo, Jing-Jing Wang, Jie Song, Jia-Qi Li, Hong-Zhe Li, Ruo-Tong Ning, Ze-Ning Yang, Hai-Tao Fan, Zhan-Long He and Long-Ding Liu
Institute of Medical Biology, Chinese Academy of Medical Sciences & Peking Union Medical College, Kunming 650118, China

Jiwei Ding, Jianyuan Zhao, Ling Ma, Zeyun Mi, Yanbing Wu, Jinmin Zhou, Xiaoyu Li, Zonggen Peng and Shan Cen
Institute of Medicinal Biotechnology, Chinese Academy of Medical Sciences and Peking Union Medical School, Beijing 100050, China

Zhijun Yang and Mei Ge
School of Pharmacy, Shanghai Jiaotong University, Shanghai 200040, China

Jiamei Guo and Ying Guo
Institute of Materia Medica, Chinese Academy of Medical Sciences and Peking Union Medical School, Beijing 100050, China

Tao Wei
Department of Food Science, Beijing Union University, Beijing 100101, China

Haisheng Yu and Liguo Zhang
Institute of Biophysics, Chinese Academy of Sciences, Beijing 100101, China

Ben X. Wang and Eleanor N. Fish
Toronto General Hospital Research Institute, University Health Network, 67 College Street, Toronto, ON M5G 2M1, Canada
Department of Immunology, University of Toronto, 1 King's College Circle, Toronto, ON M5S 1A8, Canada

Lianhu Wei and Lakshmi P. Kotra
Toronto General Hospital Research Institute, University Health Network, 67 College Street, Toronto, ON M5G 2M1, Canada
Center for Molecular Design and Preformulations, University Health Network, 101 College Street, Toronto, ON M5G 1L7, Canada
Department of Pharmaceutical Sciences, Leslie Dan Faculty of Pharmacy, University of Toronto, 144 College Street, Toronto, ON M5S 3M2, Canada

Earl G. Brown
Department of Biochemistry, Microbiology and Immunology, Faculty of Medicine, University of Ottawa, 451 Smyth Road, Ottawa, ON K1H 8M5, Canada

Wen Shi, Wenlu Fan, Jing Bai, Wen Cui, Yigang Xu, Yijing Li, Li Wang, Yanping Jiang and Lijie Tang
College of Veterinary Medicine, Northeast Agricultural University, Harbin 150030, China

Yandong Tang
Harbin Veterinary Research Institute, Chinese Academy of Agricultural Sciences, Harbin 150001, China

Min Liu
College of Animal Science and Technology, Northeast Agricultural University, Harbin 150030, China

Christine M. Livingston, Dhivya Ramakrishnan, Simon P. Fletcher and Rudolf K. Beran
Gilead Sciences, Foster City, CA 94404, USA

Michel Strubin
Department of Microbiology and Molecular Medicine, University Medical Center (CMU), 1211 Geneva, Switzerland

Chuangang Cheng, Chengfeng Lei and Xiulian Sun
Wuhan Institute of Virology, Chinese Academy of Sciences, Wuhan 430071, China

Lan Su and Congrui Xu
Wuhan Institute of Virology, Chinese Academy of Sciences, Wuhan 430071, China
University of Chinese Academy of Sciences, Beijing 100049, China

Haodi Jia, Chaoyang Chen, Fei Zhang, Cheng Li, Li Zong , Yongxiang Wang and Yumei Wen
Key Lab of Medical Molecular Virology, School of Basic Medical Sciences, Fudan University, Shanghai 200032, China

Yanli Qin and Jiming Zhang
Department of Infectious Diseases, Huashan Hospital, Fudan University, Shanghai 200032, China

Jisu Li
Liver Research Center, Rhode Island Hospital, Warren Alpert School of Medicine, Brown University, Providence, RI 02903, USA

Shuping Tong
Key Lab of Medical Molecular Virology, School of Basic Medical Sciences, Fudan University, Shanghai 200032, China
Liver Research Center, Rhode Island Hospital, Warren Alpert School of Medicine, Brown University, Providence, RI 02903, USA

Xiaozhen Guo, Mengjia Zhang, Xiaoqian Zhang, Xin Tan, Hengke Guo, Wei Zeng, Atta Muhammad Memon, Zhonghua Li, Yinxing Zhu, Bingzhou Zhang, Xugang Ku, Meizhou Wu, Shengxian Fan and Qigai He
State Key Laboratory of Agricultural Microbiology, College of Veterinary Medicine, Huazhong Agricultural University, Wuhan 430070, China
The Cooperative Innovation Center for Sustainable Pig Production, Wuhan 430070, China

Guokai Yan
The Cooperative Innovation Center for Sustainable Pig Production, Wuhan 430070, China
College of Animal Sciences and Technology, Huazhong Agricultural University, Wuhan 430070, China

Index

www.ingramcontent.com/pod-product-compliance
Lightning Source LLC
Chambersburg PA
CBHW050449200326
41458CB00014B/5122